BIOLOGICAL AND MEDICAL PHYSICS,
BIOMEDICAL ENGINEERING

BIOLOGICAL AND MEDICAL PHYSICS, BIOMEDICAL ENGINEERING

The fields of biological and medical physics and biomedical engineering are broad, multidisciplinary and dynamic. They lie at the crossroads of frontier research in physics, biology, chemistry, and medicine. The Biological and Medical Physics, Biomedical Engineering Series is intended to be comprehensive, covering a broad range of topics important to the study of the physical, chemical and biological sciences. Its goal is to provide scientists and engineers with textbooks, monographs, and reference works to address the growing need for information.

Books in the series emphasize established and emergent areas of science including molecular, membrane, and mathematical biophysics; photosynthetic energy harvesting and conversion; information processing; physical principles of genetics; sensory communications; automata networks, neural networks, and cellular automata. Equally important will be coverage of applied aspects of biological and medical physics and biomedical engineering such as molecular electronic components and devices, biosensors, medicine, imaging, physical principles of renewable energy production, advanced prostheses, and environmental control and engineering.

Editor-in-Chief:
Elias Greenbaum, Oak Ridge National Laboratory, Oak Ridge, Tennessee, USA

Editorial Board:

Masuo Aizawa, Department of Bioengineering, Tokyo Institute of Technology, Yokohama, Japan

Olaf S. Andersen, Department of Physiology, Biophysics & Molecular Medicine, Cornell University, New York, USA

Robert H. Austin, Department of Physics, Princeton University, Princeton, New Jersey, USA

James Barber, Department of Biochemistry, Imperial College of Science, Technology and Medicine, London, England

Howard C. Berg, Department of Molecular and Cellular Biology, Harvard University, Cambridge, Massachusetts, USA

Victor Bloomfield, Department of Biochemistry, University of Minnesota, St. Paul, Minnesota, USA

Robert Callender, Department of Biochemistry, Albert Einstein College of Medicine, Bronx, New York, USA

Britton Chance, Department of Biochemistry/ Biophysics, University of Pennsylvania, Philadelphia, Pennsylvania, USA

Steven Chu, Department of Physics, Stanford University, Stanford, California, USA

Louis J. DeFelice, Department of Pharmacology, Vanderbilt University, Nashville, Tennessee, USA

Johann Deisenhofer, Howard Hughes Medical Institute, The University of Texas, Dallas, Texas, USA

George Feher, Department of Physics, University of California, San Diego, La Jolla, California, USA

Hans Frauenfelder, CNLS, MS B258, Los Alamos National Laboratory, Los Alamos, New Mexico, USA

Ivar Giaever, Rensselaer Polytechnic Institute, Troy, New York, USA

Sol M. Gruner, Department of Physics, Princeton University, Princeton, New Jersey, USA

Judith Herzfeld, Department of Chemistry, Brandeis University, Waltham, Massachusetts, USA

Mark S. Humayun, Doheny Eye Institute, Los Angeles, California, USA

Pierre Joliot, Institute de Biologie Physico-Chimique, Fondation Edmond de Rothschild, Paris, France

Lajos Keszthelyi, Institute of Biophysics, Hungarian Academy of Sciences, Szeged, Hungary

Robert S. Knox, Department of Physics and Astronomy, University of Rochester, Rochester, New York, USA

Aaron Lewis, Department of Applied Physics, Hebrew University, Jerusalem, Israel

Stuart M. Lindsay, Department of Physics and Astronomy, Arizona State University, Tempe, Arizona, USA

David Mauzerall, Rockefeller University, New York, New York, USA

Eugenie V. Mielczarek, Department of Physics and Astronomy, George Mason University, Fairfax, Virginia, USA

Markolf Niemz, Klinikum Mannheim, Mannheim, Germany

V. Adrian Parsegian, Physical Science Laboratory, National Institutes of Health, Bethesda, Maryland, USA

Linda S. Powers, NCDMF: Electrical Engineering, Utah State University, Logan, Utah, USA

Earl W. Prohofsky, Department of Physics, Purdue University, West Lafayette, Indiana, USA

Andrew Rubin, Department of Biophysics, Moscow State University, Moscow, Russia

Michael Seibert, National Renewable Energy Laboratory, Golden, Colorado, USA

David Thomas, Department of Biochemistry, University of Minnesota Medical School, Minneapolis, Minnesota, USA

Samuel J. Williamson, Department of Physics, New York University, New York, New York, USA

U. Bastolla   M. Porto
H.E. Roman   M. Vendruscolo
(Eds.)

Structural Approaches to Sequence Evolution

Molecules, Networks, Populations

With 95 Figures, 1 in color and 19 Tables

Dr. Ugo Bastolla
Universidad Autonoma Madrid, Fac. de Ciencias
Cantoblanco, 28049 Madrid, Spain
E-mail: ubastolla@cbm.uam.es

Professor Dr. Markus Porto
TU Darmstadt, Institut für Festkörperphysik
Hochschulstr. 8, 64289 Darmstadt, Germany
E-mail: porto@fkp.tu-darmstadt.de

Dr. H. Eduardo Roman
Università di Milano-Bicocca, Dipartimento di Fisica
Piazza della Scienza 3, 29126 Milano, Italy
E-mail: roman@mib.infn.it

Dr. Michele Vendruscolo
University of Cambridge, Department of Chemistry
Lensfield Road, Cambridge CB2 1EW, United Kingdom
E-mail: mv245@cam.ac.uk

Library of Congress Control Number: 2007924598

ISSN 1618-7210

ISBN-10 3-540-35305-4 Springer Berlin Heidelberg New York

ISBN-13 978-3-540-35305-8 Springer Berlin Heidelberg New York

This work is subject to copyright. All rights are reserved, whether the whole or part of the material is concerned, specifically the rights of translation, reprinting, reuse of illustrations, recitation, broadcasting, reproduction on microfilm or in any other way, and storage in data banks. Duplication of this publication or parts thereof is permitted only under the provisions of the German Copyright Law of September 9, 1965, in its current version, and permission for use must always be obtained from Springer. Violations are liable to prosecution under the German Copyright Law.

Springer is a part of Springer Science+Business Media

springer.com

© Springer-Verlag Berlin Heidelberg 2007

The use of general descriptive names, registered names, trademarks, etc. in this publication does not imply, even in the absence of a specific statement, that such names are exempt from the relevant protective laws and regulations and therefore free for general use.

Typesetting by the editors and SPi using a Springer LaTeX macro package
Cover: eStudio Calamar Steinen

Printed on acid-free paper   SPIN 11417156   57/3180/SPi - 5 4 3 2 1 0

To my parents, *UB*

To Marion, *MP*

To Fulvia, *HER*

To my parents, *MV*

Preface

Soon after the first sequences of proteins and nucleic acids became available for comparative analysis, it became apparent that they can play a key role for reconstructing the evolution of life. The availability of the sequence of several proteins prompted the birth of the field of molecular evolution, which aims at both the reconstruction of the biochemical history of life and the understanding of the mechanisms of evolution at the molecular level through the analysis of the macromolecules of existing organisms. These ambitious goals can only be accomplished within a wide interdisciplinary approach that combines together experimental techniques of molecular biology, bioinformatics and mathematical modeling. Indeed, the huge amount of data made available in recent years by genome sequencing projects is demanding simultaneous skills on these three approaches.

At its beginnings, the study of molecular evolution was almost entirely based on the analysis of macromolecular sequences. More recently, progress in structural biology has opened the possibility of using also structural information in evolutionary studies. It now appears that a paradigm shift is taking place within the field of molecular evolution, from coding symbols (sequence) to coded meanings (structure and function). This book investigates such a structural approach at different levels of biological organization, i.e., of molecules, networks and populations, showing that their understanding can significantly contribute to elucidate the mechanisms of evolution and to reconstruct its course. Synergies between experimental, theoretical, computational, and statistical approaches are expected to widen our understanding of the processes and pathways of molecular evolution. However, relevant fields such as those dealing with the structure and thermodynamics of biomolecules, gene networks, and the mechanisms of molecular biology are not fully integrated into the field of molecular evolution yet, and these missings links are becoming increasingly evident.

The central goal of the present tutorial book is to stimulate this integration by bringing these different disciplines together. The idea of such a book emerged during the interdisciplinary workshop *Structural approaches to sequence evolution: Molecules, networks, populations* that we organized at the Max-Planck-Institut für Physik komplexer Systeme in Dresden (Germany) in July 2004. The book collects a series of tutorial chapters written by experts from different scientific communities, most of whom have participated in the workshop.

As this workshop was the birthplace of the present book, the editors wish to express their sincere gratitude to the Max-Planck-Institut für Physik komplexer Systeme for hosting and financing it. In particular, they would like to thank Dr. Sergej Flach (head of the conference programme) for his very helpful support and Mrs. Mandy Lochar (conference secretary) for her very efficient organization.

Madrid, *Ugo Bastolla*
Darmstadt, *Markus Porto*
Milano, *H. Eduardo Roman*
Cambridge, April 2006 *Michele Vendruscolo*

Contents

Part I Molecules: Proteins and RNA

1 Modeling Conformational Flexibility and Evolution of Structure: RNA as an Example
P. Schuster and P.F. Stadler 3
1.1   Definition and Computation of RNA Structures 3
      1.1.1   RNA Secondary Structures 4
      1.1.2   Compatibility of Sequences and Structures 8
      1.1.3   Sequence Space, Shape Space, and Conformation Space 11
      1.1.4   Computation of RNA Secondary Structures 14
      1.1.5   Mapping Sequences into Structures 15
      1.1.6   Suboptimal Structures and Partition Functions 18
1.2   Design of RNA Structures 19
      1.2.1   Inverse Folding 19
      1.2.2   Multiconformational RNAs 20
      1.2.3   Riboswitches 22
1.3   Processes in Conformation, Sequence, and Shape Space 23
      1.3.1   Kinetic Folding 23
      1.3.2   Evolutionary Optimization 25
      1.3.3   Evolution of Noncoding RNAs 30
References 32

2 Gene3D and Understanding Proteome Evolution
J.G. Ranea, C. Yeats, R. Marsden, and C. Orengo 37
2.1   Protein Family Clustering................................. 42
      2.1.1   SYSTERS.. 42
      2.1.2   ProtoNet.. 42
      2.1.3   ADDA.. 42
      2.1.4   ProDom 43

2.2	The PFscape Method		43
2.3	The NewFams		44
2.4	Describing the Proteome		45
2.5	Superfamily Evolution and Genome Complexity		46
2.6	Superfamily Evolution and Functional Relationships		48
2.7	Limits to Genome Complexity in Prokaryotes		50
2.8	The Bacterial Factory		52
2.9	Conclusions		53
References			54

3 The Evolution of the Globins: We Thought We Understood It
A.M. Lesk 57

3.1	Introduction		58
3.2	Coordinates and Calculations		59
3.3	Results		59
	3.3.1	Description of Secondary and Tertiary Structure of Full-Length (~150–Residue) Globins	59
	3.3.2	Description of Secondary and Tertiary Structure of Truncated Globins	60
	3.3.3	Alignment	60
3.4	Helix Contacts		62
	3.4.1	Geometry of Inter-Helix Contacts	62
	3.4.2	Pairs of Helices Making Contacts	63
	3.4.3	Structures of Helix Interfaces in Truncated Globins, Compared to Those in Sperm Whale Myoglobin	65
	3.4.4	The B/G Interface	65
	3.4.5	The A/H Interface	66
	3.4.6	The B/E Interface	67
3.5	Patterns of Residue–Residue Contacts at Helix Interfaces		68
	3.5.1	The G/H Interface	69
3.6	Haem Contacts		72
3.7	The Tunnel		72
3.8	Conclusions		72
References			73

4 The Structurally Constrained Neutral Model of Protein Evolution
*U. Bastolla, M. Porto, H.E. Roman,
and M. Vendruscolo* 75

4.1	Aspects of Population Genetics		76
	4.1.1	Population Size and Mutation Rate	76
	4.1.2	Natural Selection	77
	4.1.3	Mutant Spectrum	78
	4.1.4	Neutral Substitutions	80
	4.1.5	Beyond the Small $M\mu$ Regime: Neutral Networks	81

4.2	Structural Aspects of Molecular Evolution		83
	4.2.1	Neutral Theory and Protein Folding Thermodynamics	83
	4.2.2	Structural Conservation and Functional Changes in Protein Evolution	84
	4.2.3	Models of Molecular Evolution with Structural Conservation	85
4.3	The SCN Model of Evolution		87
	4.3.1	Representation of Protein Structures	88
	4.3.2	Stability Against Unfolding	88
	4.3.3	Stability Against Misfolding	89
	4.3.4	Calculation of $\alpha(\mathbf{A})$	89
	4.3.5	Sampling the Neutral Networks	91
	4.3.6	Fluctuations and Correlations in the Evolutionary Process	91
	4.3.7	Substitution Process	93
4.4	Site-Specific Amino Acid Distributions		97
	4.4.1	Vectorial Representation of Protein Sequences	98
	4.4.2	Vectorial Representation of Protein Folds	99
	4.4.3	Relation Between Sequence and Structure	99
	4.4.4	The PE as a Structural Determinant of Evolutionary Conservation	100
	4.4.5	Site-Dependent Amino Acid Distributions	101
	4.4.6	Sequence Conservation and Structure Designability	104
	4.4.7	Site-Specific Amino Acid Distributions in the PDB	105
	4.4.8	Mean-Field Model of Mutation plus Selection	107
4.5	Conclusions		109
References			109

5 Towards Unifying Protein Evolution Theory
N.V. Dokholyan and E.I. Shakhnovich 113

5.1	Two Views on Protein Evolution	113
5.2	Challenges in Functionally Annotating Structures	113
5.3	The Importance of the Tree of Life	115
5.4	Building the PDUG	116
5.5	Properties of the PDUG: Power Laws on Very Different Evolutionary Scales	117
5.6	Functional Flexibility Score: Calculating Entropy in Function Space	118
5.7	Lattice Proteins and Its Random Subspaces: Structure Graphs	119
5.8	Divergence and Convergence Explored: What Power Laws Tell Us about Evolution	120
5.9	Context Is Important	122
5.10	Not All Functions Are Created Equal and Neither Are Structures	122

5.11   Concluding Remarks 124
References 124

Part II Molecules: Genomes

6 A Twenty-First Century View of Evolution: Genome System Architecture, Repetitive DNA, and Natural Genetic Engineering
J.A. Shapiro 129
6.1   Introduction: Cellular Computation and DNA
       as an Interactive Data Storage Medium 129
6.2   Genome System Architecture and Repetitive DNA 130
6.3   Genomes and Cellular Computation: *E. coli lac* Operon ... 132
6.4   New Principles of Evolution:
       The Lessons of Sequenced Genomes 135
6.5   Natural Genetic Engineering 136
6.6   Conclusions: A Twenty-First Century View of Evolution ... 141
6.7   Twenty-First Century Directions in Evolution Research ... 143
References 144

7 Genomic Changes in Bacteria: From Free-Living to Endosymbiotic Life
F.J. Silva, A. Latorre, L. Gómez-Valero, and A. Moya 149
7.1   Introduction 149
7.2   Genetic and Genomic Features
       of Endosymbiotic Bacteria 153
       7.2.1   Sequence Evolution in Endosymbionts 153
       7.2.2   Reductive Evolution: DNA Loss and Genome
               Reduction in Obligate Bacterial Mutualists 158
       7.2.3   Chromosomal Rearrangements
               Throughout Endosymbiont Evolution 160
7.3   Conclusions and Prospects 162
References 163

Part III Phylogenetic Analysis

8 Molecular Phylogenetics: Mathematical Framework and Unsolved Problems
X. Xia 169
8.1   Introduction 169
8.2   Substitution Models 170
       8.2.1   Nucleotide-Based Substitution Models
               and Genetic Distances 171

	8.2.2	Amino Acid-Based and Codon-Based Substitution Models 176
8.3	Tree-Building Methods 178
	8.3.1	Distance-Based Methods 178
	8.3.2	Maximum Parsimony Methods 181
	8.3.3	Maximum Likelihood Methods 182
	8.3.4	Bayesian Inference 185
8.4	Final Words... 187
References 187

9 Phylogenetics and Computational Biology of Multigene Families
P. Liò, M. Brilli, and R. Fani 191

- 9.1 Introduction 191
- 9.2 How Do Large Gene Families Arise?........................... 193
- 9.3 The Classical Model of Gene Duplication 193
- 9.4 Subfunctionalization Model 194
- 9.5 Subneofunctionalization 195
- 9.6 Tests for Subfunctionalization 196
- 9.7 Tests for Functional Divergence After Duplication.............. 196
  - 9.7.1 Case Study 1: Chemokine Receptors Expansion in Vertebrates............................ 197
  - 9.7.2 Case Study 2: The Evolution of TIM Barrel Coding Genes 199
- References 204

10 SeqinR 1.0-2: A Contributed Package to the R Project for Statistical Computing Devoted to Biological Sequences Retrieval and Analysis
D. Charif and J.R. Lobry 207

- 10.1 Introduction 207
  - 10.1.1 About R and CRAN............................... 207
  - 10.1.2 About this Document............................. 208
  - 10.1.3 About Sequin and `seqinR` 208
  - 10.1.4 About Getting Started............................ 208
  - 10.1.5 About Running R in Batch Mode 208
  - 10.1.6 About the Learning Curve 209
- 10.2 How to Get Sequence Data..................................... 213
  - 10.2.1 Importing Raw Sequence Data from Fasta Files 213
  - 10.2.2 Importing Aligned Sequence Data 214
  - 10.2.3 Complex Queries in ACNUC Databases 218
- 10.3 How to Deal with Sequence 220
  - 10.3.1 Sequence Classes 220
  - 10.3.2 Generic Methods for Sequences 220
  - 10.3.3 Internal Representation of Sequences 221

10.4	Multivariate Analyses	225
	10.4.1 Correspondence Analysis	225
	10.4.2 Synonymous and Nonsynonymous Analyses	230
References		232

Part IV Networks

11 Evolutionary Genomics of Gene Expression
I.K. Jordan and L. Mariño-Ramírez 235

11.1	Sequence Divergence	236
	11.1.1 Ortholog Identification	236
	11.1.2 Sequence Alignment	237
	11.1.3 Sequence Distance Calculation	237
11.2	Gene Expression Divergence	240
	11.2.1 Database Sources	241
	11.2.2 Probe-to-Gene Mapping	241
	11.2.3 Structure of the Data	242
	11.2.4 Transformation and Normalization	242
	11.2.5 Measuring Divergence	243
	11.2.6 Clustering and Visualization	245
11.3	Integrated Analysis	246
	11.3.1 Sequence vs. Expression Divergence	246
	11.3.2 Neutral Changes in Gene Expression	247
	11.3.3 Evolutionary Conservation of Gene Expression	250
References		251

12 From Biophysics to Evolutionary Genetics: Statistical Aspects of Gene Regulation
M. Lässig 253

12.1	Introduction	253
12.2	Biophysics of Transcriptional Regulation	254
	12.2.1 Factor-DNA Binding Energies	255
	12.2.2 Energy Distribution in the Genome	257
	12.2.3 Search Kinetics	258
	12.2.4 Thermodynamics of Factor Binding	258
	12.2.5 Sensitivity and Genomic Design of Regulation	260
	12.2.6 Programmability and Evolvability of Regulatory Networks	260
12.3	Bioinformatics of Regulatory DNA	261
	12.3.1 Markov Model for Background Sequence	261
	12.3.2 Probabilistic Model for Functional Sites	262
	12.3.3 Bayesian Model for Genomic Loci	263
	12.3.4 Dynamic Programming and Sequence Analysis	264
12.4	Evolution of Regulatory DNA	266
	12.4.1 Deterministic Population Dynamics and Fitness	267

	12.4.2	Stochastic Dynamics and Genetic Drift	268
	12.4.3	Mutation Processes and Evolutionary Equilibria	270
	12.4.4	Substitution Dynamics	271
	12.4.5	Neutral Dynamics in Sequence Space, Sequence Entropy	273
	12.4.6	Dynamics Under Selection, the Score-Fitness Relation	274
	12.4.7	Measuring Selection for Binding Sites	275
	12.4.8	Nucleotide Frequency Correlations	276
	12.4.9	Stationary Evolution of Binding Sites	276
	12.4.10	Adaptive Evolution of Binding Sites	278
12.5	Toward a Dynamical Picture of the Genome		278
	12.5.1	Evolutionary Interactions Between Sites	279
	12.5.2	Site–Shadow Interactions	280
	12.5.3	Gene Interactions	280
	12.5.4	Evolutionary Innovations	281
References			281

Part V Populations

13 Drift and Selection in Evolving Interacting Systems
T. Ohta 285
13.1 Hierarchy of Networks 286
13.2 Drift and Selection, a Historical Perspective 287
13.3 Molecular Clock and Near-Neutrality 288
13.4 Mutants' Effects on Fitness 291
13.5 Evolution of Form and Shape: Cooption 294
References 296

14 Adaptation in Simple and Complex Fitness Landscapes
K. Jain and J. Krug 299
14.1 Basic Concepts and Models 300
  14.1.1 Fitness, Mutations, and Sequence Space 300
  14.1.2 Mutation–Selection Models 304
14.2 Simple Fitness Landscapes 307
  14.2.1 The Error Threshold: Preliminary Considerations 307
  14.2.2 Error Threshold in the Sharp Peak Landscape 308
  14.2.3 Exact Solution of a Sharp Peak Model 311
  14.2.4 Modifying the Shape of the Fitness Peak 312
  14.2.5 Beyond the Standard Model 317
14.3 Complex Fitness Landscapes 321
  14.3.1 An Explicit Genotype–Phenotype Map for RNA Sequences 322
  14.3.2 Uncorrelated Random Landscapes 322
  14.3.3 Correlated Landscapes 323

		14.3.4	Neutrality	326
14.4			Dynamics of Adaptation	327
		14.4.1	Peak Shifts and Punctuated Evolution	328
		14.4.2	Evolutionary Trajectories for the Quasispecies Model	328
		14.4.3	Dynamics in Smooth Fitness Landscapes	332
14.5			Evolution in the Laboratory	333
		14.5.1	RNA Evolution In Vitro	333
		14.5.2	Quasispecies Formation in RNA Viruses	334
		14.5.3	Dynamics of Microbial Evolution	334
14.6			Conclusions	335
References				336

15 Genetic Variability in RNA Viruses: Consequences in Epidemiology and in the Development of New Strategies for the Extinction of Infectivity
E. Lázaro 341

15.1	Introduction	341
15.2	Replication of RNA Viruses and Generation of Genetic Variability	343
15.3	Structure of Viral Populations	344
15.4	Viral Quasi-Species and Adaptation	345
15.5	Population Dynamics of Host–Pathogen Interactions	348
15.6	The Limit of the Error Rate	350
	15.6.1 Increases in the Error Rate of Replication. Lethal Mutagenesis As a New Antiviral Strategy	352
	15.6.2 Evolution of Viral Populations Through Successive Bottlenecks	355
15.7	Conclusions	359
References		360

Index 363

List of Contributors

Ugo Bastolla
Centro de Biologia Molecular
"Severo Ochoa"
Campus UAM, Cantoblanco
28049 Madrid, Spain
ubastolla@cbm.uam.es

Matteo Brilli
Dipartimento di Biologia
Animale e Genetica
Universitá di Firenze
Via Romana 17-19
50125 Firenze, Italy
matteo.brilli@dbag.unifi.it

Delphine Charif
Université Claude Bernard
Villerbanne Cedex
Lyon 1, France
Charif@biomserv.univ-lyon1.fr

Nikolay V. Dokholyan
Department of Biochemistry
and Biophysics
The University of North Carolina
at Chapel Hill, School of Medicine
NC 27599 Chapel Hill, USA
dokh@med.unc.edu

Renato Fani
Dipartimento di Biologia
Animale e Genetica
Universitá di Firenze
Via Romana 17-19
50125 Firenze, Italy
renato.fani@unifi.it

Juan Garcia Ranea
Department of Biochemistry
and Molecular Biology
University College London
Gower Street
WC1E 6BT London, UK
ranea@biochemistry.ucl.ac.uk

Laura Gómez-Valero
Institut Cavanilles de Biodiversitat
i Biologia Evolutiva
and Departament de Genètica
University of Valencia
Apartado Postal 22085
46071 Valencia, Spain
laura.gomez@uv.es

Kavita Jain
Department of Physics
of Complex Systems
Weizmann Institute of Science
PO Box 26
76100 Rehovot, Israel
kavita.jain@weizmann.ac.il

I. King Jordan
National Center
for Biotechnology Information
National Institutes of Health
8600 Rockville Pike
MD 20894 Bethesda, USA
jordan@ncbi.nlm.nih.gov

Joachim Krug
Institut für Theoretische Physik
Universität zu Köln
Zülpicher Str. 77
50937 Köln, Germany
krug@thp.uni-koeln.de

Michael Lässig
Universität zu Köln
Institut für Theoretische Physik
Zülpicher Str. 77
50937 Köln, Germany
lassig@thp.uni-koeln.de

Amparo Latorre
Institut Cavanilles de Biodiversitat
i Biologia Evolutiva
and Departament de Genètica
University of Valencia
Apartado Postal 22085
46071 Valencia, Spain
amparo.latorre@uv.es

Ester Lázaro
Centro de Astrobiologia
(INTA-CSIC)
Ctra de Ajalvir Km 4
Torrejón de Ardoz
28850 Madrid, Spain
lazarolm@inta.es

Arthur M. Lesk
Department of Biochemistry
and Molecular Biology
and the Huck Institutes
of the Life Sciences:
Genomics, Proteomics,
and Bioinformatics Institute
The Pennsylvania State University
512A Wartik Laboratory
PA 16802 University Park, USA
aml25@psu.edu

Pietro Lió
Computer Laboratory
University of Cambridge
15 JJ Thomson Avenue
CB3 0FD Cambridge, UK
pl219@cl.cam.ac.uk

Jean R. Lobry
Laboratoire de Biométrie et Biologie
Evolutive (UMR 5558), CNRS
Univ. Lyon 1, 43 bd 11 nov, 69622
Villeurbanne Cedex
France
HELIX, Unité de recherche INRIA.
Lobry@biomserv.univ-lyon1.fr

Leonardo Mariño-Ramírez
National Center
for Biotechnology Information
National Institutes of Health
8600 Rockville Pike
MD 20894 Bethesda, USA
marino@ncbi.nlm.nih.gov

Russell Marsden
Department of Biochemistry
and Molecular Biology
University College London
Gower Street
WC1E 6BT London, UK
marsden@biochemistry.ucl.ac.uk

Andres Moya
Institute Cavanilles for Biodiversity
and Evolutionary Biology
University of Valencia
Apartado Postal 22085
46071 Valencia, Spain
andres.moya@uv.es

Tomoko Ohta
National Institute of Genetics
Department of Population Genetics
Yata Str. 1111
411-8540 Mishima, Japan
tohta@lab.nig.ac.jp

Christine Orengo
Department of Biochemistry
and Molecular Biology
University College London
Gower Street
WC1E 6BT London, UK
orengo@biochemistry.ucl.ac.uk

Markus Porto
Technische Universität Darmstadt
Institut für Festkörperphysik
Hochschulstr. 8
64289 Darmstadt, Germany
porto@fkp.tu-darmstadt.de

H. Eduardo Roman
Universitá di Milano - Bicocca
Dipartimento di Fisica
Piazza della Scienza 3
20126 Milano, Italy
eduardo.roman@mib.infn.it

Peter Schuster
Institut für Theoretische Chemie
Universität Wien
Währingerstrasse 17
1090 Wien, Austria
pks@tbi.univie.ac.at

Eugene I. Shakhnovich
Department of Chemistry
and Chemical Biology
Harvard University
MA 02138, Cambridge, USA
eugene@belok.harvard.edu

James A. Shapiro
University of Chicago
Department of Biochemistry
and Molecular Biology
920 E. 58th Str.
IL 60637 Chicago, USA
jsha@uchicago.edu

Francisco J. Silva
Institut Cavanilles de Biodiversitat
i Biologia Evolutiva
and Departament de Genètica
University of Valencia
Apartado Postal 22085
46071 Valencia, Spain
francisco.silva@uv.es

Peter F. Stadler
Institut für Informatik
Universität Leipzig
Härtelstraße 16-18
04107 Leipzig, Germany
studla@bioinf.uni-leipzig.de

Michele Vendruscolo
University of Cambridge
Department of Chemistry
Lensfield Road
CB2 1EW Cambridge, UK
mv245@cam.ac.uk

Xuhua Xia
University of Ottawa, CAREG
and Biology Department
150 Louis Pasteur
P.O. Box 450, Station A, Ottawa
K1N 6N5 Ontario, Canada
xxia@uottawa.ca

Corin Yeats
Department of Biochemistry
and Molecular Biology
University College London
Gower Street
WC1E 6BT London, UK
yeats@biochem.ucl.ac.uk

Part I

Molecules: Proteins and RNA

1

Modeling Conformational Flexibility and Evolution of Structure: RNA as an Example

P. Schuster and P.F. Stadler

In this chapter, RNA secondary structures are used as an appropriate toy model to illustrate an application of the landscape concept to understand the molecular basis of structure formation, optimization, adaptation, and evolution in simple systems. Two classes of landscapes are considered (1) conformational landscapes mapping RNA conformations into free energies of formation and (2) sequence–structure mappings assigning minimum free energy structures to sequences. Even without referring to suboptimal conformations, optimization of RNA structures by mutation and selection reveals interesting features on the population level that can be interpreted by means of sequence–structure maps. The full power of the RNA model unfolds when sequence–structure maps and conformational landscapes are merged into a more advanced mapping that assigns a whole spectrum of conformations to the individual sequence. The scenario is complicated further – but at the same time made more realistic – by considering kinetic effects that allow for the assignment of two or more long-lived conformations, together with their suboptimal folds, to a single sequence. In this case, molecules can be designed, which fulfil multiple functions by switching back and forth from one stable conformation to the other or by changing conformation through allosteric binding of effectors. The evolution of noncoding RNAs is presented as an example for the application of landscape-based concepts.

1.1 Definition and Computation of RNA Structures

RNA sequences form structures under appropriate conditions consisting of aqueous solution at sufficiently low temperatures, approximately neutral pH, and ionic strength. In most of the sufficiently well studied examples RNA folding occurs in two steps [1, 2] (1) the formation of a flexible so-called secondary structure requiring monovalent counterions and (2) the folding of the secondary structure into a rigid 3D-structure in the presence on divalent ions, especially $Mg^{2\oplus}$ [3] (for an exception see [4]). Experimental determination of full spatial RNA structures is a hard task for crystallographers and NMR spectroscopists [5,6]. Prediction of 3D-structures is also an enormously complex problem and at least as demanding as in the case of proteins [7]. RNA secondary structures, however, in contrast to protein secondary structures, have a physical meaning as folding intermediates and are useful tools in the interpretation and prediction of RNA function. In addition, conventional RNA secondary structures (Sect. 1.1.1) can be represented as (restricted) strings over a three-letter alphabet and they are accessible, therefore, to combinatorial analysis and other techniques of discrete mathematics [8–10]. The discreteness of secondary structures allows for straightforward comparisons of the spaces of sequences, structures, and conformations and provides the insights into flexibility and robustness of RNA molecules. Moreover, RNA secondary structures and lattice protein models are at present the only biological objects for which conformational landscapes and sequence–structure maps can be computed and analyzed in complete detail. Therefore, this contribution will be exclusively dealing with them.

1.1.1 RNA Secondary Structures

A conventional RNA secondary structure[1] is a listing of base pairs that can be visualized by a planar graph. The nodes of the graph are nucleotides of the RNA molecule, $i \in \{1, 2, \ldots, n\}$ numbered consecutively along the chain (Fig. 1.1). The edges of the graph represent bonds between, nodes which fall into two classes: (1) the backbone, $\{i-(i+1) \; \forall i = 1, \ldots, n-1\}$, and (2) the base pairs. The two ends of the sequence (5'- and 3'-end) are chemically different. The backbone is completely defined for known n and hence a secondary structure is completely determined by a listing of base pairs, S, where a pair between i and j will be denoted by $i - j$. For a conventional secondary structure, the base pairs fulfil three conditions:

1. *Binary interaction restriction.* An individual nucleotide is either involved in one base pair or it is a single nucleotide forming no base pair.
2. *No nearest neighbor pair restriction.* Base pairs to nearest neighbors, $i - j$ with $j = i - 1$ or $j = i + 1$ are excluded.
3. *No pseudoknot restriction.* Two base pairs $i - j$ and $k - l$ with $i < j, i < k$ and $k < l$ are only accepted if either $i < k < l < j$ or $i < j < k < l$ are fulfilled – the second base pair is either enclosed by the first base pair or lies completely outside (Fig. 1.1).

Condition 1 forbids the formation of base triplets or higher interactions between nucleotides. Condition 2 is required for steric reasons because stereochemistry does not allow for pairing geometries between neighboring nucleotides. As we shall mention later, this condition is even more stringent in the

---

[1] "Conventional" means here that the structure is free of pseudoknots (Condition 3). Some other definitions include certain or all classes of pseudoknots.

1 Flexibility and Evolution of Structure 5

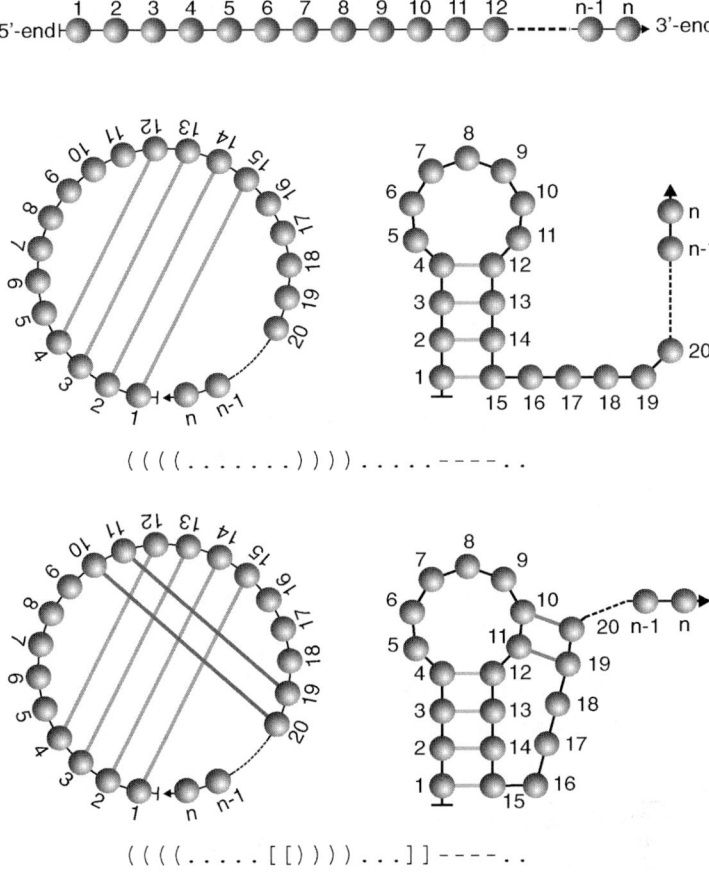

Fig. 1.1. Definition of RNA secondary structures. Each nucleotide inside the sequence forms two backbone bonds to its neighbors, the two nucleotides at the ends, 1 and n, are connected to one neighbor (topmost drawing: nucleotides are shown as *spheres*, the 3'-end is represented by an *arrow*). Each nucleotide can stay unpaired or form one (and only one) base pair to another nucleotide. In the circular representation of structures (left-hand side of the drawings in the *middle* and at the *bottom*), base pairs appear as *lines* crossing the circle. The upper secondary structures has no pseudoknot. The structure at the bottom contains a pseudoknot, which is easily recognized by crossings of *lines* in the circular representation. On the right-hand side of the two structures, we show the conventional drawings of secondary structures as they are used by biochemists and molecular biologists. Parentheses representations (see text) are shown below the two structures

sense that hairpin loops with less than three single nucleotides do not occur in real structures. Condition 3 is mainly a technical constraint, because the explicit consideration of pseudoknots impedes mathematical analysis of structures substantially and makes actual computations much more time consuming [11].

Throughout this chapter, it will be convenient to identify a secondary structure by its set of base pairs Ω. More abstractly, we consider Ω as an arbitrary matching on $\{1, \ldots, n\}$. In other words, we shall sometimes relax the conventional no-pseudoknot Condition 3 and insist only that each nucleotide takes part in at most one base pairs (Condition 1).[2] Furthermore, let Υ be the set of unpaired bases, which is the subset of $\{1, \ldots, n\}$ that is not met by the matching Ω.

The graphic representation of secondary structures is fully equivalent to other representations that we shall not discuss here except two, the adjacency matrix[3]

$$A = \left\{ a_{ij} = a_{ji} = \begin{cases} 1 & \text{if } i, j \in \Omega, \\ 0 & \text{otherwise}, \end{cases} \quad i, j = \{1, \ldots, n\} \right\}, \qquad (1.1)$$

and the parentheses notation, which will be used later on to calculate base pair probabilities and compute distances between structures, respectively. In this notation, single nucleotides, $i \in \Upsilon$, are represented by dots and base pairs by parentheses (Fig. 1.1). Structures are strings of length n over the three-letter alphabet, $\{., (,)\}$ with the restrictions that the number of left parentheses, "(," has to match exactly the number of right parentheses, ")," and no parenthesis must be closed before it had been opened. The no-pseudoknot restriction guarantees that left and right parentheses are assigned according to the rules of mathematics. Colored parentheses are required for the correct assignment in the presence of pseudoknots (bottom plot in Fig. 1.1).

Three classes of elements occur in structures (1) stacks, (2) various kinds of loops, and (3) external elements (Fig. 1.2). Stacks are arrays of consecutive base pairs in which the two strands run in opposite direction:

$$\begin{array}{cccccc} 5'\text{-end} \cdots & - & i & - & i+1 & - & i+2 & - & \cdots & 3'\text{-end} \\ & & | & & | & & | & & & \\ 3'\text{-end} \cdots & - & j & - & j-1 & - & j-2 & - & \cdots & 5'\text{-end} \,. \end{array}$$

Loops are commonly classified by the number of closing base pairs:[4]

(1) A loop of degree one has one closing base pair and is commonly called a hairpin loop.
(2) Loops of degree two are bulges or internal loops depending on the positioning of the two closing pairs. In bulges, the closing pairs are neighbors

---

[2] Wherever confusion is possible we shall be precise and use S for conventional secondary structures and Ω for the generalization.
[3] Here the backbone is excluded from the adjacency matrix but its makes no difference when it is considered too because the backbone does not change in superpositions of the structures discussed here.
[4] Each stack neighboring the loop ends in a pair is called a *closing pair* of the loop. The number of closing base pairs is easily determined: Imagine the loop as a circle and count all base pairs whose nucleotides are members of this circle.

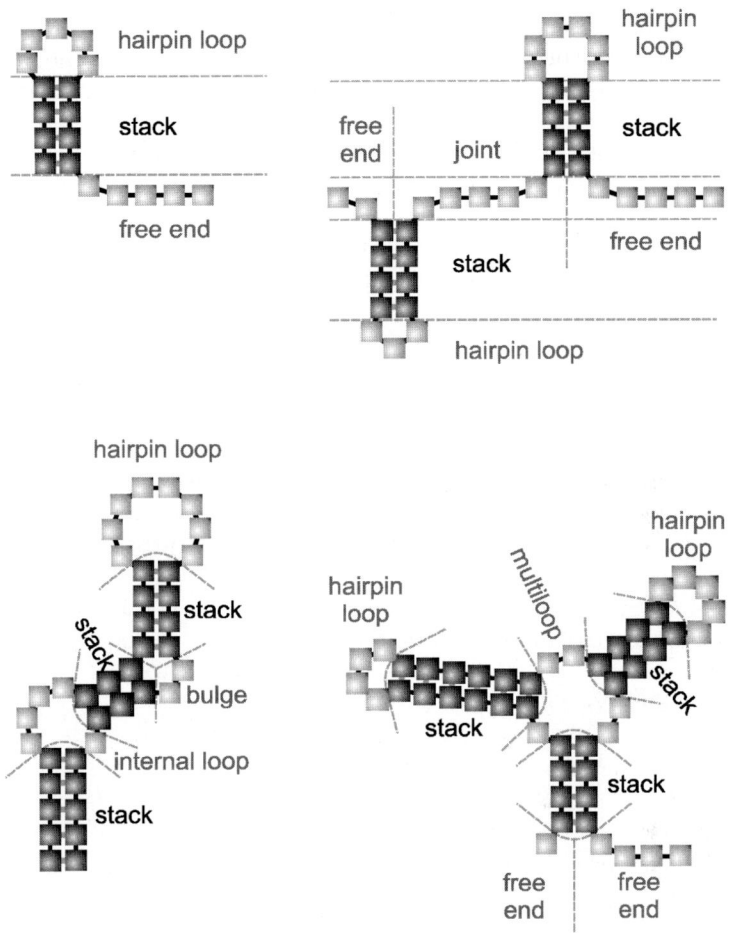

Fig. 1.2. Elements of RNA secondary structures. Three classes of structural elements are distinguished: (1) stacks (indicated by nucleotides in dark color), (2) loops, and (3) external elements being joints and free ends. Loops fall into several subclasses: Hairpin loops have one base pair, called the closing pair, in the loop. Bulges and internal loops have two closing pairs, and loops with three or more closing pairs are called multiloops

without a single nucleotide in between while they are separated by single bases on both sides in internal loops. Algorithmically, two stacked adjacent base pairs are treated as an interior loop without unpaired bases. Higher degree loops have three or more closing pairs and are called multiloops.
(3) Flexible substructures are free ends and parts of the nucleotide chain that join two modules of structure.

As indicated in Fig. 1.2 it is important for calculations of free energies that the individual substructures are independent in the sense that the free energy of a substructure is not changed by changes in the pairing pattern of another substructure.

It will turn out useful to introduce the notion of acceptable structures, which are a subset of the conventional structures [12]. Two restrictions are introduced that eliminate structures of high free energies, which are commonly well above the energy of the open chain (a) Condition 2 in the definition of secondary structures is made more stringent in the sense that base pairs to next nearest neighbors are also excluded, and hence the base pairs with the shortest distance along the sequence are $i - i+3$, and (b) isolated base pairs are excluded implying that the shortest stacking regions consists of at least two base pairs formed by neighboring bases.

1.1.2 Compatibility of Sequences and Structures

A sequence $X = (x_1 x_2 \cdots x_n)$ over an alphabet $\mathcal{A}$ with κ letters is *compatible* with the matching Ω if $\{i-j\} \in \Omega$ implies that $x_i x_j$ is an allowed base pair. This situation is expressed by $x_i x_j \in \mathcal{B}$. For natural RNAs, we have $\mathcal{A} = \{\alpha_i\} = \{A, C, G, U\}$ (or $\{A, T, G, C\}$ for DNA) and $\mathcal{B} = \{\beta_{ij} = \alpha_i - \alpha_j\} = \{AU, UA, GC, CG, GU, UG\}$. We denote the set of all sequences that are compatible with a structure Ω by

$$\mathbf{C}[\Omega] = \{X \mid \{i-j\} \in \Omega \implies x_i x_j \in \mathcal{B}\} \ . \tag{1.2}$$

Clearly, for each $i \in \Upsilon$ we may choose an arbitrary letter from the nucleic acid alphabet $\mathcal{A}$, while for each pair we may choose any of the ϱ base pairs contained in $\mathcal{B}$. For a given structure we have, therefore,

$$|\mathbf{C}[\Omega]| = \kappa^{|\Upsilon|} \varrho^{|\Omega|} \ , \tag{1.3}$$

compatible sequences.

The problem has a relevant inverse too: How many structures are compatible with a given sequence X? The set of these structures comprises all possible conformations, i.e., the minimum free energy structure together with the suboptimal structures. The computation of this number is rather involved and has to use a recursion that has some similarity to the computation of the minimum free energy structure (Sect. 1.1.4). It can be also obtained as the partition function [13] in the limit of infinite temperature, $T \to \infty$ (Sect. 1.1.6). A simpler estimate is possible in terms of the stickiness of the sequence,

$$p(X) = 2 \sum_{\beta_{ij} \in \mathcal{B}} p_i(X) p_j(X) \quad \text{with} \quad p_i(X) = \frac{n_i(X)}{n} \quad \text{and} \quad p_j(X) = \frac{n_j(X)}{n} \ ,$$

$$\tag{1.4}$$

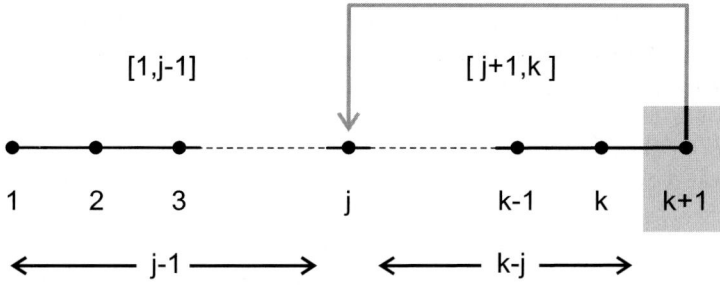

Fig. 1.3. Basic principle of recursions for secondary structures. The property of a sequence with chain length n is built up recursively from the properties of smaller segments under the assumption that the contributions are additive: The property for the segment $[1, k+1]$ is identical with that of the segment $[1, k]$ if the nucleotide x_{k+1} forms no base pair. If it forms a base pair with the nucleotide x_j the segment $[1, k+1]$ is bisected into two smaller fragments $[1, j-1]$ and $[j+1, k]$. The solution of a problem can be found by starting from the smallest segments and progressing successively to larger segments. This procedure leads either to a recursion formula (1.6, 1.7) or it can be converted into a dynamic programming algorithm as in the case of minimum free energy structure determination

where $n_i(X)$ and $n_j(X)$ are the numbers of nucleotides α_i and α_j in the sequence X, respectively, and $n = \sum_{\alpha_i \in \mathcal{A}} n_i(X)$, the chain length of the molecule.

On the basis of the assumption of additive contributions from structure elements, the properties associated with secondary structures can be computed in recursive manner from smaller to larger segments (Fig. 1.3). It is straightforward to enumerate, for example, all possible secondary structures for a given chain length n, s_n, by means of a recursion [14, 15]. For a minimal length for hairpin loops, $n_{\mathrm{lp}} \geq \lambda$, one finds [12, 16]:

$$s_{m+1} = s_m + \sum_{j=1}^{m-\lambda} s_{j-1} \cdot s_{m-j} = s_m + \sum_{j=\lambda}^{m-1} s_j s_{m-j-1}$$
$$\text{with} \quad s_0 = s_1 = \cdots = s_\lambda = 1 \,. \tag{1.5}$$

For a (random) sequence X with nucleotide composition $(p_1, \ldots, p_\kappa)$, the probability that two nucleotides form a base pair is given by the stickiness $p(X)$. Insertion into the recursion leads to [17]:

$$s_{m+1}(p) = s_m(p) + p \sum_{j=1}^{m-\lambda} s_{j-1}(p) \cdot s_{m-j}(p)$$
$$\text{with} \quad s_0(p) = s_1(p) = \cdots = s_\lambda(p) = 1 \,, \tag{1.6}$$

and $s_n(p)$ yields a rough estimate of the number of structures that are compatible with the sequence X. The recursion and the estimate can be extended to a restriction of the length of stacks, $n_{\text{st}} \geq \sigma$ [12]:

$$s_{m+1}(p) = \Xi_{m+1}(p) + \phi_{m-1}(p),$$

$$\Xi_{m+1}(p) = s_m(p) + \sum_{k=\lambda+2\sigma-2}^{m-2} \phi_k(p) \cdot s_{m-k-1}(p)$$

$$\phi_{m+1}(p) = p \sum_{k=\sigma-1}^{\lfloor (m-\lambda+1)/2 \rfloor} \Xi_{m-2k+1}(p) \cdot p^k \qquad (1.7)$$

with $s_0 = s_1 = \cdots = s_{\lambda+2\sigma-1} = 1$, $\phi_0 = \phi_1 = \cdots = \psi_{\lambda+2\sigma-3} = 0$, and $\Xi_0 = \Xi_1 = \cdots = \Xi_{\lambda+2\sigma-1} = 1$. Performing the recursion up to $m+1 = n$ provides us with a rough estimate for the numbers of secondary structures.

Physically acceptable suboptimal structures exclude hairpin loops with one or two single nucleotides and hence $\lambda = 3$. Since suboptimal conformations need not fulfil the criterion of negative free energies, no restriction on stack lengths is appropriate. For a minimum hairpin loop length of $\lambda = 3$ and $\sigma = 1$ we find the numbers collected in Table 1.1. The numbers of suboptimal structures become very large at moderate chain length n already. The expressions given here become asymptotically correct for long sequences. In order to provide a test for smaller chain lengths, we refer to one particular case where the number of suboptimal structures has been determined by exhaustive enumeration: The sequence

AAAGGGCACAGGGUGAUUUCAAUAAUUUUA

with $n = 30$ and $p = 0.4067$ has $1,416,661$ configurations and the estimate by means of the recursion (1.7) yields a value $s_{30}(0.4067) = 1.17 \times 10^6$ for $\lambda = 3$ and $\sigma = 1$ that is fairly close to the exact number.

Table 1.1. Estimates on the numbers of suboptimal structures, $s_n(p)$ with $\lambda = 3$ and $\sigma = 1$ and $p(X)$ being the stickiness of sequence X

Chain length	Stickiness $p(X)$			
(n)	1.0	0.5	0.375	0.25
10	65	21.4	14.3	8.6
20	1.07×10^5	7,403	2,778	787.8
50	1.82×10^{15}	1.27×10^{12}	8.52×10^{10}	2.57×10^9
100	6.32×10^{32}	2.09×10^{26}	8.05×10^{23}	5.81×10^{20}
200	2.07×10^{68}	1.55×10^{55}	1.95×10^{50}	8.06×10^{43}

1.1.3 Sequence Space, Shape Space, and Conformation Space

The analysis of relations between sequences and structures is facilitated by means of three formal discrete spaces (1) the sequence space being the space of all sequences of chain length n, (2) the shape space meant here as the space of all secondary structures that can be formed by sequences of chain length n, and (3) a conformation space containing all structures that can be formed by one particular sequence of chain length n.

Sequence Space

The sequence space is a metric space of cardinality κ^n with κ being the size of the alphabet. In addition to natural molecules built from the four-letter alphabet, $\{A, T, G, C\}$ for DNA and $\{A, U, G, C\}$ for RNA, sequences over three-letter, $\{A, U, G\}$ [18] and two letter, $\{D, U\}^5$ [19], alphabets were found to form perfect catalytic RNA molecules. Accordingly, we shall discuss also non-natural alphabets. The Hamming distance $d_H(X_1, X_2)$, defined as the number of positions in which two aligned sequences differ,[6] fulfills the three requirements of a metric on sequence space:

$$d_H(X_1, X_1) = 0 , \quad (1.8a)$$
$$d_H(X_1, X_2) = d_H(X_2, X_1) , \quad \text{and} \quad (1.8b)$$
$$d_H(X_1, X_3) \leq d_H(X_1, X_2) + d_H(X_2, X_3) . \quad (1.8c)$$

The Hamming metric corresponds to choosing the single point mutation as the elementary move in sequence space.

Shape Space

The shape space comprises all possible secondary structures of chain length n. The number of structures is given by recursion (1.6) with $p = 1$, or the recursion (1.7) with $p = 1$, in case physically meaningful restrictions are applied to the lengths of hairpin loops (n_{lp}) or stacks (n_{st}). It is also straightforward to define a distance between structures. Several choices are possible (Sect. 1.2.1), we shall make use of two of them because they correspond to move sets that are important in kinetic folding of RNA (1) the base pair distance, $d_P(S_j, S_j)$, and (2) the Hamming distance between the parentheses notations of structures, $d_H(S_j, S_j)$, (Fig. 1.4). The Hamming distance between structures is simply the number of positions in which the two strings representing the secondary structures differ whereas the base pair distance is twice the minimal number

---

[5] Because of weak bonding in the **A – U** pair adenine has been replaced by **D** being 2,6-diamino-purine in these studies.

[6] Unless stated otherwise we shall consider here binary end-to-end alignments of sequences with equal lengths.

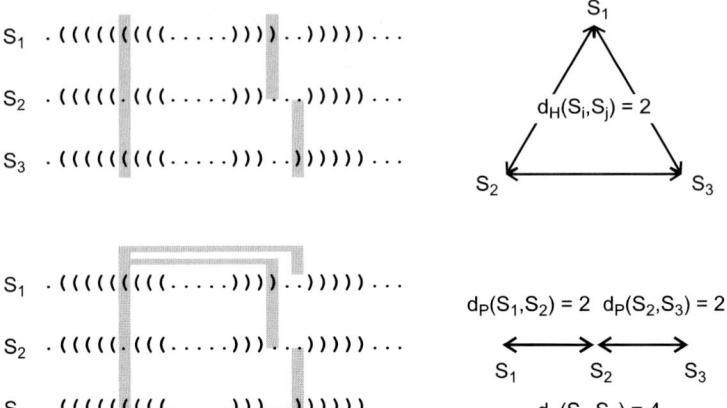

Fig. 1.4. Two measures of distances between secondary structures. The Hamming distance between parentheses notations of secondary structures is shown in the upper plot. Base pair opening and base pair closure contribute $d_H = 2$, but simultaneous opening and closing, corresponding to a shift of one or more nucleotides, leads also to the same distance and the three structures are equidistant in shape space with Hamming metric. If we use the base pair distance instead, we find also $d_P = 2$ for opening or closing of a base pair, but now the shift move is not in the move set and the two contributions for opening and closing add up to $d_P = 4$

of base pairs that have to be erased and formed to convert one structure into the other.[7] Figure 1.4 shows the difference between the two distances in a sequence of two consecutive steps (1) a base pair is removed in going from S_1 to S_2, and (2) a base pair is closed, which involves one of the two nucleotides that formed the pair in S_1, in the step from S_2 to S_3. In base pair distance, we have $d_P(S_1, S_3) = d_P(S_1, S_2) + d_P(S_2, S_3) = 4$, but in Hamming distance we find $d_H(S_1, S_3) = d_H(S_1, S_2) = d_H(S_2, S_3) = 2$. The interpretation is straightforward: The base pair distance corresponds to a set of two moves, base pair opening and base pair closure, whereas the Hamming distance corresponds to a larger move set that involves, in addition to single base pair operations, (synchronous) shifts of one or more base pairs resulting in the migration of a bulge, internal loop or other structural element.

Conformation Space

The conformational space refers to a single sequence (X) and contains all structures that are compatible with X. Accordingly, it is a subspace of shape space:

$$\mathbf{C}[X] = \{\Omega \,|\, X \in \mathbf{C}[\Omega_i]\} \ . \tag{1.9}$$

---

[7] To make the two measures of distance comparable base pair distances are multiplied by factor 2 since base pair involves two nucleotides.

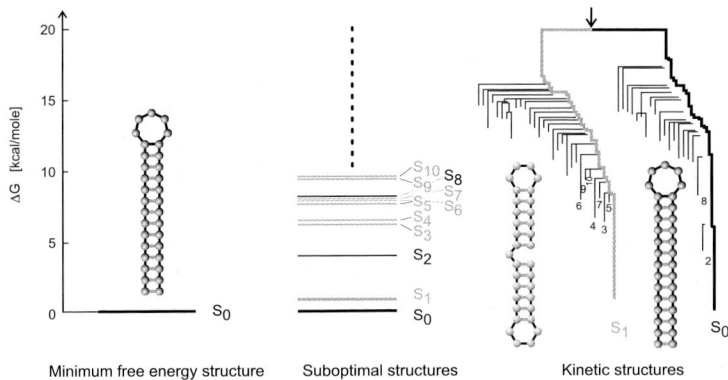

Fig. 1.5. Three notions of structures. The mfe-structure is shown as the only relevant conformation on the left-hand side corresponding in a formal sense to the zero temperature limit ($\lim T \to 0$). In the middle, we show the set of suboptimal structures as it is considered at equilibrium and temperature T in form of the partition function. The notion of the equilibrium structure implies the limit of infinite time ($\lim t \to \infty$). On the right-hand side, we show the barrier-tree of a molecule which exemplifies a situation that is encountered, for example, in RNA switches. At finite time we may find one or more long-lived conformations in addition to the mfe-structure

The conformation space is of particular importance for kinetic folding of RNA. In addition, it represents the structural diversity of conformations that is accessible from the ground state Ω_0 on excitation. The two move sets discussed in the context of a measure of distance on shape space are also relevant for conformational space since are tantamount to elementary moves in kinetic folding of RNA [20–22]. In Fig. 1.5, we show by means of a real example how the notion of RNA structure is extended to account for suboptimal foldings and kinetic effects. Conventional RNA folding assigns the minimum free energy (mfe) structure to the sequence. As we have seen above many suboptimal structures accompany the mfe-structure and contribute to the molecular properties in the sense of a Boltzmann ensemble. The partition function is the proper description of the RNA molecule at thermodynamic equilibrium or in the limit of infinite time. At finite time (Fig. 1.5; energy diagram on the right-hand side showing an RNA switch) the situation might be different and the RNA molecule may have one or more long-lived metastable conformation in addition to the mfe-structure. Then the actual molecular structure depends also on initial conditions and on the time window of the observation. The transitions between long-lived states are determined by the activation energies, which are shown in the construct of a barrier tree.[8]

---

[8] The barrier tree is a simplification of the conformational energy landscape and will be discussed in Sect. 1.2.2.

1.1.4 Computation of RNA Secondary Structures

Computation of secondary structures with minimum free energies [23] is based on the same principle as shown for counting the numbers of structures (Fig. 1.3). First, the free energies of the smallest possible substructures are taken or computed from a list of parameters, then a dynamic programming table of free energies is progressively completed by proceeding from smaller to larger segments until the minimum free energy of the whole molecule is obtained. Backtracking reveals the structure. The conventional approach is empirical and uses the free energies and enthalpies of RNA model compounds to derive the parameters for the individual structural elements. These elements correspond to the substructures shown in Fig. 1.2 at sufficiently high resolution for sequence specific contributions.

As an example, we show the free stacking energy of a cluster of **GC**-pairs in Fig. 1.6, which is obtained from three free stacking energy parameters for the **GC**-pairs interacting at different geometries. On total, 21 different free stacking energy parameters are required for the six base pairs. To be able to compute the temperature dependence, 21 stacking enthalpy parameters are required in addition. Loops are taken into account with loop size dependent parameters and hairpin loops, bulges, internal loops, and multiloops are treated differently. Other parameters consider nucleotides stacking on top of regular stacks, especially stable configurations, for example tetraloops[9] with specific sequences, end-on-end stacking of stacks, etc. Stacks are (almost) the only structure stabilizing elements, because base pair stacking is a contribution with substantial negative free energy. Further structure stabilization comes from single bases stacking on stacks called "dangling ends" and some

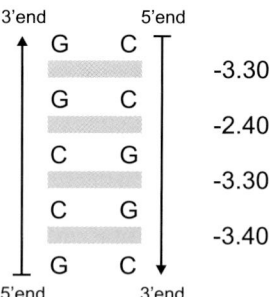

Fig. 1.6. The stacking parameters for the interaction between **GC** base pairs. Free energies of stacking are given for the three different interaction geometries (the first and the third paired pairs are identical). Values are given in kcal mol^{-1}. Additivity is assumed and therefore, we obtain a free energy of interaction of $\Delta G = -12.40$ kcal mol^{-1} for the stack of five pairs

---

[9] It is common to indicate the size of small hairpin loops by special wording: "triloops" are hairpin loops with three single nucleotides in the loop, "tetraloops" have four, and "pentaloops" five singles bases.

other sequence specific contributions. Loops are almost always destabilizing because of the entropic effect of the ring closure that freezes degrees of internal rotation.

Listings of parameters, which are updated every few years, can be found in the literature [24–27]. These parameters enter an energy function $E(X;\Omega)$ that assigns a unique free energy value to every substructure and provides the tool for completing the entries in the dynamic programming table. Several software packages are available and web servers make secondary structure calculations easily accessible for everybody (see, for example, the Vienna RNA package and the Vienna RNA server [28, 29]).

1.1.5 Mapping Sequences into Structures

The numbers of physically accessible structures obtained from the recursion (1.7) are compared in Table 1.2 with the actual numbers of minimum free energy structures computed by means of a folding routine. To this end, all sequences of a chain length n were folded, grouped with respect to structures, and enumerated. The numbers refer to structures without single base pairs. Exhaustive folding of entire sequence spaces was performed for five different alphabets: **GC**, **UGC**, **AUGC**, **AUG**, and **AU**. As follows directly from the table, the mapping $\Omega = f(X)$ is many-to-one in all five alphabets. The set of sequences that form a given matching Ω, the preimage of Ω in sequence space

$$\mathbf{G}[\Omega] = f^{-1}(\Omega) \doteq \{X | f(X) = \Omega\}, \quad (1.10)$$

is turned into a graph, the neutral network G, by connecting all pairs of nodes with Hamming distance one by an edge. Global properties of neutral networks are derived by means of random graph theory [30]. The characteristic quantity for a neutral network is the degree of neutrality $\bar{\lambda}$, which is obtained by averaging the fraction of Hamming distance one neighbors that form the same minimum free energy structure, $\lambda_X = N_{\text{ntr}}^{(1)} / (n \cdot (\kappa - 1))$ with $N_{\text{ntr}}^{(1)}$ being the number of neutral one-error neighbors, over the whole network, $\mathbf{G}[\Omega]$:

$$\bar{\lambda}[\Omega] = \frac{1}{|\mathbf{G}(\Omega)|} \sum_{X \in \mathbf{G}[\Omega]} \lambda_X. \quad (1.11)$$

Connectedness of neutral networks is, among other properties, determined by the degree of neutrality [31]:

$$\text{With probability one network is} \begin{cases} \text{connected} & \text{if } \bar{\lambda} > \lambda_{\text{cr}} \\ \text{not connected} & \text{if } \bar{\lambda} < \lambda_{\text{cr}}, \end{cases} \quad (1.12)$$

$$\text{where } \lambda_{\text{cr}} = 1 - \kappa^{-1/(\kappa - 1)}.$$

Computations yield $\lambda_{\text{cr}} = 0.5$, 0.423, and 0.370 for the critical value in two-, three-, and four-letter alphabets, respectively.

Table 1.2. Comparison of exhaustively folded sequence spaces

Chain length	Number of sequences			Number of structures				
(n)	2^n	4^n	$s_n(1)$	GC	UGC	AUGC	AUG	AU
7	128	1.64×10^4	2	1	1	1	1	1
8	256	6.55×10^4	4	3	3	3	2	1
9	512	2.62×10^5	8	7	7	7	3	1
10	1,024	1.05×10^6	14	13	13	13	5	3
12	4,096	1.68×10^7	37	35	35	36	14	8
14	1.64×10^4	2.68×10^7	101	83	89	93	31	20
16	6.55×10^4	4.29×10^9	304	214	246	260	72	44
18	2.62×10^5	6.87×10^{10}	919	582	735		180	96
20	1.05×10^6	1.10×10^{12}	2,741	1,599	2,146		504	232
25	3.36×10^7	1.13×10^{15}	44,695	18,400				1,471
30	1.07×10^9	1.15×10^{18}	760,983	218,318				21,315

The values are derived through exhaustive folding of all sequences of chain length n from a given alphabet. The numbers refer to actually occurring minimum free energy structures (open chain included) without isolated base pairs and are directly comparable to the total numbers of acceptable structures $s_n(1)$ with $\lambda = 3$ and $\sigma = 2$ as computed from the recursion (1.7) [12]. The parameters are taken from [25]

Random graph theory predicts a single largest component for nonconnected networks, i.e. networks below threshold, that is commonly called the "giant component." Real neutral networks derived from RNA secondary structures may deviate from the prediction of random graph theory in the sense that they have two or four equally sized largest components. This deviation is readily explained by nonuniform distribution of the sequences belonging to **G**[S_k] in sequence space caused by specific structural properties of S_k [32,33]. In particular, sequences that fold into structures, which allow for closure of additional base pairs at the ends of the stacks, are more probable to be formed by sequences that have an excess of one of the two bases forming a base pair than by those with the uniform distribution: $x_G = x_C$ and $x_A = x_U$. In case of **GC**-sequences, the neutral network is then depleted from sequences in the middle of sequence space and we find two largest components, one at excess **G** and one at excess **C**.

In Table 1.3 we show, as an example, computed values of the degree of neutrality, $\bar{\lambda}[S]$ in neutral networks derived from tRNA-like cloverleaf structures with different stack lengths of the hairpin loops. The most striking feature of the data is the weak structure dependence of $\bar{\lambda}[S]$ with a family: For a given alphabet the cloverleafs S_1, S_2, S_3, and S_4, have almost the same $\bar{\lambda}$ values irrespective of the stability of the corresponding folds. Because of the shorter stack lengths in S_1, S_2 and S_3 and the weakness of the **AU** pair no

Table 1.3. Degree of neutrality in different nucleotide alphabets

Structure[a]	Nucleotide alphabet				
	GC	UGC	AUGC	AUG	AU
S_1	0.05 ± 0.03	0.26 ± 0.07	0.28 ± 0.06	–	–
S_2	0.06 ± 0.03	0.26 ± 0.07	0.28 ± 0.06	0.22 ± 0.05	–
S_3	0.06 ± 0.03	0.25 ± 0.07	0.29 ± 0.06	0.21 ± 0.06	–
S_4	0.07 ± 0.03	0.25 ± 0.06	0.31 ± 0.06	0.20 ± 0.06	0.07 ± 0.03

The values for the degree of neutrality, $\bar{\lambda}$, were obtained by sampling 1,000 random sequences folding into the four cloverleaf structures with different stack sizes[a] using the inverse folding routine [28]. [a] The following cloverleaf structures were used:

```
S1: ((((((...((((.......)))).((((......)))).....((((......)))).))))))....
S2: ((((((...((((.....)))).((((......)))).....((((......)))).))))))....
S3: ((((((...((((.....)))).((((......)))).....(((((.....)))))).))))))....
S4: ((((((...(((((....))))).((((((.....)))))).....(((((.....)))))).))))))....
```

AU-sequences forming these structures were obtained by inverse folding. The same was found for S_1 in case of **AUG**-sequences. Considering the fact that λ_{cr} decreases from two to four-letter alphabets, we see that neutral networks in two-letter sequence spaces ($\bar{\lambda} \approx 0.06$ and $\lambda_{\text{cr}} = 0.5$) and four-letter sequence spaces ($\bar{\lambda} \approx 0.3$ and $\lambda_{\text{cr}} = 0.37$) must have very different extensions, the former being certainly non connected and whereas the latter come close to threshold.

The extension of neutral networks can be visualized also by evaluating the lengths of neutral path. A neutral path connects pairs of neighboring neutral sequences of Hamming distance $d_{\text{H}} = 1$ for single nucleotide exchanges or $d_{\text{H}} = 1, 2$ for base pair exchange with the condition that the Hamming distance from a reference sequence increases monotonously along the path. The path ends when it reaches a sequence, which has only neutral neighbors that are closer to the reference sequence. Table 1.4 compares the degree of neutrality and the length of neutral path for **GC** and **AUGC** sequences of chain length $n = 100$ with the expected result: Networks in **AUGC** space extend through whole sequence space whereas **GC** networks sustain neutral path of roughly only half of this length. The table also contains comparisons with constrained molecules that were cofolded with one or two fixed sequences. The three values demonstrate the influence of multiple constraints on neutrality, which lead to a decrease in both, degree of neutrality and length of neutral path, and provide an explanation why the (almost unconstrained) ribozymes of Schultes and Bartel [35] stay functional along very long neutral paths whereas functional tRNAs, which have to fulfil multiple constraints, tolerate only very limited variability in their sequences.

Table 1.4. The lengths of neutral paths through sequence space

Molecule	Alphabet	Degree of neutrality ($\bar{\lambda}$)	Neutral path length $\bar{d}_\mathrm{H}(X_0, X_\mathrm{f})$
Single fold	**GC**	0.08	≈45
Single fold	**AUGC**	0.33	>95
Cofold with one sequence	**AUGC**	0.32	75
Cofold with two sequences	**AUGC**	0.18	40

The degree of neutrality, $\bar{\lambda}$, and the mean lengths of neutral paths through sequence space, $\bar{d}_\mathrm{H}(X_0, X_\mathrm{f})$ (with X_0 being the initial and X_f the last sequence), is compared for three examples (1) folding of (stand alone) **AUGC** sequences of chain lengths $n = 100$, (2) cofolding of **AUGC** sequences of chain lengths $n = 100$ with a single fixed sequence, and (3) cofolding of **AUGC** sequences of chain lengths $n = 100$ with two single fixed sequences. The values represent averages over samples of 1,200 random sequences. The value for the path length in **GC** sequence space with $n = 100$ is an estimate from Fig. 10 in [34].

The existence of neutral networks and neutral paths in real RNA molecules has been demonstrated by several experimental studies on selection of RNA molecules with predefined properties (e.g., [36, 37]). Several theoretical investigations were also dealing with random pools of RNA sequences [38–41] and showed, for example, that natural RNA molecules have lower free folding energies than the average of random energies thus demonstrating the effect of evolutionary selection for stable structures.

1.1.6 Suboptimal Structures and Partition Functions

Algorithms for the computation of suboptimal conformations have been developed and two of them are frequently used [42, 43]. As we have already seen from our estimate, the numbers of suboptimal states are very large and, moreover, they increase exponentially with chain length n. The latter of the two algorithms [43] has been designed for the calculation of all conformations within a given energy band above the mfe and adopts a technique originally proposed for suboptimal alignments of sequences [44]. The algorithm starts from the same dynamic programming table as the conventional mfe conformation but considers all backtracking results within the mentioned energy band. As indicated in Fig. 1.5, the set of structures, mfe and suboptimal conformations $\{S_0, S_1, S_2, \ldots\}$, is ordered since their free energies, $\{\varepsilon_0, \varepsilon_1, \varepsilon_2, \ldots\}$ fulfill the relation $\varepsilon_0 \leq \varepsilon_1 \leq \varepsilon_2 \ldots$.

At equilibrium and temperature T, the individual conformations form a Boltzmann ensemble that contains a structure S_j with the Boltzmann weight

$\gamma_j = g_j \exp\bigl(-(\varepsilon_j - \varepsilon_0)/RT\bigr)/Q(T)$, where R is the Boltzmann constant for one mole, $R = N_\mathrm{L} \cdot k_\mathrm{B}$, and $Q(T)$ is the partition function[10]

$$Q(T) = \sum_i g_i \exp\bigl(-(\varepsilon_i - \varepsilon_0)/RT\bigr). \tag{1.13}$$

Instead of having a structure with a set of defined base pairs, the ground state is now described by a temperature-dependent linear combination of states where the weighted superposition of base pairs gives rise to base pairing probabilities $p_{ij}(X,T)$ which are the elements of the matrix

$$P(X,T) = \sum_k \gamma_k A(S_k) \quad \text{or} \quad p_{ij}(X,T) = \sum_k \gamma_k a_{ij}(S_k), \tag{1.14}$$

which is a Boltzmann weighted superposition of the adjacency matrices (1.1) of the individual structures with the following properties: In the limit $T \to 0$, the base pairing probabilities converge to the base pairing pattern of S_0 (for a nondegenerate ground state, $\varepsilon_0 < \varepsilon_1$) as described by the adjacency matrix $A(S_0)$ and in the limit $T \to \infty$ all (micro)states have equal weights and the partition function converges to the total number of all conformations of the sequence X. An elegant algorithm that computes the partition function $Q(T)$ directly by dynamic programming is found in [13]. It has been incorporated into the Vienna RNA package [28].

1.2 Design of RNA Structures

The design of RNA molecules boils down to finding sequences that fold into molecules with predefined structures and properties. Consequently, an algorithm is needed that computes sequences that fold into predefined mfe structures. The required procedure thus corresponds to an inversion of the conventional folding procedure.

1.2.1 Inverse Folding

Given a sequence X, the *folding problem* consists in finding a matching Ω that minimizes an energy function $E(X;\Omega)$ and (if desired) satisfies other constraints, such as the no-pseudoknot condition. In Sect. 1.1.4, we have seen that the folding problem for pseudoknot-free secondary structures is easily solved by means of dynamic programming.

In the *inverse folding problem*, we have the same energy function E and the same constraints, but we are given the structure Ω and search for a sequence

---

[10] Sometimes different microstates S_i with the same free energy ε_j are lumped together to form one "mesoscopic" state in the partition function and then the factor g_j accounts for this degeneracy.

X that has Ω as an optimal structure. We denote the set of solutions of the inverse folding problem by $f^{-1}(\Omega)$. Note that $f^{-1}(\Omega)$ may be empty, since there are logically possible secondary structures that are not formed as minimum energy structures of any sequence.

Just as the folding problem can be regarded as an optimization problem on the energy landscape of a given sequence, we can also rephrase the inverse folding problem as a combinatorial optimization problem. To this end, we consider a measure $D(\Omega_1, \Omega_2)$ for the structural dissimilarity of two RNA secondary structures Ω_1, Ω_2. A variety of such distance measures have been described in the literature [28, 45–48]. Since we will be interested here only in the sequences of equal length, we may simply use the cardinality of the symmetric difference of Ω_1 or in Ω_2:

$$D(\Omega_1, \Omega_2) = \big|(\Omega_1 \cup \Omega_2) \setminus (\Omega_1 \cap \Omega_2)\big|. \tag{1.15}$$

Clearly, sequence X folds into structure Ω, if and only if $\Xi(X) = D(\Omega, f(X)) = 0$. Hence, inverse folding translates into minimizing D over all sequences. We know a priori that solutions to the inverse folding problem must be compatible with the structure:

$$f^{-1}(\Omega) \subseteq \mathbf{C}[\Omega]. \tag{1.16}$$

It is straightforward to modify this approach to search, for instance, for sequences in which the ground state is much more stable than any structural alternative [28]: Let $E(X; \Omega)$ be the energy of structure Ω for sequence X, and let $G(X)$ be the ensemble free energy of sequence X, which can be computed by McCaskill's algorithm [13]. Sequences with the desired property minimize

$$\Xi(X) = E(X; \Omega) - G(X) = -RT \ln \gamma_X(\Omega), \tag{1.17}$$

where $\gamma_X(\Omega)$ is the probability of structure Ω in the Boltzmann ensemble of sequence X.

It has been found empirically [28] that this combinatorial optimization problem is easily solvable by means of adaptive walks. Starting from a randomly chosen initial sequence X_0, we produce mutants by exchanging a nucleotide at the unpaired positions Υ or by replacing one of the six pairing combinations by another one in a pair in Ω. A mutant is accepted if the cost function $\Xi(X)$ decreases. In a more sophisticated version, implemented in the program RNAinverse, a significant speedup is achieved by optimizing parts of the structure individually. This reduces the number of evaluations of the folding procedure for long sequences. A more sophisticated stochastic local search algorithm is used in the RNA-SSD software [49].

1.2.2 Multiconformational RNAs

Figure 1.5 indicates that the energy surface of a typical RNA sequence has a large number of local minima with often high energy barriers separating different basins of attraction. Thus non-native conformations can have energies comparable to the ground state, and they can be separated from the native state by very high energy barriers. Stable alternative conformations have been observed experimentally for a variety of RNA molecules [50–53].

Alternative conformations of the same RNA sometimes determine completely different functions [54, 55]. SV11, for instance, is a relatively small molecule that is replicated by Qβ replicase [56, 57]. It exists in two major conformations, a metastable multicomponent structure and a rod-like conformation, constituting the stable state, separated by a huge energy barrier. While the metastable conformation is a template for Qβ replicase, the ground state is not. By melting and rapid quenching the molecule can be reverted from the inactive stable to the active metastable form [58]. Another, particularly impressive, example is a designed sequence that can satisfy the base-pairing requirements of both the hepatitis delta virus self-cleaving ribozyme and an artificially selected self-ligating ribozyme, which have no base pairs in common. This *intersection sequence* displays catalytic activity for both cleavage and ligation reactions [35].

To deal with multiple conformations, we consider a collection of structures (matchings) $\Omega_1, \Omega_2, \ldots, \Omega_k$ on the same sequence X. The fundamental question in this context is whether there is a sequence in

$$\mathbf{C}[\Omega_1, \Omega_2, \ldots, \Omega_k] = \bigcap_{j=1}^{k} \mathbf{C}[\Omega_j] \tag{1.18}$$

and if so, what is the size of this intersection of sets of compatible sequences. To answer this question, it is useful to consider the graph Ψ with vertex set $\{1, \ldots, n\}$ and edge set $\bigcup_{j=1}^{k} \Omega_j$.

Generalized Intersection Theorem

Suppose $\mathcal{B} \subseteq \mathcal{A} \times \mathcal{A}$ contains at least one symmetric pair, i.e., xy $\in \mathcal{B}$ implies yx $\in \mathcal{B}$. Then

(1) $\mathbf{C}[\Omega_1, \ldots, \Omega_k] \neq \emptyset$ if Ψ is bipartite.
   For $k = 2$, Ψ is a disjoint union of paths and cycles with even length, and hence always bipartite.
(2) The number of sequences that are compatible with all structures can be written in the form

$$|\mathbf{C}[\Omega_1, \Omega_2, \ldots, \Omega_k]| = \prod_{\text{components } \psi \text{ of } \Psi} F(\psi), \tag{1.19}$$

where $F(\psi)$ is the number of sequences that are compatible with the connected component ψ.

(3) For the biophysical alphabet holds: $\bigcap_j \mathbf{C}[\Omega_j] \neq \emptyset$ if and only if Ψ is a bipartite graph.
   In particular, for the case of bistable sequences, $k = 2$, we can express the size of the intersection explicitly in terms of Fibonacci numbers

$$F(P_k) = 2(\text{Fib}(k) + \text{Fib}(k+1)) = 2\text{Fib}(n+2) \quad (1.20)$$
$$F(C_k) = 2(\text{Fib}(k-1) + \text{Fib}(k+1)), \quad (1.21)$$

where P_k and C_k are path and cycle components of Ψ with k vertices.

For a proof of these propositions see [31, 59]. Interestingly, for two structures there is always a nonempty intersection $\mathbf{C}[\Omega_1] \cap \mathbf{C}[\Omega_2]$. In contrast, the chance that the intersection of three randomly chosen structures in nonempty decreases exponentially with sequence length [60]. Recently, an alternative attempt has been made to extend the design aspect of the intersection theorem to three or more sequences [61].

Given a collection of alternative secondary structures, we can again ask the *inverse folding* or *sequence design* question. For simplicity, we restrict ourselves to two structures Ω_1 and Ω_2 here. For example, one might be interested in sequences that have two prescribed structures Ω_1 and Ω_2 as stable local energy minima with roughly equal energy, and for which the energy barrier between these two minima is roughly ΔE. It is not hard to design a cost function $\Xi(X)$ for this problem. In [59], the following ansatz has been used successfully:

$$\Xi(X) = E(X, \Omega_1) + E(X, \Omega_2) - 2G(X) + \xi \left(E(X, \Omega_1) - E(X, \Omega_2)\right)^2$$
$$+ \zeta \left(B(X, \Omega_1, \Omega_2) - \Delta E\right)^2. \quad (1.22)$$

Here, $B(X, \Omega_1, \Omega_2)$ is the energy barrier between the two conformations Ω_1, Ω_2, which can be readily computed from the barrier tree of the sequence X.

1.2.3 Riboswitches

The capability of RNA molecules to form multiple (meta)-stable conformations with different function is used in nature to implement so called *molecular switches* that regulate and control the flow of a number of biological processes. Gene expression, for example, can be regulated when the two mutually exclusive structural alternatives correspond to an active and in-active conformation of the transcript [62]. Mechanistically, one fold of the mRNA, the repressing conformation, contains a terminator hairpin or some other structural element, which conceals the translation initiation site, whereas in the alternative conformation the gene can be expressed [63]. The switching between two competing RNA conformations can be triggered by molecular events such as the binding of a target metabolite.

The best-known example of such a behavior are the riboswitches [64]. These are autonomous structural elements primarily found within the 5'-UTRs of bacterial mRNAs, which, upon direct binding of small organic molecules, can trigger conformational changes, leading to an alteration of the expression for the downstream located gene. Their general architecture shows two modular units [65], a "sensor" for a small metabolite and a unit which "interprets" the signal from the "sensor" unit and interfaces to those RNA elements involved in gene expression regulation. The size of the "sensor"-unit ranges typically from 70 to 170 nucleotides, which is unexpectedly large compared with artificial aptamers obtained by in vitro directed evolution experiments. Riboswitches regulate several key metabolic pathways [66, 67] in bacteria including those leading to coenzyme B_{12}, thiamine, pyrophosphate, flavin monophosphate, S-adenosylmethionine, and a couple of important amino acids. The search for additional elements is ongoing, e.g., [68, 69]. Riboswitches and engineered allosteric ribozymes [70, 71] demonstrate impressively that RNA is indeed capable of maintaining and regulating a complex metabolic state without the help of proteins.

1.3 Processes in Conformation, Sequence, and Shape Space

Kinetic folding and evolutionary optimization of RNA molecules are considered as stochastic processes, in particular as constrained walks in conformation and sequence and/or shape space. We present a brief overview of the basic concepts and then consider the evolution of noncoding RNA molecules as one actual and particular interesting example.

1.3.1 Kinetic Folding

Kinetic folding of RNA molecules can be understood and modeled as a stochastic process in RNA conformation space. The process corresponds to a time-ordered series of secondary structures, a trajectory

$$\Omega_0 \to \Omega_1 \to \Omega_2 \to \cdots \to \Omega_T , \qquad (1.23)$$

where initial and target structures, Ω_0 and Ω_T, may be chosen at will. Commonly, $\Omega_0 = \mathbf{O}$ and $\Omega_T = S_0$ are used corresponding to the open chain and the mfe-structure, respectively. Individual trajectories (1.23) may contain loops, i.e., the same structure may be visited two or more times. In general, it is of advantage to define the target conformation as an absorbing state. Leaving the target state unconstrained causes the trajectory to approach a thermodynamic ensemble in the sense that it visits the individual conformations with frequencies according to the Boltzmann weights. For practical purposes, the time required to fulfil the condition of ergodicity, however, is prohibitively long. Basic to the stochastic process is a set of moves that defines the allowed transitions between conformations. In the simplest case, it

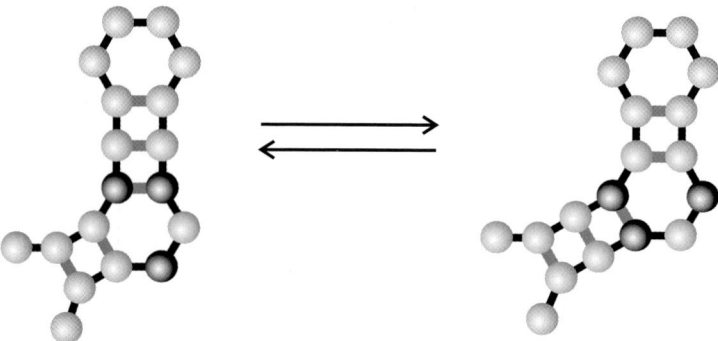

Fig. 1.7. The shift move in kinetic RNA folding. The shift move is a combination of base pair opening and base pair closure that occurs simultaneously. The requirement for an allowed shift move is that it takes place within one substructure element, bulge, internal loop or multiloop. Shifts involving free ends are also considered legitimate

contains base pair closure and base pair opening according to the conventional secondary structure rules (Conditions 1–3). Such a move set corresponds to the base pair distance, d_P, as metric in shape space (Fig. 1.4). It turned out to be important to introduce also a shift move (Fig. 1.7) since the trajectories approach the target much faster then [20]. If the move set is extended to simultaneous shifts of as many nucleotides as possible within a given substructure element, the set has the Hamming metric between parentheses notation of structures, $d_H(S_i, S_j)$ (Fig. 1.4), as proper measure of distance.

The stochastic process (1.23) can also be described by a master equation for the probabilities of the ensemble: $P_k(t)$ is the probability to observe the conformation S_k at time t. The time derivatives fulfil the equation

$$\frac{dP_k}{dt} = \sum_{i=0}^{m+1}(P_{ik}(t) - P_{ki}(t)) = \sum_{i=0}^{m+1} k_{ik} P_i - P_k \sum_{i=0}^{m+1} k_{ik}$$

with $\quad k = 0, 1, \ldots, m+1 \quad$ and $\quad i \to k \in$ move set, $\hfill (1.24)$

where we assume that the open chain conformation **O** is not part of the suboptimal conformations, $S_1, \ldots, S_m$. The transition probabilities are computed from the free energies of the conformations

$$P_{ik}(t) = k_{ik} P_i(t) = P_i(t)\, e^{-(g_k - g_i)/(2RT)}/\Sigma_i , \hfill (1.25)$$
$$P_{ki}(t) = k_{ki} P_k(t) = P_k(t)\, e^{-(g_i - g_k)/(2RT)}/\Sigma_k , \hfill (1.26)$$

with $\quad \Sigma_j = \sum_{i=0, i\neq j}^{m+1} \exp\left(-(g_j - g_i)/(2RT)\right) .$

To avoid the necessity of additional parameters the free energies are taken from the suboptimal foldings. Calibration of the time scale occurs through

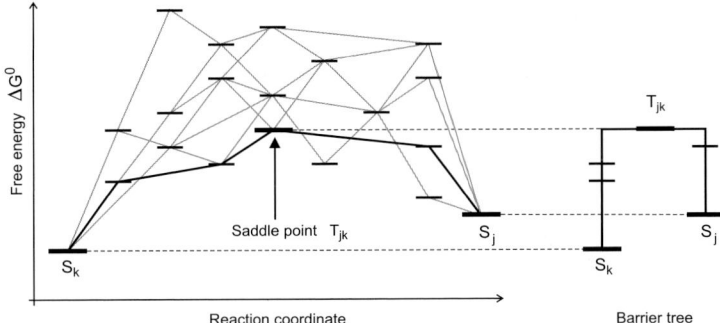

Fig. 1.8. Construction of barrier trees. The set of suboptimal conformations is related by a move set as shown in the left-hand part of the sketch. The barrier tree is derived from the set of suboptimal structures by eliminating all conformations except local minima of the free energy surface and minima connecting saddle points of lowest free energy. We remark that the set of local minima depends on the choice of the move set, although important local minima are very unlikely to be changed on physically meaningful alterations of the move set

adjusting the folding kinetics of a model system to the experimental data. Although it is straightforward to solve the master equation (1.24) by means of an eigenvalue problem, practical difficulties arise from the enormously high number of suboptimal conformations determining the dimensionality of the system [72].

A simplification of full kinetic folding is introduced in the form of "barrier trees" (Fig. 1.8). All suboptimal conformations that do neither represent a local minimum of the conformational energy landscape nor a lowest energy transition state between two local minima are neglected. The remaining barrier tree can be used to simulate kinetic folding by means of conventional Arrhenius kinetics. The results are often in astonishingly good agreement with the exact computations based on (1.24). Cases of less satisfactory agreement can be predicted [72].

1.3.2 Evolutionary Optimization

Evolution of RNA molecules based on replication, mutation, and selection in constant environment can be described by an ODE [73]:

$$\frac{dx_i}{dt} = \sum_{k=1}^{m} f_k Q_{ki} x_k - x_i \phi(t), \quad i = 1, \ldots, m,$$

$$\phi(t) = \sum_{k=1}^{m} f_k x_k(t). \tag{1.27}$$

Herein the concentrations of individual RNA sequences are denoted by $x_i = [X_i]$ and Q_{ij} are the elements of a mutation matrix whose elements, in the simplest case of the uniform error rate assumption, can be expressed by an (average) error rate p per site and replication.

$$Q_{ij} = p^{d_H(X_i,X_j)} \cdot (1-p)^{n-d_H(X_i,X_j)}. \tag{1.28}$$

The mutation probability thus is only a function of the error rate and the Hamming distance $d_H(X_i, X_j)$ between the two sequences involved. The results of the analysis of replication–mutation kinetics have been presented and discussed extensively [74–77] and we dispense here from repeating them. Kinetic differential equations refer to infinite population size and accordingly, a different description is required for the study of finite size effects on evolutionary optimization. In addition, population dynamics is considered as a process taking place exclusively in sequence space and structural properties enter the model as parameters only.

Replication and mutation of RNA molecules leading to selection in confined populations have indeed been studied also in finite populations. The best-suited stochastic methods for modeling the system are multitype branching processes [78]. A simplified version of the branching trajectories in replication and mutation is shown in Fig. 1.9. As expected, the mean value of the stochastic process coincides with the deterministic solution [80]. The standard deviation, however, can be enormous as we shall see in detail later.

To simulate the interplay between mutation acting on the RNA sequence and selection operating on phenotypes, here RNA structures, the sequence–structure map has to be an integral part of the model [81–83]. The simulation tool starts from a population of RNA molecules and simulates chemical reactions corresponding to replication and mutation in a continuous stirred flow reactor (CSTR) by using Gillespie's algorithm [84, 85]. In target search problems, the replication rate of a sequence X_k is chosen to be a function of the Hamming distance between the mfe-structure formed by the sequence, $S_k = f(X_k)$ and the target structure S_T,

$$f_k(S_k, S_T) = \frac{1}{\alpha + d_H(S_k, S_T)/n}, \tag{1.29}$$

which increases when S_k approaches the target (α is an adjustable parameter that was commonly chosen to be 0.1). A trajectory is completed when the population reaches a sequence that folds into the target structure. Accordingly, the simulated stochastic process has two absorbing barriers, the target and the state of extinction. For sufficiently large populations ($N > 30$ molecules), the probability of extinction is very small, for population sizes reported here, $N \geq 1,000$ it has been never observed.

A typical trajectory is shown in Fig. 1.10. The mean distance to target of the population decreases in steps until the target is reached [82,83,86]. Individual (short) adaptive phases are interrupted by long quasi-stationary epochs. To reconstruct the optimization dynamics, a time-ordered series of structures was determined that leads from an initial structure S_I to the target structure

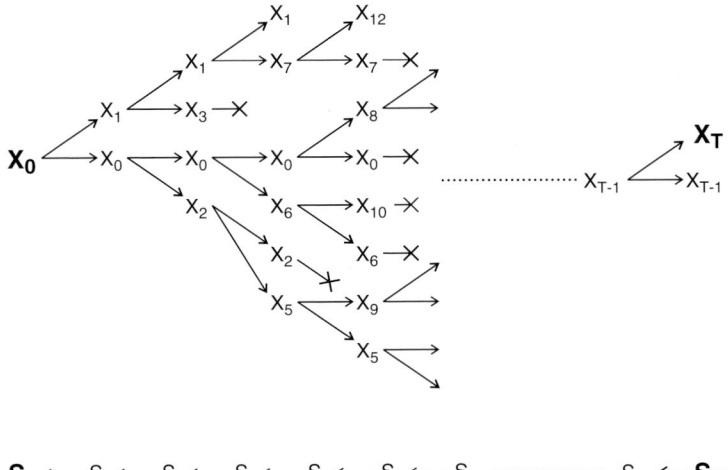

Fig. 1.9. Evolutionary optimization as a multitype branching process. The sketch in the upper part shows only replication acts that lead to mutation. A full genealogy is a time ordered series, which records all individual replication acts, for example $X_0, \ldots, X_0, X_a, \ldots, X_a, X_b, \ldots, \ldots, X_{T-1}, X_T$ leading to target. The population size is either constant (Moran model [79]) or it fluctuates around a constant value (flow reactor: $N \pm \sqrt{N}$), and hence every replication act has to be compensated by the elimination of one molecules that is tantamount to the end of some trajectory in the system. The sketch on the *bottom* illustrates the reconstruction of the optimization run by means of a "relay series"

S_T. This series, called the *relay series*, is a uniquely defined and uninterrupted sequence of shapes. It is retrieved through backtracking, that is in opposite direction from the final structure to the initial shape (see the lower part of Fig. 1.9). The procedure starts by highlighting the final structure and traces it back during its uninterrupted presence in the flow reactor until the time of its first appearance. At this point, we search for the parent shape from which it descended by mutation. Now we record time and structure, highlight the parent shape, and repeat the procedure. Recording further backwards yields a series of shapes and times of first appearance, which ultimately ends in the initial population.[11] Usage of the relay series and its theoretical background allows for classification of transitions [83, 87]. Inspection of the relay series on the quasistationary plateaus allows for a distinction of two scenarios:

(1) The structure is constant and we observe neutral evolution in the sense of Kimura's theory of neutral evolution [88]. In particular, the number of

---

[11] It is important to stress two facts about relay series (1) the same shape may appear two or more times in a given relay series. Then, it was extinct between two consecutive appearances. (2) A relay series is not a genealogy, which is the full recording of parent–offspring relations a time-ordered series of genotypes.

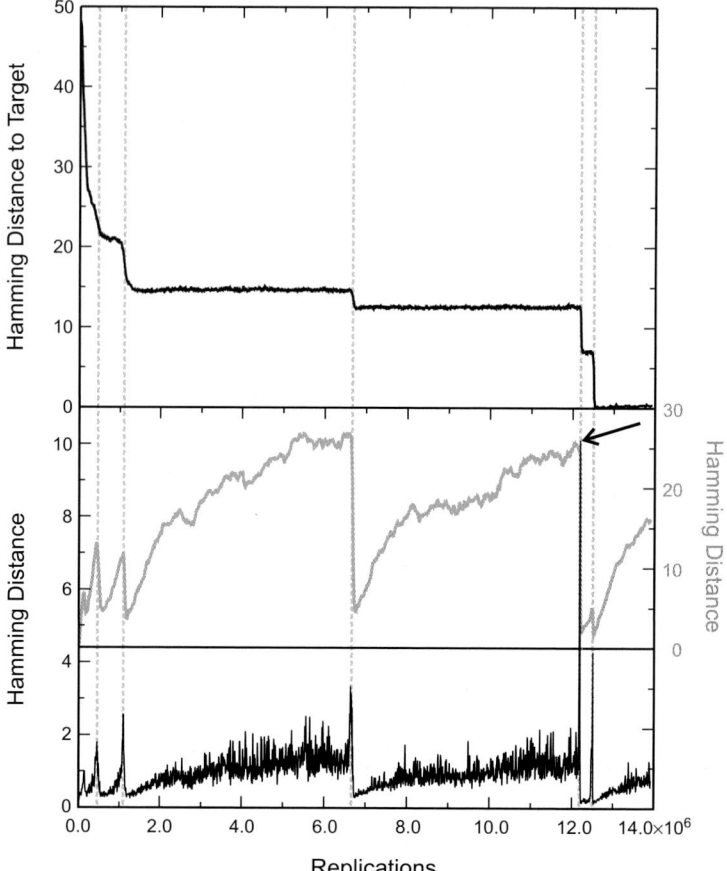

Fig. 1.10. A trajectory of evolutionary optimization. The *topmost* plot presents the mean distance to the target structure of a population of 1,000 molecules. The plot in the middle shows the width of the population in Hamming distance between sequences and the plot at the *bottom* is a measure of the velocity with which the center of the population migrates through sequence space. A remarkable synchronization is observed: At the end of a quasi-stationary plateau an adaptive phase of the migration to target is initiated that is accompanied by a drastic shrinking of the population width and a jump in the population center. A mutation rate of $p = 0.001$ was chosen, the replication rate parameter is defined in (1.29), and initial as well as target structure is shown in Table 1.5

neutral mutations accumulated is proportional to the number of replications in the population, and the evolution of the population can be understood as a diffusion process on the corresponding neutral network [89].
(2) The process during the stationary epoch involves several structures with identical replication rates and the relay series reveal a kind of random walk in the space of these neutral structures.

The diffusion of the population on the neutral network is illustrated by the plot in the middle of Fig. 1.10 that shows the width of the population as a function of time [86, 90]. The population width increases during the quasi-stationary epoch and sharpens almost instantaneously after a sequence had been formed that allows for a continuation of the optimization process. The scenario at the end of the plateau corresponds to a *bottle neck* of evolution. The lower part of the figure shows a plot of the migration rate or drift of the population center and confirms this interpretation: The drift is almost always very slow unless the population center "jumps" from one point in sequence space to the other point where the sequence initiating the new adaptive phase had appeared. A closer look at the figure reveals the coincidence of the three events (1) beginning of a new adaptive phase, (2) collapse-like narrowing of the population spread, and (3) jump-like migration of the population center.

Table 1.5 collects some numerical data obtained from repeated evolutionary trajectories under identical conditions.[12] Individual trajectories show enormous scatter in the time or the number of replications required to reach

Table 1.5. Statistics of the optimization trajectories

Alphabet	Population size (N)	Number of runs (n_R)	Real time from start to target		Number of replications (10^7)	
			Mean value	σ	Mean value	σ
AUGC	1,000	120	900	$+1,380 - 542$	1.2	$+3.1 - 0.9$
	2,000	120	530	$+880 - 330$	1.4	$+3.6 - 1.0$
	3,000	1,199	400	$+670 - 250$	1.6	$+4.4 - 1.2$
	10,000	120	190	$+230 - 100$	2.3	$+5.3 - 1.6$
	30,000	63	110	$+97 - 52$	3.6	$+6.7 - 2.3$
	100,000	18	62	$+50 - 28$	–	–
GC	1,000	46	5,160	$+15,700 - 3,890$	–	–
	3,000	278	1,910	$+5,180 - 1,460$	7.4	$+35.8 - 6.1$
	10,000	40	560	$+1,620 - 420$	–	–

The table shows the results of sampled evolutionary trajectories leading from a random initial structure S_I to the structure of tRNA^{phe}, S_T as target.[a] Simulations were performed with an algorithm introduced by Gillespie [84,85,91]. The time unit is here undefined. A mutation rate of $p = 0.001$ per site and replication was used. The mean and standard deviation were calculated under the assumption of a log-normal distribution that fits well the data of the simulations [a] The structures S_I and S_T were used in the optimization:

S_I: ((.(((((((((((((..........(((.....)))......)))))).)))))).))...(((......)))
S_T: ((((((...((((........)))).((((.......)))))......(((((.......))))).))))))....

---

[12] Identical means here that everything was kept constant except the seeds for the random number generators.

the target. The mean values and the standard deviation were obtained from statistics of trajectories under the assumption of a log-normal distribution. Despite the scatter three features are unambiguously detectable:

(1) The search in **GC** sequence space takes about five time as long as the corresponding process in **AUGC** sequence space in agreement with the difference in neutral network structure discussed above.
(2) The time to target decreases with increasing population size.
(3) The number of replications required to reach the target increases with population size.

Combining items (2) and (3) allows for a clear conclusion concerning time and material requirements of the optimization process: Fast optimization requires large populations whereas economic use of material suggests to work with small population sizes.

1.3.3 Evolution of Noncoding RNAs

In recent year, there has been mounting evidence that noncoding RNAs in fact dominate the regulatory networks of the cell (see, e.g., [92–96] for reviews). Unlike protein coding genes, noncoding RNA (ncRNA) gene sequences do not exhibit a strong *common* statistical signal that separates them from their genomic context. Consequently, a reliable general purpose computational gene-finder for noncoding RNA genes has remained elusive, see e.g., [97]. Most classes of the currently known noncoding RNAs, however, are characterized by a common, evolutionarily very well conserved, secondary structure, while at the same time their sequence is rather variable. This feature can be understood as a consequence of stabilizing selection acting (predominantly) on the secondary structure, while the sequence remains (mostly) free to diffuse on the neutral network.

Diffusion in sequence space, i.e., Kimura's *neutral theory* [88], in fact, forms the conceptual basis of phylogenetic inference. It is important to notice, however, that substitution rates differ dramatically between unpair regions and base-paired regions, since sequence positions that form conserved base pairs are highly correlated. This effectively restricts the diffusion process to the neutral network [89]. Corresponding stochastic models of sequence evolution are described, e.g., in [98–101]. The phase package [102,103] implements such a model and is specifically designed to infer phylogenies from RNAs that have a conserved secondary structure, including rRNAs.

Structural conservation in the presence of sequence variation is also the basis of recent comparative genomics approaches toward RNA gene finding. The first tool of this type, qrna [104] is based upon an SCFG approach to asses the probability that a pair of aligned sequences evolved under a constraint for preserving a secondary structure. The program RNAz [105] uses two independent criteria for classification: a z-score measuring thermodynamic stability of individual sequences, and a *structure conservation index* obtained by comparing folding energies of the individual sequences with the predicted consensus folding. Both quantities measure different aspects of stabilizing selection for RNA structure.

In the remainder of this section, we give a brief overview of the evolutionary patterns of the most prominent RNA families. For a recent, much more detailed review, we refer to [106]. Similar to protein-coding genes, most ncRNAs appear in multiple paralogous copies in the genome. Unlike protein coding genes, however, some classes of ncRNAs appear to be associated with a large number of pseudogenes, this is in particular true for tRNAs and small nuclear RNAs.

Ribosomal RNA sequences are probably the most widely used source of data in molecular phylogenetics: rRNAs are abundant, very well conserved, and therefore easy to access experimentally. Because of concerted evolution, usually, there are no divergent paralogues despite the fact that rRNA genes, in higher eukaryotes at least, typically are arranged in large tandem-repeated clusters. It may not come as a surprise, however, that divergent paralogues of both SSU [107, 108] and LSU [109] do occur in some lineages.

Multiple copies of functional tRNA genes, the existence of numerous pseudogenes, and tRNA-derived repeats are general characteristics of tRNA evolution [110]. Comparative sequence analysis of transfer RNA by means of statistical geometry provides strong evidence that transfer RNA sequences diverged long before the divergence of archaea and eubacteria [111]. Indeed, in a sample of tRNAs for very diverse organisms, those with the same anticodon rather than those from the same organism form coherent subtrees. Models for the origin of tRNA from even simpler components are discussed, e.g., in [112–114].

Like rRNAs and tRNAs, there are typically multiple genomic copies of the spliceosomal snRNAs. Surprisingly, the copy numbers in the genome vary significantly between even closely related species. The mechanism generating this pattern remains unclear at present.

The absence of small nucleolar RNAs (snoRNAs) from bacterial genomes suggests that snoRNPs arose in the archaeal and eukaryotic branch after the divergence of the bacteria. SnoRNAs fall into two structurally distinct classes, box C/D and H/ACA snoRNAs, that guide two different types of chemical modifications of rRNAs and some other ncRNAs, see e.g., [115] for a review. The numerous box C/D and H/ACA snoRNAs of Archaea and Eukarya are likely to have arisen through duplication and variation of the guide sequence [116]. A recent case study of the evolution of the vertebrate U17/E1, E2, and E3 snoRNAs [106] shows that divergent paralogues of snoRNAs have been produced throughout vertebrate evolution. Most vertebrate snoRNAs are encoded in introns. Interestingly, paralogues often reside in adjacent introns of the same gene. In some cases at least, these copies appear to be subject to concerted evolution.

MicroRNA evolution follows a pattern on its own. The mature microRNA is only about 22nt long. It is processed from a thermodynamically very stable stem-loop structure of about 70–80nt in length. Frequently, tandem duplications seem to lead to poly-cistronic transcripts [117]. In contrast to rRNA, tRNAs, and snRNAs, divergent paralogues appear to be the rule rather than the exception for microRNAs. Consequently, most microRNAs that can be traced back to the vertebrate ancestor are present in 2–4 paralogues copies that are remnants of the vertebrate-specific genome duplications. Interestingly, it has been found that tandem-duplications typically predate the nonlocal duplication events [118]. The origin of microRNAs remains unknown. As yet, no microRNA with homologues in both animals and plants has been described so far, although the microRNA processing machinery in animals and plants is clearly homologous. In [119] it has been argued that microRNA could easily arise *de novo* since stem-loop structures resembling pre-miRNAs are very abundant secondary structures in genomic sequences. A recent study on the evolution of animal miRNAs showed that a large number of novel microRNAs appeared in early vertebrates and in placental mammals, while the rate of annotation is otherwise much lower.

Acknowledgments

This work has been supported financially by the Austrian "Fonds zur Förderung der wissenschaftlichen Forschung" (FWF), Project Nos. P-13093, P-13887, and P-14898. Part of the work has been carried out during a visit at the Santa Fe Institute within the External Faculty Program. The support is gratefully acknowledged.

References

1. D. Thirumalai, Proc. Natl. Acad. Sci. **95**, 11506 (1998)
2. D. Thirumalai, N. Lee, S.A. Woodson, D.K. Klimov, Annu. Rev. Phys. Chem. **52**, 751 (2001)
3. D.E. Draper, RNA **10**, 335 (2004)
4. M. Wu, I. Tinoco, Jr., Proc. Natl. Acad. Sci. USA **95**, 11555 (1998)
5. S.R. Holbrook, Curr. Opt. Struct. Biol. **15**, 302 (2005)
6. G. Varani, I. Tinoco, Jr., Q. Rev. Biophys. **24**, 479 (1991)
7. S. Louise-May, P. Auffinger, E. Westhof, Curr. Opin. Struct. Biol. **6**, 289 (1996)
8. P.F. Stadler, J. Math. Chem. **20**, 1 (1996)
9. P. Schuster, P.F. Stadler, in *Discrete Models of Biopolymers*. ed. by M.J.C. Crabbe, M. Drew, A. Konopka. Handbook of Computational Chemistry (Marcel Dekker, New York, 2004) pp. 187–222
10. C.M. Reidys, P.F. Stadler, SIAM Rev. **44**, 3 (2002)
11. E. Rivas, S.R. Eddy, J. Mol. Biol. **285**, 2053 (1999)
12. I.L. Hofacker, P. Schuster, P.F. Stadler, Discr. Appl. Math. **89**, 177 (1998)
13. J.S. McCaskill, Biopolymers **29**, 1105 (1990)

14. M.S. Waterman, *Introduction to Computational Biology: Maps Sequences and Genomes* (Chapman and Hall/CRC, London/Boca Raton, 2000)
15. M.S. Waterman, T.F. Smith, Math. Biosci. **42**, 257 (1978)
16. M. Tacker, P.F. Stadler, E.G. Bornberg-Bauer, I.L. Hofacker, P. Schuster, Eur. Biophys. J. **25**, 115 (1996)
17. M. Zuker, D. Sankoff, Bull. Math. Biol. **46**, 591 (1984)
18. J. Rogers, G. Joyce, Nature **402**, 323 (1999)
19. J.S. Reader, G.F. Joyce, Nature **420**, 841 (2002)
20. C. Flamm, W. Fontana, I.L. Hofacker, P. Schuster, RNA **6**, 325 (1999)
21. W. Zhang, S.J. Chen, J. Chem. Phys. **118**, 3413 (2003)
22. W. Zhang, S.J. Chen, J. Chem. Phys. **119**, 8716 (2003)
23. M. Zuker, P. Stiegler, Nucleic Acids Res. **9**, 133 (1981)
24. D.H. Turner, N. Sugimoto, Annu. Rev. Biophys. Chem. **17**, 167 (1988)
25. A.E. Walter, D.H. Turner, J. Kim, M.H. Lyttle, P. Müller, D.H. Mathews, M. Zuker, Proc. Natl. Acad. Sci. USA **91**, 9218 (1994)
26. D.H. Mathews, J. Sabina, M. Zuker, D.H. Turner, J. Mol. Biol. **288**, 911 (1999)
27. D.H. Mathews, M.D. Disney, J.L. Childs, S.J. Schroeder, M. Zuker, D.H. Turner, Proc. Natl. Acad. Sci. USA **101**, 7287 (2004)
28. I.L. Hofacker, W. Fontana, P.F. Stadler, L.S. Bonhoeffer, M. Tacker, P. Schuster, Mh. Chemie **125**, 167 (1994)
29. I.L. Hofacker, Nucleic Acids Res. **31**, 3429 (2003)
30. B. Bollobás, *Random Graphs* (Academic, London, 1985)
31. C. Reidys, P.F. Stadler, P. Schuster, Bull. Math. Biol. **59**, 339 (1997)
32. W. Grüner, R. Giegerich, D. Strothmann, C. Reidys, J. Weber, I.L. Hofacker, P. Schuster, Mh. Chemie **127**, 355 (1996)
33. W. Grüner, R. Giegerich, D. Strothmann, C. Reidys, J. Weber, I.L. Hofacker, P. Schuster, Mh. Chemie **127**, 375 (1996)
34. P. Schuster, J. Biotechnol. **41**, 239 (1995)
35. E. Schultes, D. Bartel, Science **289**, 448 (2000)
36. D.M. Held, S.T. Greathouse, A. Agrawal, D.H. Burke, J. Mol. Evol. **57**, 299 (2003)
37. Z. Huang, J.W. Szostak, RNA **9**, 1456 (2003)
38. W. Fontana, D.A.M. Konings, P.F. Stadler, P. Schuster, Biopolymers **33**, 1389 (1993)
39. P.G. Higgs, J. Phys. I (France) **3**, 43 (1993)
40. J. Gevertz, H.H. Gan, T. Schlick, RNA **11**, 853 (2005)
41. P. Clote, F. Ferré, E. Kranakis, D. Krizanc, RNA **11**, 578 (2005)
42. M. Zuker, Science **244**, 48 (1989)
43. S. Wuchty, W. Fontana, I.L. Hofacker, P. Schuster, Biopolymers **49**, 145 (1999)
44. M.S. Waterman, T.H. Byers, Math. Biosci. **77**, 179 (1985)
45. B.A. Shapiro, K. Zhang, Comput. Appl. Biosci. **6**, 309 (1990)
46. C. Reidys, P.F. Stadler, Comput. Chem. **20**, 85 (1996)
47. V. Moulton, M. Zuker, M. Steel, R. Pointon, D. Penny, J. Comput. Biol. **7**, 277 (2000)
48. M. Höchsmann, T. Töller, R. Giegerich, S. Kurtz, *Proceedings of the Computational Systems Bioinformatics Conference*, vol. 159 (Stanford, CA, CSB 2003)
49. M. Andronescu, A.P. Fejes, F. Hutter, H.H. Hoos, A. Condon, J. Mol. Biol. **336**, 607 (2004)

50. J.R. Fresco, A. Adains, R. Ascione, D. Henley, T. Lindahl, Cold Spring Harb. Symp. Quant. Biol. **31**, 527 (1966)
51. E.R. Hawkins, S.H. Chang, W.L. Mattice, Biopolymers **16**, 1557 (1977)
52. V.L. Emerick, S.A. Woodson, Biochemistry **32**, 14062 (1993)
53. R. Micura, C. Höbartner, Chembiochem **4**, 984 (2003)
54. T. Baumstark, A.R. Schroder, D. Riesner, EMBO J. **16**, 599 (1997)
55. A.T. Perrotta, M.D. Been, J. Mol. Biol. **279**, 361 (1998)
56. C.K. Biebricher, S. Diekmann, R. Luce, J. Mol. Biol. **154**, 629 (1982)
57. C.K. Biebricher, R. Luce, EMBO J. **11**, 5129 (1992)
58. H. Zamora, R. Luce, C.K. Biebricher, Biochemistry **34**, 1261 (1995)
59. C. Flamm, I.L. Hofacker, S. Maurer-Stroh, P.F. Stadler, M. Zehl, RNA **7**, 254 (2000)
60. I. Abfalter, C. Flamm, P.F. Stadler, in *Design of Multistable Nucleic Acid Sequences*. ed. by H.W. Mewes, V. Heun, D. Frishman, S. Kramer. Proceedings of the German Conference on Bioinformatics (GCB 2003), vol. 1 (Belleville Verlag Michael Farin, München, 2003) pp.1–7
61. P. Clote, L. Gąsieniec, R. Kolpakov, E. Kranakis, D. Krizanc, J. Theor. Biol. **236**, 216 (2005)
62. E. Merino, C. Yanofsky, in *Regulation by Termination-Antitermination: A Genomic Approach*. ed. by A.L. Sonenshein, J.A. Hoch, R. Losick. *Bacillus subtilis and its Closest Relatives: From Genes to Cells* (ASM, Washington, DC, 2002) pp. 323–336
63. T.M. Henkin, C. Yanofsky, Bioessays **24**, 700 (2002)
64. A.G. Vitreschak, D.A. Rodionov, A.A. Mironov, M.S. Gelfand, Trends Genet. **20**, 44 (2004)
65. W.C. Winkler, R.R. Breaker, Chembiochem **4**, 1024 (2003)
66. S. Brantl, Trends Microbiol. **12**, 473 (2004)
67. E. Nudler, A.S. Mironov, Trends Biochem. Sci. **29**, 11 (2004)
68. J.E. Barrick, K.A. Corbino, W.C. Winkler, A. Nahvi, M. Mandal, J. Collins, M. Lee, A. Roth, N. Sudarsan, I. Jona, J.K. Wickiser, R.R. Breaker, Proc. Natl. Acad. Sci. USA **101**, 6421 (2004)
69. E.A. Lesnik, G.B. Fogel, D. Weekes, T.J. Henderson, H.B. Levene, R. Sampath, D.J. Ecker, Biosystems **80**, 145 (2005)
70. R.R. Breaker, Curr. Opin. Biotechnol. **13**, 31 (2002)
71. S.K. Silverman, RNA **9**, 377 (2003)
72. M.T. Wolfinger, W.A. Svrcek-Seiler, C. Flamm, I.L. Hofacker, P.F. Stadler, J. Phys. A: Math. Gen. **37**, 4731 (2004)
73. M. Eigen, Naturwissenschaften **58**, 465 (1971)
74. M. Eigen, P. Schuster, Naturwissenschaften **64**, 541 (1977)
75. M. Eigen, P. Schuster, Naturwissenschaften **65**, 7 (1978)
76. J. Swetina, P. Schuster, Biophys. Chem. **16**, 329 (1982)
77. M. Eigen, J. McCaskill, P. Schuster, Adv. Chem. Phys. **75**, 149 (1989)
78. P. Jagers, *Branching Processes with Biological Applications* (Wiley, London, 1975)
79. P.A.P. Moran, *The Statistical Processes of Evolutionary Theory* (Clarendon, Oxford, UK, 1962)
80. L. Demetrius, P. Schuster, K. Sigmund, Bull. Math. Biol. **47**, 239 (1985)
81. W. Fontana, P. Schuster, Biophys. Chem. **26**, 123 (1987)
82. W. Fontana, W. Schnabl, P. Schuster, Phys. Rev. A **40**, 3301 (1989)

83. W. Fontana, P. Schuster, Science **280**, 1451 (1998)
84. D.T. Gillespie, J. Comput. Phys. **22**, 403 (1976)
85. D.T. Gillespie, J. Phys. Chem. **81**, 2340 (1977)
86. P. Schuster, in *Molecular Insight into the Evolution of Phenotypes*. ed. by J.P. Crutchfield, P. Schuster. Evolutionary Dynamics: Exploring the Interplay of Accident, Selection, Neutrality, and Function (Oxford University Press, New York, 2003) pp. 163–215
87. B.R.M. Stadler, P.F. Stadler, G.P. Wagner, W. Fontana, J. Theor. Biol. **213**, 241 (2001)
88. M. Kimura, *The Neutral Theory of Molecular Evolution* (Cambridge University Press, Cambridge, UK, 1983)
89. M.A. Huynen, P.F. Stadler, W. Fontana, Proc. Natl. Acad. Sci. USA **93**, 397 (1996)
90. K. Grünberger, U. Langhammer, A. Wernitznig, P. Schuster, *RNA evolution in Silico* (Technical Report, Institut für Theoretische Chemie, Universität Wien, 2005)
91. D.T. Gillespie, J. Stat. Phys. **16**, 311 (1977)
92. D.P. Bartel, C.Z. Chen, Nat. Genet. **5**, 396 (2004)
93. O. Hobert, Trends Biochem. Sci. **29**, 462 (2004)
94. J.S. Mattick, Bioessays **25**, 930 (2003)
95. J.S. Mattick, Nat. Genet. **5**, 316 (2004)
96. M. Szymański, M.Z. Barciszewska, M. Żywicki, J. Barciszewski, J. Appl. Genet. **44**, 1 (2003)
97. S.R. Eddy, Nat. Genet. **2**, 919 (2001)
98. M. Schöninger, A. von Haeseler, J. Mol. Evol. **49**, 691 (1999)
99. B. Knudsen, J.J. Hein, Bioinformatics **15**, 446 (1999)
100. N.J. Savill, D.C. Hoyle, P.G. Higgs, Genetics **157**, 399 (2001)
101. J. Otsuka, N. Sugaya, J. Theor. Biol. **222**, 447 (2003)
102. H. Jow, C. Hudelot, M. Rattray, P.G. Higgs, Mol. Biol. Evol. **19**, 1591 (2002)
103. C. Hudelot, V. Gowri-Shankar, H. Jow, M. Rattray, P.G. Higgs, Mol. Phylogenet. Evol. **28**, 241 (2003)
104. E. Rivas, R.J. Klein, T.A. Jones, S.R. Eddy, Curr. Biol. **11**, 1369 (2001)
105. S. Washietl, I.L. Hofacker, P.F. Stadler, Proc. Natl. Acad. Sci. USA **102**, 2454 (2005)
106. A.F. Bompfünewerer, C. Flamm, C. Fried, G. Fritzsch, I.L. Hofacker, J. Lehmann, K. Missal, A. Mosig, B. Müller, S.J. Prohaska, B.M.R. Stadler, P.F. Stadler, A. Tanzer, S. Washietl, C. Witwer, Theor. Biosci. **123**, 301 (2005)
107. S. Carranza, J. Baguñà, M. Riutort, J. Mol. Evol. **49**, 250 (1999)
108. A.P. Rooney, Mol. Biol. Evol. **21**, 1704 (2004)
109. M.J. Telford, P.W.H. Holland, J. Mol. Evol. **44**, 135 (1997)
110. F.E. Frenkel, M.B. Chaley, E.V. Korotkov, K.G. Skryabin, Gene **335**, 57 (2004)
111. M. Eigen, B.F. Lindemann, M. Tietze, R. Winkler-Oswatitsch, A.W.M. Dress, A. von Haeseler, Science **244**, 673 (1989)
112. M. Eigen, R. Winkler-Oswatitsch, Naturwissenschaften **68**, 282 (1981)
113. S. Rodin, S. Ohno, A. Rodin, Proc. Natl. Acad. Sci. USA **90**, 4723 (1993)
114. M. Di Giulio, J. Theor. Biol. **226**, 89 (2004)
115. M.P. Terns, R.M. Terns, Gene Expr. **10**, 17 (2002)

116. D. Lafontaine, D. Tollervey, Trends Biochem. Sci. **23**, 383 (2002)
117. Y. Lee, K. Jeon, J.T. Lee, S. Kim, V.N. Kim, EMBO J. **21**, 4663 (2002)
118. J. Hertel, M. Lindemeyer, K. Missal, C. Fried, A. Tanzer, C. Flamm, I.L. Hofacker, P.F Stadler, The Students of Bioinformatics Computer Labs 2004 and 2005. BMC Genomics **7**, 25 (2006)
119. A. Tanzer, P.F. Stadler, J. Mol. Biol. **339**, 327 (2004)

2

Gene3D and Understanding Proteome Evolution

J.G. Ranea, C. Yeats, R. Marsden and C. Orengo

Gene3D is a database of protein sequence families. The families have been created through clustering the proteome sequences of over 200 species, including more than 15 eukaryotes, and totalling over 750,000 proteins (as derived from UniProt [1]). Each family is then further subclustered on the basis of sequence similarity. Using remote homologue detection methods, we have been able to assign structures – based on the CATH [2] and Pfam [3] databases – to a majority of these sequences. This allows a high resolution view of the functions and evolution of specific protein families, as well as the evolution of a species' gene content.

As can be seen from the numbers of proteins involved, the scale of the protein annotation problem is far greater than can be dealt with per manu. To describe all these proteins with the highest possible accuracy, it is necessary to accurately characterise and comprehend the evolution of variant functions from single ancestral sequences.

Central to this effort is the modular theory of protein evolution. As the first structures of proteins were solved in the late 1960s, it became swiftly apparent that several of these structures were made up from smaller substructures [4]; furthermore, these substructures appeared in varying contexts and in proteins of varying function [5] (see Table 2.1). It was proposed that these substructures ('domains') are the true units of protein evolution.

The use of such units in evolution allows two methods of generating novel functions. First, the traditional concept of stepwise mutation still applies to the individual domains. Second, multidomain proteins can obtain new functions through the recombination, subtraction or addition of new domains. Recent work by various researchers (i.e. Ponting, Teichmann, Koonin and Aravind to name a few) has indicated that domain shuffling is one of the driving forces behind evolution, and can underlie processes such as the generation of antigenic diversity in pathogens [6] or the diversification of transcription regulation [7].

Table 2.1. Top ten most frequent domain assignments in each kingdom and in total

Archea	Bacteria	Eukaryota	In All
P-loop containing nucleotide triphosphate hydrolases	P-loop containing nucleotide triphosphate hydrolases	Classic zinc finger	P-loop containing nucleotide triphosphate hydrolases
NAD(P)-binding Rossman-like	NAD(P)-binding Rossman-like	Immunoglobulins	Classic zinc finger
Transferase (methyltransferase)	Winged helix repressor DNA binding	P-loop containing nucleotide triphosphate hydrolases	Immunoglobulins
Winged helix repressor DNA binding	Periplasmic-binding protein-like I	Transferase (phosphotransferase) domain I	NAD(P)-binding Rossman-like
CBS domain	Homeodomain-like	Laminin	Winged helix repressor DNA binding
Electron transport	BPD transport system inner membrane component	Fibronectin type I	Homeodomain-like
Oxidoreductase	Transferase (methyltransferase)	Phosphorylase kinase domain I	Transferase (phosphotransferase) domain I
Photoreceptor	Binding proteins	Membrane spanning alpha helix pairs	Phosphorylase kinase domain I
ABC transporter	Oxidoreductase	Nuclear protein	Periplasmic-binding protein-like
Spore coat protein polysaccharide biosynthesis protein	ABC transporter	Cadherins	Laminin

Further support has come from investigations into whole genome sequences, for instance by Teichmann et al. [8] and by Gerstein [9]. These suggest that around 70% of all proteins are multidomain and that this figure is only slightly higher in eukaryotes over prokaryotes. We have extended these

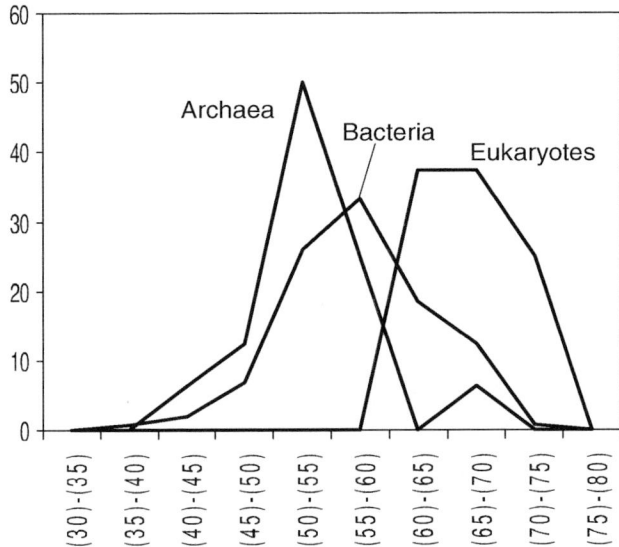

Fig. 2.1. Approximation of the percentage of multidomain proteins in approximately 200 genomes. The x-axis indicates the proportion of multidomain proteins found in a genome; the y-axis indicates the proportion of species (expressed as a percentage) that have the proportion of multidomain proteins. The data have been split into three sets – archaea (16 species), bacteria (162 species) and eukaryotes (16 species) – and binned into ranges of 5%. Although each specific value may not be correct it can be assumed that the overall graph shows the correct shape and range (unpublished work by C. Yeats, UCL)

observations to over 200 species through using the data available in Gene3D (see Fig. 2.1). Furthermore, the Gene3D data shows (see Fig. 2.2) that around 50% of identified domains in these genomes belong to around 219 universal (found in eukaryotes, prokaryotes and archaea) domain families [10]. In contrast, only about 10% of proteins are universal. This provides a strong indication that much of the variation between species comes from the shuffling of common components to provide new functions, rather than the introduction of novel components.

Examination of the distribution of identified domains has shown that they follow a power-law behaviour; i.e. a few families account for most domains (see Fig. 2.3), while many families account for the remainder. Other estimates have concluded that half of all discovered domain sequences will belong to around 1,500 structural families (reviewed by Grant et al. [11]). Hence, by careful characterisation of the individual families, as well as understanding the effects of their various recombinations, it should be possible to predict the functions of most proteins from sequence.

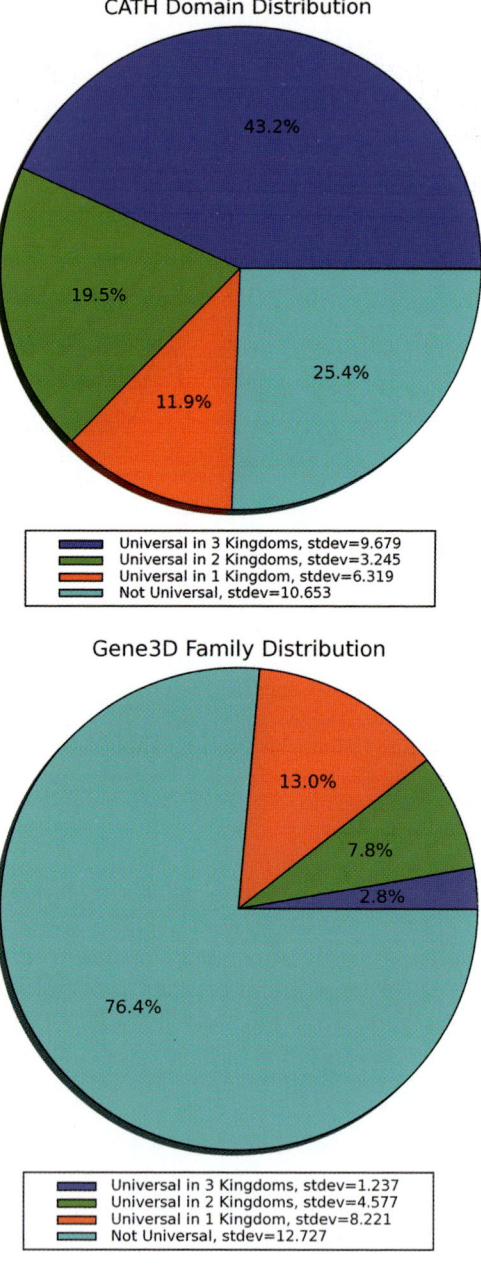

Fig. 2.2. (*Top*) Kingdom distribution of domain families in Gene3D. A domain is regarded as universal in a kingdom if it is found in over 70% of species. (*Bottom*) Kingdom distribution of protein families in Gene3D. These statistics are calculated in the same manner as in (*Top*), except that the units measured are entire proteins rather than domains. As can be seen protein families are far less likely to be common or widespread than protein domains

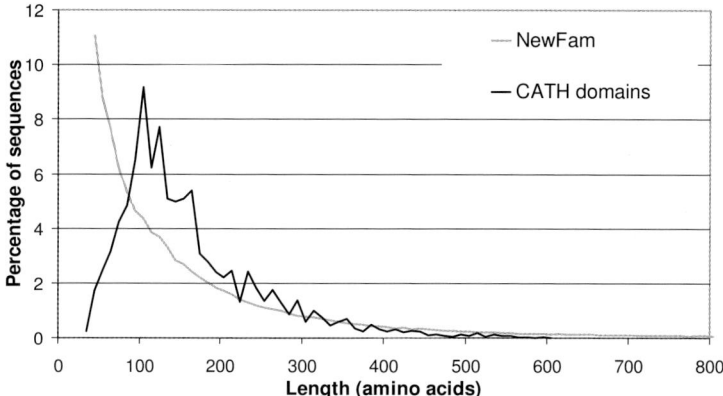

Fig. 2.3. The length distribution of NewFam families compared to CATH. The length distributions of CATH domains and NewFam domains are binned and plotted on the same axis. The size distributions are very similar, suggesting that NewFams do approximately represent single domains

Hand-in-hand with understanding the modular nature of proteins has been the creation of tools for the identification of related domains and the transfer of functional information. On top of this information, databases are being built that catalogue and describe domain families and attempt to predict the function of newly identified proteins. Gene3D is one of these databases; others are described later. At its heart, Gene3D uses a protocol called PFscape (see later). PFscape is a benchmarked automated process that derives and assigns information from two manually curated databases to an automatically clustered set of proteins. Novel sequences can then be annotated through comparison to these described clusters.

As mentioned earlier, a major issue in the analysis of genomic sequence is the sheer volume of data. A typical bacterial genome will consist of around 4,000 genes, whereas eukaroytic genomes can range from 5,000 to at least 40,000 genes. Currently there are over 200 genomes in Gene3D, comprising a total of over 750,000 unique proteins. Hence, PFscape has been designed to have the advantages of a high-throughput automated procedure and the accuracy of annotation of a manual expert-based procedure, and enables the production of regular updates of the genome annotation.

This information is presented through the Gene3D website.[1] The web resource allows browsing of individual genomes, and viewing domain and functional annotations for the families. These include links to CATH [12] or Pfam [3] domains that have been assigned to these families.

---

[1] www.biochem.ucl.ac.uk/bsm/cath/Gene3D

2.1 Protein Family Clustering

Several protocols have recently been developed to attempt to classify proteins into families and to determine their domain structure. These all aim to deal with large numbers of proteins in a fully automated fashion. Examples are: SYSTERS [13], ProtoNet [14], ADDA (automatic domain decomposition algorithm) [15] and ProDom [16]. A short description is provided of each so as to allow comparison to PFscape.

2.1.1 SYSTERS

Underlying SYSTERS is a two-step algorithm that clusters sequences according to their similarities as determined by exhaustive Smith–Waterman searches [17]. This algorithm takes advantage of the observation that the tree structure of single-linkage clustered proteins tends to show several distinct groupings towards the leaves that rapidly merge into 'superfamilies' at a certain point towards the trunk of the tree. This is believed to capture the evolutionary radiation of domain families.

The first step of the algorithm is to cluster the proteins into a single linkage tree using E-values. All sequence similarities with an E-value of above 0.05 were regarded as being of infinite distance. The unrelated subtrees are connected using an artificial overall root node with weight infinity. Then an automatic approach is used to determine the maximum change in family size while moving down from leaf to root. This has been shown to often accurately capture the largest grouping of genuinely related sequences. Step two employs a graph theory-based algorithm, derived from Hartuv et al. [18], to parse the superfamily tree into subclusters. This method generates 158,153 protein families, of which approximately 40,000 contain more than one sequence, from a starting set of around 1.1 million sequences.

2.1.2 ProtoNet

ProtoNet uses a bottom-up clustering method based on E-values from an all-against-all BLAST search. At each step, the two most related clusters are merged together to create a hierarchical tree of clusters. This is an extremely large tree, which contains many biologically incorrect or irrelevant clusters. So a second algorithm, which determines the most stable size of a cluster, is employed to condense these clusters into a realistic set. Using this approach on approximately 110,000 sequences created a tree of 28,000 clusters.

2.1.3 ADDA

ADDA (automatic domain decomposition algorithm) is based around a global maximum likelihood model for the likely domain composition of a protein based on its pairwise alignments to other proteins. These are determined using BLAST. Having delineated the likely domains, these are clustered using a method that carries out pairwise profile–profile comparisons of nearest neighbours. If two subclusters are found to be unrelated, an edge to the family is formed, and otherwise they are merged.

2.1.4 ProDom

ProDom is based on an algorithm originally developed by Sonnhammer and Kahn [19] and extended to become mkdom2 by Gouzy et al. [20]. It is a greedy algorithm that assumes that the smallest protein present in a sequence database consists of a single domain. It then identifies all homologous regions and removes these from the sequence database. Hence each fragment is dealt with in turn until all have been clustered or removed. The problem with this method is that it assumes that the sequence database is very clean and does not contain fragment sequences or mispredictions. This approach clustered 750,000 sequences into 186,000 clusters of at least two proteins.

2.2 The PFscape Method

PFscape uses the TribeMCL of Enright et al. [21] algorithm to create the initial multilinkage clusters. Using the genomes of 90 bacteria, 14 eukaryota and 16 archaea, 112,464 gene families were created, of which 50,219 included more than one member. This is in line with the results obtained from other clustering methods. The clustering results were benchmarked against a manually validated dataset of structurally characterised proteins obtained from CATH; this allowed the derivation of optimum parameters for TribeMCL. The consistency of the results was confirmed by examining families for which each sequence's domain architecture could be fully annotated using Pfam and CATH. For at least 70% of the families, over 90% of the members show identical domain architectures. For the remainder, the variances are usually slight. The clusters were then functionally annotated using information extracted from COG [22], GO [23] and KEGG [24].

The PFUpdate protocol [10] is employed for the inclusion of new genomes. Each protein from the new genome is searched against all the sequences previously in Gene3D. The protein is then assigned to the cluster that contains the sequence that showed the highest similarity, provided that the overlap of the two proteins is above 80% and the E-value is below 0.001. At the end of this process, any clusters that have been altered (whether expanded or reduced by error correction) are blasted against each other again and re-clustered using multilinkage clustering. At present, around 75% of the protein sequences from new bacterial genomes and 60% from archaeal genomes can be assigned to existing clusters in Gene3D.

2.3 The NewFams

Whilst Pfam and CATH combined manage to typically assign at least one domain to around 60–80% of proteins within a genome they only describe around 40–60% of amino acids encoded [10]. Hence, it is clear that there are many remaining regions of suitable size and composition as to encode unidentified domains.

Using the assumption that a region of undescribed complex amino acid sequence greater than 40 residues in length is likely to encode a domain, a protocol was developed [10] to identify these regions and to cluster them together. This has created a set of 40,000 putative domain families, termed 'NewFams,' that are likely to represent protein structures; work is currently underway to validate and describe as many of these as possible. It is important to note that the protocol for building these families are likely to miss homologues and hence that several NewFams may represent a single domain.

Incorporating the NewFams into the annotation process should allow a more highly resolved view of protein evolution and functional adaptation. This can be achieved, for example, by identifying GO terms that associate with particular NewFams, or by identifying protein familes that have diverged through the introduction of an undescribed region. Of note, the 6,000 biggest CATH, Pfam and NewFam domain families account for over 80% of sequenced residues (see Fig. 2.4), discounting protein sequences with no apparent homology to any other sequence ('singletons').

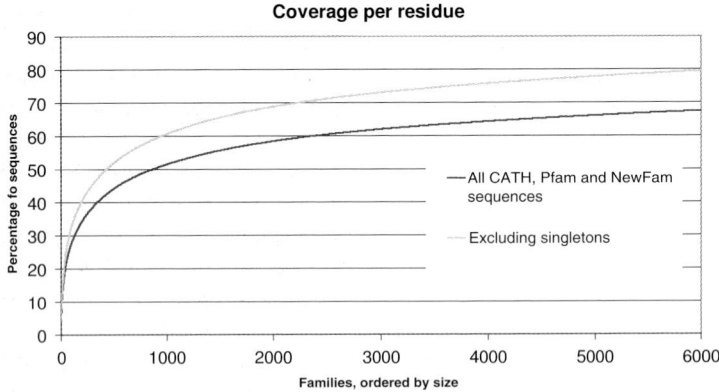

Fig. 2.4. Coverage of the protein universe by Gene3D: the percentage of residues in Gene3D that are matched, at least once, by CATH, Pfam and the NewFams. For both graphs the cumulative coverage provided each family is shown, with the families ordered by size on the x-axis

2.4 Describing the Proteome

As discussed in the Introduction to this chapter, the genome sequencing projects of the last ten years have created a large and diverse dataset. With over 200 genomes sequenced, the variation between whole genomes can now be approached in a generalised statistical manner. In this section, we will describe some examples of the ways that this data can be exploited through Gene3D.

Since Gene3D is based around whole genome sequences, it is possible to generate a frequency of occurrence profile for each superfamily. The profiles can then be compared and correlated with each other or with other extant data. These investigations have clarified the evolutionary restraints on the development of bacterial genome complexity and provide models that can allow us to predict the effect of increasing or decreasing bacterial genome size.

Furthermore, since the Gene3D clusters have been annotated through structural domain assignment, it is possible to detect more remote relationships than are possible through sequence comparison. Indeed, it has often been shown that two very similar structures can show little more than random sequence similarity to each other. Hence, using Gene3D has the additional benefit that it is possible to derive relationships between clusters due to shared domain architectures, despite a lack of sequence evidence. This also allows better curation and validation of the Gene3D clusters.

The first study presented focuses on the identification of the genetic determinants of genome complexity variation in prokaryotes. Genome complexity can be measured simply by counting the number of ORFs [25]. In prokaryotes, a strong correlation is found between their genome and proteome sizes, and the genomes tend to consist of only a few genetic elements; hence, prokaryotes strongly conform to this assumption and should show easily discernible trends linking their protein content and genome complexity [26]. Additionally, from a simple statistical point of view, the high number of sequenced prokaryotic species, as compared to eukaryotes, increases the significance of observed phenomena.

The genome complexity of an organism reflects the balance of disparate selective pressures working over time. Gene duplication and lineage-specific gene loss seem to be the key processes determining bacterial genome size, followed by horizontal gene transfer [27–29]. All these genetic changes that finally shape the genome size and content are caused by the sum of more primary processes, which have taken place at the lower level of gene family rearrangements. Therefore, it is reasonable to suppose that gene family evolution (expansions and contractions of the families) and genome size variation have been interdependent events in evolution [30]. The genome size is not related to phenotype or lineage in prokaryotes. For example, bacteria with a broad range of phenotypes and lifestyles can have similar genome sizes, and diversity in genome size has been observed for bacteria belonging to relatively narrow phylogenetic groups [31]. Furthermore, the smallest genomes are usually derived from bacteria with larger genomes through a process called evolution by reduction [26, 32]. Therefore, contrary to what is generally thought, the most complex genomes do not correspond with the most evolved prokaryotes. All these facts point to other evolutionary mechanisms being needed to explain the relationship between superfamily evolution and variation in genome complexity.

2.5 Superfamily Evolution and Genome Complexity

To determine general genetic mechanisms involved in the development of bacterial complexity, the occurrence of each CATH domain superfamily was calculated for 100 prokaryotic species (85 eubacteria and 15 archaea) and assembled into trans-species occurrence profiles. The correlation between genome sizes, measured in number of ORFs, and superfamily occurrences was then calculated using Spearman's rank correlation coefficient [33]. Superfamilies for which domain occurrence was highly correlated with genome size (Spearman coefficient ≥ 0.7) were selected, and are referred to as size-dependent superfamilies (Table 2.2). The superfamilies for which occurrence had very low correlation with genome size (Spearman coefficient ≤ 0.2) were also selected, and referred to as size-independent superfamilies (Table 2.2).

This approach allowed us to distinguish those superfamily domains that exhibit a frequency of occurrence that is dependent on genome size. We also wished to identify which of these domains represented general molecular mechanisms or functions present in the majority of organisms. For this, we selected all those domains that occurred in greater than 70% of bacterial species; these were termed the universal domains (Table 2.2).

The universal domains were divided into two groups, depending on their genome-size correlation coefficient: the size-independent and the size-dependent sets. The sum of the size-independent superfamily domains occurrences shows a flat slope with respect to the increasing genome size (Fig. 2.5a) [34]. In the size-dependent group, three different trends or correlation models were distinguishable. These were the linearly (Fig. 2.5b), the power-law-like (Fig. 2.5c), and the logarithmically distributed superfamilies (Fig. 2.5d) [30].

All these groups were functionally annotated using the COG database, PFAM domain annotations and literature searches. This revealed significant functional tendencies for the size-independent superfamily group and two of the three size-dependent superfamily subgroups. The size-independent superfamilies are mainly involved in information storage and processing functions, with a significant proportion of the functional annotations associated with translation, ribosomal structure and protein synthesis (Fig. 2.6) [34]. In the size-dependent set, the linearly distributed superfamilies are primarily associated with metabolism. While the power-law-like distributed superfamilies are, for a significant percentage, involved in basic and ancestral mechanisms of gene regulation and signal transduction (Fig. 2.6). In the

Table 2.2. Statistics from size correlation and universal percentage analysis

Superfamily Groups	All superfamilies				Universal superfamilies			
	N. Sup.	% Sup.	N. domains	% domains	Mean (Dom./Sup.)	N. Sup.	% Sup.	N. domains	% domains	Mean (Dom./Sup.)
Size-dependent (Sper. Coeff $\geq$ 0.7)	116	12%	114,244	60%	985	85	9%	106,774	56%	1,256
Size-dependent (Sper. Coeff $\leq$ 0.2)	308	33%	12,877	7%	42	76	8%	8,474	4%	112
Rest (Sper. Coeff < 0.2)	516	55%	63,506	33%	123	188	20%	43,224	23%	230
Total	940	100%	190,627	100%	203	349	37%	158,472	83%	454

Statistics of the 940 annotated superfamilies (all superfamilies box, left hand of the table), and the universal set of 349 superfamilies (universal superfamilies box, right hand of the table). These former groups are divided into three different Superfamily Groups: Size-dependent superfamilies with a Spearman's coefficient $\geq$ 0.7 (first row in the table), Size-independent superfamilies with a Spearman's coefficient $\leq$ 0.2 (second row in the Table), and the rest of the superfamilies (third row in the table). Inside of each box, from left to right, columns give: N. Sup., Number of Superfamilies; Sup., percentage of superfamilies considering all superfamilies; N. domains, total number of domains; domains, percentage of domains regarding all domains; and Mean (N. Dom./N. Sup.), domains average per superfamily

Four major types of superfamily distributions:

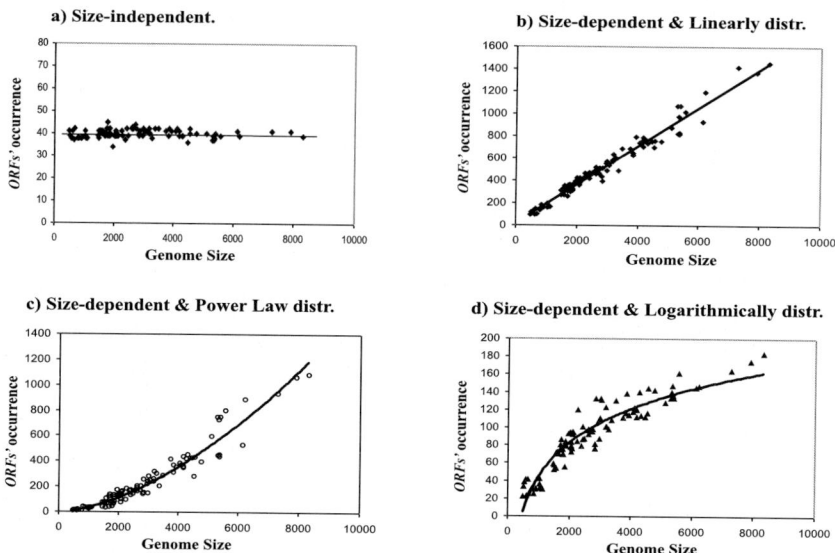

Fig. 2.5. Four major types of superfamilies distributions: (**a**) size-independent superfamilies; (**b**) size-dependent, linearly distributed superfamilies; (**c**) size-dependent, power law distributed; and (**d**) size-dependent, logarithmically distributed. ORFs occurrence (y-axis) is plotted against genome size (x-axis, also expressed in number of ORFs)

remaining size-dependent set, the logarithmically distributed superfamilies, no single functional tendency was found and, since they represented a small proportion of the superfamilies, they are not shown in Fig. 2.6 [30].

The analysis of common superfamily domains shared by all bacterial species allows us to identify their general features, which are common to all the prokaryotes. As discussed above, domain shuffling is prevalent in bacteria, and so universal genes are few in number. Furthermore, as first suggested by Chothia [35], half of all domains will be accounted for by a small number of domain families. Previously, studies have tended to focus on clustering genes by function in order to investigate the rules governing genome complexity [36, 37]. In Sect. 2.6, we will further show why considering universal superfamily domains as the basic units for both evolution and analysis is a key paradigm for the identification and characterisation of the general determinants of genome complexity.

2.6 Superfamily Evolution and Functional Relationships

It can be argued that these universal domain superfamilies represent common functions that are used by all organisms. In this case, it might be expected that

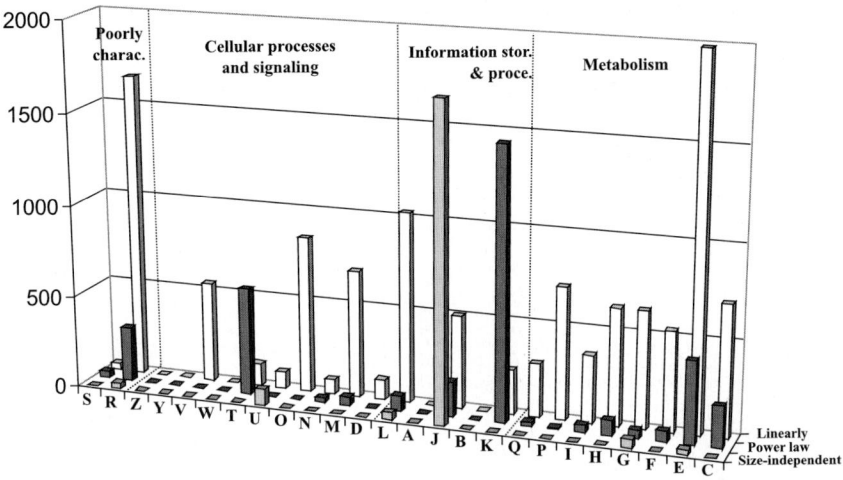

Fig. 2.6. Functional distribution of the (a), (b) and (c) types of superfamilies (see Fig. 2.5) in the 25 functional subcategories in the COG database. The COG categories one letter codes are as follows: J, translation, ribosomal structure and biogenesis; A, RNA processing and modification; K, transcription; L, replication, recombination and repair; B, chromatin structure and dynamics; D, cell cycle control, cell division, chromosome partitioning; Y, nuclear structure; V, defense mechanisms; T, signal transduction mechanisms; M, cell wall/membrane/envelope biogenesis; N, cell motility; Z, cytoskeleton; W, extracellular structures; U, intracellular trafficking, secretion and vesicular transport; O, post-translational modification, protein turnover, chaperones; C, energy production and conversion; G, carbohydrate transport and metabolism; E, amino acid transport and metabolism; F, nucleotide transport and metabolism; H, coenzyme transport and metabolism; I, lipid transport and metabolism; P, inorganic ion transport and metabolism; Q, secondary metabolites biosynthesis, transport and catabolism; R, general function prediction only; S, function unknown. The broader classes in which the 25 functional categories are grouped are also indicated

grouping superfamilies by molecular function would give clearer trends than grouping by genome size. However, it is important to note that two unrelated genes in separate lineages with a similar functional annotation should be considered as separate molecular technologies. This is because they will be fundamentally different in the details, even though the overall biochemical reaction may be highly similar. This will lead to the two proteins behaving differently in different physicochemical environments. In contrast, when two genes or domains belong to the same superfamily, it indicates that they share structural, sequence and functional characteristics [38]. Hence, it would be expected that the members of a superfamily show shared and related behaviours. In contrast, two unrelated technologies that have been grouped on the bases of a similar function may show no shared behaviour.

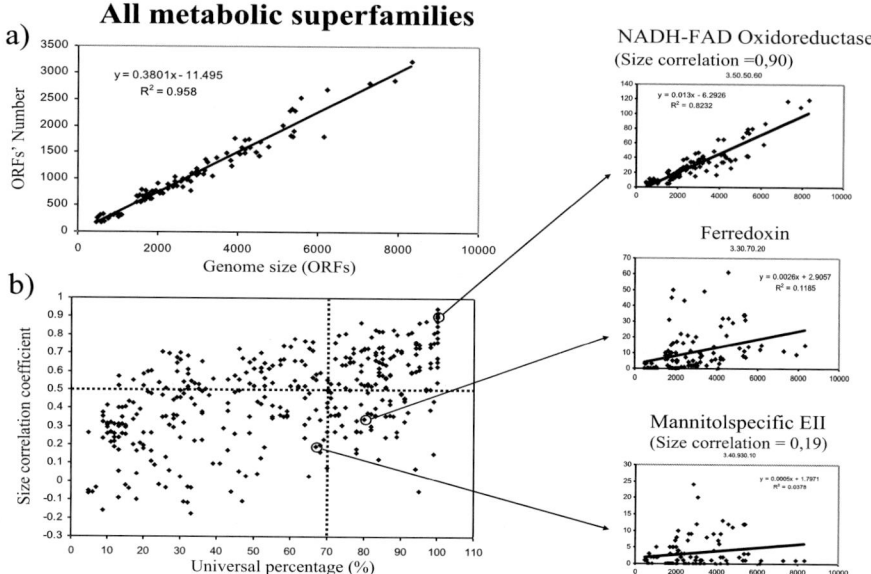

Fig. 2.7. Analysis of 385 superfamily domains annotated as metabolic domains in the COG database: (**a**) total number of metabolic genes (y-axis) versus genome size (x-axis) in 100 bacteria species; (**b**) size correlation coefficient values (y-axis) for each of 385 superfamily domains displayed against their universal percentages (x-axis). Subplots in the left side show three examples of superfamilies with different size correlation and universal values

To prove and illustrate this assumption, we selected all the superfamilies annotated as metabolic domains in the latest COG database (385 superfamilies), and displayed their respective size correlation coefficient values against their universal distribution percentages throughout all species (see Fig. 2.7). This plot shows that an important percentage of metabolic super families have low correlation with size (206, 54%, superfamilies have a size correlation below 0.5). For example, the Mannitol specific EII domain and the Ferredoxin domain show correlations with size of 0.19 and 0.34, respectively. If size-dependent superfamilies are grouped with the size-independent ones, the resulted sum also shows a size-dependent profile (Fig. 2.7a). In this case, the size-dependent superfamilies hide the neutral behaviour with respect to genome size variation of the rest of size-independent domains. Hence, as argued above, proteins grouped by function do not necessarily show shared behaviour.

2.7 Limits to Genome Complexity in Prokaryotes

The regulatory and metabolic subdivisions, the power-law and linearly distributed superfamilies respectively, represent an important percentage of all

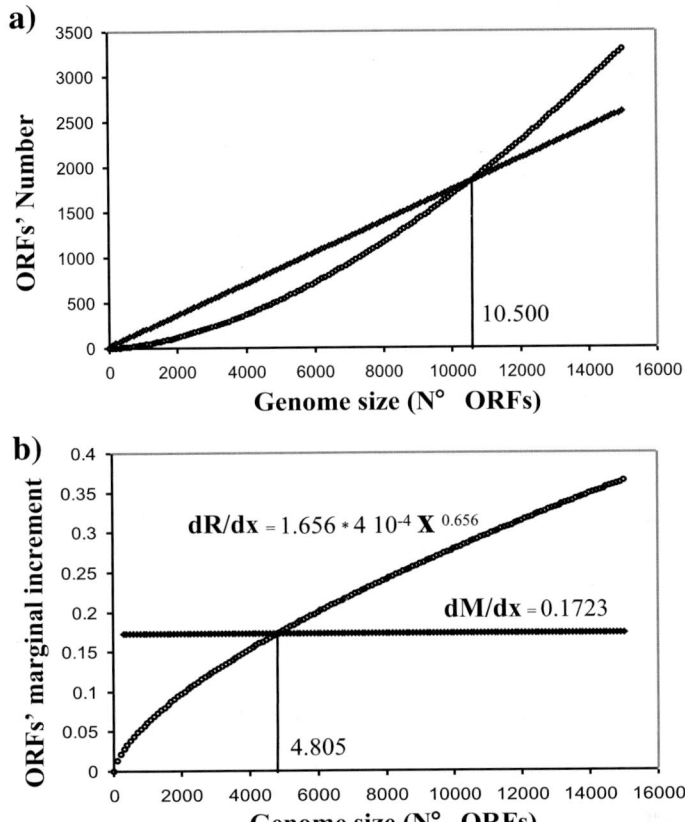

Fig. 2.8. Optimum size determination and bacterial size distribution comparison. (a) Number of ORFs (*y*-axis) calculated from metabolic (*black line*) and regulatory (*white circle line*) distribution models versus genome size (*x*-axis). (b) ORFs marginal increment (*y*-axis) estimated from the derivatives of metabolic (*black line*) and regulatory (*white circle line*) functions calculated in function of the genome size (*x*-axis)

size-dependent superfamilies (90% of all size-dependent and universal domains). Therefore, they are the subdivisions most likely to contribute to general trends in the whole prokaryote sample. Fitting the distributions of these two types of superfamilies to regression lines showed that the functions cross when the genome size reaches 10,500 ORFs (Fig. 2.8a). This suggests that when bacterial size increases above this value, regulatory complexity exceeds metabolic complexity.

These results indicate that regulation could be a primary restriction to increasing complexity in bacterial genomes, an observation in agreement with

Bird's 1995 hypothesis [25], which proposes that increasing complexity is limited by the increasing logistical problems in distinguishing signal from noise [30].

2.8 The Bacterial Factory

To a certain extent, the cell can be considered to be like a factory. Simplistically, the cell can be seen as a manufacturer of functional elements (normally proteins), which then carry out the metabolic reactions to maintain and reproduce the cell; this is equivalent to the production of goods to generate income. We can also consider the means of regulating production (i.e. management of production) as being costs for the cell. Through this analogy a similarity can be seen between optimum bacterial complexity and the optimum production level, giving maximum profit, in a factory.

In a factory, the total profit is the difference between total revenue and total cost, and the marginal profit is the additional profit derived from the production of one additional unit. Consequently, the optimum size of a factory, giving maximum total profit, is reached at a production level where the marginal revenue from producing an additional unit is equal to its marginal cost. Above this point, the marginal cost to produce an additional unit is higher than its marginal revenue [39]. One major reason for the existence of an optimum size in productive systems is the fact that any linear increment in production complexity is usually associated with a larger increment in the associated management cost [40, 41]. This behaviour can be used as an analogous model to describe the behaviour of the universal size-dependent superfamilies in bacteria.

An increase in metabolic complexity provides bacteria with new enzymes to exploit the environment and thus it is reasonable to equate metabolism with bacteria factory revenue, and the increasing regulatory system with the associated management cost. We can compare the effects that increases in genome complexity have on the marginal increases in metabolic and regulatory systems by calculating the derivatives of metabolism and regulation as a function of genome complexity (Fig. 2.8b). The point at which these two marginal increments match (metabolism/regulation) identifies a statistical optimum for bacterial genome size (Fig. 2.6b). In other words, a point at which 'maximum profit' is realised since bacteria get the maximum metabolic variability for minimal regulation cost [30]. As indicated, the optimum size is hence calculated to be roughly 4,800 ORFs – remarkably similar to the genome size of the model prokaryote *Escherichia coli*.

Overall, these results clearly suggest that increasing complexity is not free of cost in bacteria; the more complex a genome becomes the more difficult it is to manage it. Because this regulatory cost increases more steeply than the linear increment in metabolic revenue it finally offsets any advantage gained by increasing the number of metabolic genes. It is reasonable to believe that bacteria achieve an optimum reproductive balance before the increasing regulatory complexity overburdens the cellular system [30].

2.9 Conclusions

Not all the protein superfamilies have contributed in the same way to genome expansion in prokaryotes. In fact, we have identified four different behaviours relative to genome size. First, there are the superfamilies that do not contribute to, and are unaffected by, genome size – the size-independent superfamilies. Second, there are three types of size-dependent behaviour displayed: the linearly, the power-law, and the logarithmically distributed superfamilies. The functional analysis of these four groups unravelled specific functional trends for the size-independent superfamilies and two of the three size-dependent groups. It was observed that the size-independent superfamilies are mainly involved in translation and protein biosynthesis. And within the size-dependent set, the linearly distributed domains are involved in metabolism, while the power-law distributed superfamilies are involved in genetic regulation [30, 34].

The universal size-dependent superfamilies represent universal molecular technology shared by all prokaryotes to perform their metabolic and regulatory processes. Comparison between the regulatory and metabolic subdivisions from this universal and size-dependent set led to models that provide rational explanations for such concepts as the 'genome complexity limit' or 'genome optimum size' in a prokaryote. These findings imply that all prokaryotes have used similar molecular technology to optimise their reproductive efficiency. This common 'molecular technology' defines common limits to prokaryote genome expansion, since the regulatory cost expands much quicker than the metabolic revenue in these organisms.

However, achieving maximum complexity is not a selective pressure. Rather there is a balance between selection for maximum metabolic diversity, which makes energy available to the cell, against selection for a minimal regulatory system, which costs the cell energy. These forces work within the framework of maximally exploiting the environment whilst maintaining minimum cellular doubling time [30, 42].

The investigations discussed above were all based on using the proteome structure information captured by the clustering mechanisms underlying Gene3D. An advantage of Gene3D is that the protein data is based on whole genome sets, allowing analysis of the natural protein universe. Clear advances have been made in describing the evolution of the bacterial protein repertoire and the forces that drive it. Gene3D also describes eukaryotes and archaea, and we hope to extend these investigations to them. Gene3D is simple to view and interrogate through its website, and the underlying data is freely available.

References

1. R. Apweiler, A. Bairoch, C.H. Wu, W.C. Barker, B. Boeckmann, S. Ferro, E. Gasteiger, H. Huang, R. Lopez, M. Magrane, M.J. Martin, D.A. Natale, C. O'Donovan, N. Redaschi, L.S. Yeh, Nucleic Acid Res. **32**, D115 (2004)
2. F. Pearl, A. Todd, I. Sillitoe, M. Dibley, O. Redfern, T. Lewis, C. Bennett, R. Marsden, A. Grant, D. Lee, A. Akpor, M. Maibaum, A. Harrison, T. Dallman, G. Reeves, S. Diboun, S. Addou, S. Lise, C. Johnston, A. Sillero, J. Thornton, C. Orengo, Nucleic Acid Res. **33**, D247 (2005)
3. A. Bateman, L. Coin, R. Durbin, R.D. Finn, V. Hollich, S. Griffiths-Jones, A. Khanna, M. Marshall, S. Moxon, E.L.L. Sonnhammer, D.J. Studholme, C. Yeats, S.R. Eddy, Nucleic Acid Res. **32**, D138 (2004)
4. G.M. Edelman, W.E. Gall, Annu. Rev. Biochem. **38**, 415 (1969)
5. A.G. Rossman, A. Liljas, J. Mol. Biol. **85**, 177 (1974)
6. S.D. Bentley, M. Maiwald, L.D. Murphy, M.J. Pallen, C. Yeats, L.G. Dover, H.T. Norbertczak, G.S. Besra, M.A. Quail, D.E. Harris, A. von Herbay, A. Goble, S. Rutter, R. Squares, S. Squares, B.G. Barrell, J. Parkhill, D.A. Relman, Lancet **361**, 637 (2003)
7. B. Morgenstern, W.R. Atchley, Mol. Biol. Evol. **16**, 1654 (1999)
8. S.A. Teichmann, J. Park, C. Chothia, Proc. Natl. Acad. Sci. USA **95**, 14658 (1998)
9. M. Gerstein, Prot. Struct. Func. Genet. **33**, 518 (1998)
10. D. Lee, A. Grant, R.L. Marsden, C. Orengo, Proteins **59**, 603 (2005)
11. A. Grant, D. Lee, C. Orengo, Genome Biol. **5**, 107 (2004)
12. A.E. Todd, R.L. Marsden, J.M. Thornton, C.A. Orengo, J. Mol. Biol. **348**, 1235 (2005)
13. T. Meinel, A. Krause, H. Luz, M. Vingron, E. Staub, Nucleic Acid Res. **33**, D226 (2005)
14. N. Kaplan, O. Sasson, U. Inbar, M. Friedlich, M. Fromer, H. Fleischer, E. Portugaly, N. Linial, M. Linial, Nucleic Acid Res. **33**, D216 (2005)
15. A. Heger, L.U. Holm, J. Mol. Biol. **328**, 749 (2003)
16. C. Bru, E. Courcelle, S. Carrére, Y. Beausse, S. Dalmar, D. Kahn, Nucleic Acid Res. **33**, D212 (2005)
17. T.F. Smith, M.S. Waterman, J. Mol. Biol. **147**, 195 (1981)
18. E. Hartuv, A. Schmitt, J. Lange, S. Meier-Ewert, H. Lehrach, R. Shamir, Proc. Recomb. **99**, 188 (1999)
19. E.L.L. Sonnhammer, D. Kahn, Prot. Sci. **3**, 482 (1994)
20. J. Gouzy, F. Corpet, D. Kahn, Comput. Chem. **23**, 333 (1999)
21. A. Enright, S. van Dongen, C.A. Ouzounis, Nucleic Acid Res. **30**, 1575 (2002)
22. R.L. Tatusov, N.D. Fedorova, J.D. Jackson, A.R. Jacobs, B. Kiryutin, E.V. Koonin, D.M. Krylov, R. Mazumder, S.L. Mekhedov, A.N. Nikolskaya, B.S. Rao, S. Smirnov, A.V. Sverdlov, S. Vasudevan, Y.I. Wolf, J.J. Yin, D.A. Natale, BMC Bioinform. **4**, 41 (2003)
23. M. Ashburner, C.A. Ball, J.A. Blake, D. Botstein, H. Butler, J.M. Cherry, A.P. Davis, K. Dolinski, S.S. Dwight, J.T. Eppig, M.A. Harris, D.P. Hill, L. Issel-Tarver, A. Kasarskis, S. Lewis, J.C. Matese, J.E. Richardson, M. Ringwald, G.M. Rubin, G. Sherlock, Nat. Genet. **25**, 25 (2000)
24. M. Kanehisa, S. Goto, Nucleic Acid Res. **28**, 27 (2000)
25. A.P. Bird, Trends Genet. **11**, 94 (1995)

26. A. Mira, H. Ochman, N.A. Moran, Trends Genet. **17**, 589 (2001)
27. E.V. Koonin, Y.I. Wolf, G.P. Karev, Nature **420**, 218 (2002)
28. C. Chothia, J. Gough, C. Vogel, S.A. Teichmann, Science **300**, 1701 (2003)
29. H. Ochman, J.G. Lawrence, E.A. Groisman, Nature **405**, 299 (2000)
30. J.A. Ranea, A. Grant, J.M. Thornton, C.A. Orengo, Trends Genet. **21**, 21 (2005)
31. U. Dobrindt, J. Hacker, Curr. Opin. Microbiol. **4**, 550 (2001)
32. N.A. Moran, Cell **108**, 583 (2002)
33. M. Kendall, J.D. Gibbons, *Rank Correlation Methods*, 5th edn. (Oxford University Press, New York, 1990), ISBN 0195208374
34. J.A. Ranea, D.W. Buchan, J.M. Thornton, C.A. Orengo, J. Mol. Biol. **336**, 871 (2004)
35. C. Chothia, Nature **357**, 543 (1992)
36. I. Cases, V. de Lorenzo, C.A. Ouzounis, Trends Microbiol. **11**, 248 (2003)
37. E. Nimwegen, Trends Genet. **19**, 479 (2003)
38. A.E. Todd, C.A. Orengo, J.M. Thornton, J. Mol. Biol. **307**, 1113 (2001)
39. N.G. Mankiw, *Principles of Microeconomics*, 2nd edn. (Dryden, Chicago, 1998), ISBN 0030776422
40. G. Frizelle, *The Management of Complexity in Manufacturing*, (Business Intelligence press, London, 1998), ISBN 1898085366
41. H.A. Orr, Evolution Int. J. Org. Evol. **54**, 13 (2000)
42. A. Dufresne, L. Garczarek, F. Partensky, Genome Biol. **6**, 1 (2005)

3

The Evolution of the Globins: We Thought We Understood It

A.M. Lesk

Protein crystallography achieved its first results in the late 1950s with the structure determinations of sperm whale myoglobin and human haemoglobin. These gave us our first glimpse of the structural changes that take place during protein evolution. Many other structures of proteins in the globin family have continued to reveal interesting and important details of the coordinated divergence during evolution of amino acid sequences and protein structures and functions.

The history of investigations of globin evolution can be divided into stages that correspond to the availability of structures of progressively more highly diverged sets of molecules. The earliest work dealt exclusively with mammalian globins [1]. (Because the divergence of myoglobin from haemoglobin occurred early in the vertebrate lineage, it would be more appropriate to regard this early work as dealing exclusively with vertebrate globins – a relatively restricted group nevertheless.) When additional structures, including invertebrate and plant globins, became available, Lesk and Chothia [2] analysed their architecture, asking the question of how such different amino acid sequences could be compatible with the same basic folding pattern. All the structures then available are now classified as 'full-length' globins, typically ~150 residues long.

Recently, a new family of globins has been discovered that are substantially shorter, containing as few as 109 residues. Globins from *Chlamydomonas eugametos*, *Paramecium caudatum* and *Mycobacterium tuberculosis* are members of the new sub-family of truncated globins. They require a reopening of the questions, supposedly "answered" in previous studies limited to full-length globins, of what are the variable and conserved features of globin structures.

Sperm whale myoglobin, a typical full-length globin, contains 9 helices, labelled A to H. Truncated globins retain most but not all of the helices of the standard globin fold. A notable exception is the loss of the N-terminus of the F helix. The F helix contains the crucial iron-linked histidine. Truncated globins show a shortening of the A helix and of the region between the C and D helices. Of the 59 sites involved in conserved helix-to-helix or helix-to-haem contacts in full-length globins, 41 of them appear, with conserved contacts, in all three truncated globins studied. The structures of the helix–helix interfaces, conserved in full-length globins, are in most cases similar in truncated globins as well, with the notable exception of the interface between B and E helices. This interface has an unusual crossed-ridge structure in full-length globins but a more common structure in truncated globins. The globin structures continue to offer an illuminating case study in protein evolution.

3.1 Introduction

Once upon a time, Lesk and Chothia [2] wrote a paper describing the structural variations in the globin structures then known. That study treated nine full-length globins, all of ∼150 residues in length. All were from metazoa, including vertebrates, invertebrates and plants. The basic conclusions were that

1. The principal determinants of the three-dimensional structure of the globins are the approximately 59 residues that are involved in the packing of helices and in the interactions between the helices and the haem group.
2. Although mutations of the buried residues keep the sidechains non-polar, these residues vary both in amino acid *identity* and *size*.
3. With the exception of some residues near the ends of the helices, most of the helices in globins have geometries close to that of a standard α-helix.
4. In the nine globins studied, the total volume of the residues that form the interfaces between homologous helices varies by up to 57%.
5. Shifts in relative position and orientation of homologous pairs of packed helices may be as much as 7 Å and 30°. (At the time, these changes in residue volume and geometry of pairs of helices in contact were large, relative to the general expectation, which had been based on the much greater similarity of structures of mammalian globins.)
6. Five helix packings occur, with extensive interfacial contact, in all globins studied: A/H, B/E, B/G, F/H and G/H.
7. Differences in the residues at homologous helix interfaces can be related qualitatively to differences in relative geometry of the helices.
8. The geometries of the nine haem pockets are very similar: the shifts in the packings produced by mutations are coupled so as to maintain the same relative geometry for the residues that form the haem pocket.
9. Despite the change in volume of residues at helix interfaces, and the relative shifts and rotations of the helices packed, there is substantial conservation of the *reticulation* of the residues; that is, homologous residues tend to make homologous contacts.

The results of this paper remained valid for additional globin structures that emerged over the years – until recently. In the new century, a novel class of shorter globin structures has appeared, with sizes as small as 109 residues [3–5]

Table 3.1. Truncated globins of known structure

Species of origin	PDB code	Number of residues	Ligand	Resolution Å	Ref.
Chlamydomonas eugametos	1dly	164	CN$^-$	1.8	[3]
Paramecium caudatum	1dlw	116	–	1.54	[3]
Mycobacterium tuberculosis	1idr	136	O$_2$	1.90	[7]

(see Table 3.1). Known as truncated globins (although the residues lost are not exclusively from the chain termini), they are distant relatives of globins from multicellular organisms. Truncated globins have been implicated in diverse functions, including detoxification of NO and photosynthesis.

The purpose of this article is to reopen the question of what fundamentally characterises the globin fold, and specifically to ask to what extent the conclusions of previous work retain validity.

3.2 Coordinates and Calculations

The coordinates of three truncated globins (see Table 3.1) are available from the Protein Data Bank [6].

3.3 Results

3.3.1 Description of Secondary and Tertiary Structure of Full-Length (~150–Residue) Globins

Globins show a characteristic folding pattern based on α-helices that creates a pocket into which the haem binds (see Fig. 3.1). Typical full-length globins, such as sperm whale myoglobin, contain 8 or 9 helices: A, B, C (a 3_{10} helix), D, E, F (split in mammalian globins into two consecutive helices, F' and F), G and H. The F and E helices contain the proximal and distal histidines, respectively – the histidines that interact with the haem group. These two helices form the mouth of the haem pocket. Many globins lack the D helix.

The structural framework is created and stabilised by contacts between pairs of helices. Full-length globin structures share five major common helix contacts: A/H, B/E, B/G, F'F/H and G/H. Many contain some or all of the following minor helix contacts: A/E, E/H, C/G and B/D. These are minor both in the sense of the extent of buried surface area in the contact and in the extent of the conservation of the appearance of the contacts in globins from different species.

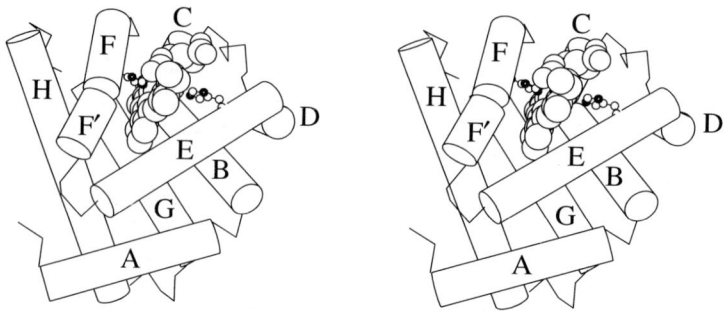

Fig. 3.1. The structure of sperm whale myoglobin, a typical full-length globin. The labels of the helices, in alphabetic order in appearance in the chain, are shown. The atoms of the haem group are shown as *large spheres*. The sidechains of the proximal and distal histidines are shown in *ball-and-stick* representation

Table 3.2. Helix assignments in sperm whale myoglobin and truncated globins

	A'	A	B	C	D	E	F'	F	G	H'	H
1mbo	–	3–18	20–36	37–43	51–58	58–78	82–87	87–96	100–119	124–149	
1dlw	–	1–7	8–27	29–34	–	37–52	–	63–69	75–92	96–107	109–114
1dly	–	1–7	8–25	29–32	–	37–52	–	63–69	76–94	97–110	111–117
1idr	2–10	14–20	21–39	42–46	–	50–67	–	76–82	87–106	108–120	120–126

1mbo = sperm whale myoglobin; 1dlw = *P. caudatum* globin; 1dly = *C. eugametos* globin; 1idr = *M. tuberculosis* globin, A chain

3.3.2 Description of Secondary and Tertiary Structure of Truncated Globins

Table 3.2 contains the helix assignments of the truncated globins and, for comparison, sperm whale myoglobin.

None of the three truncated globins contains a D helix. Surprisingly, the F helix is shortened in the truncated globins, although the proximal histidine retains its position to bind the iron (see later). The A helix is also shortened. In *C. eugametos* and *P. caudatum* globins (1dly and 1dlw), the chain begins at the residue corresponding to residue 10 in sperm whale myoglobin. In contrast, *M. tuberculosis* globin (1dlr) contains an N-terminal extension. Although the A helix in *M. tuberculosis* globin (1idr) is no longer than that in *P. caudatum* globin (1dlw) and *C. eugametos* globin (1dly), the N-terminal extension contains a separate additional helix called the A' helix. In the truncated globins, the H helix is split into two parts, H' and H.

3.3.3 Alignment

It is relatively easy to align the three truncated globins with one another. Figure 3.2 shows that the conformations of *P. caudatum* globin (1dlw) and *C.*

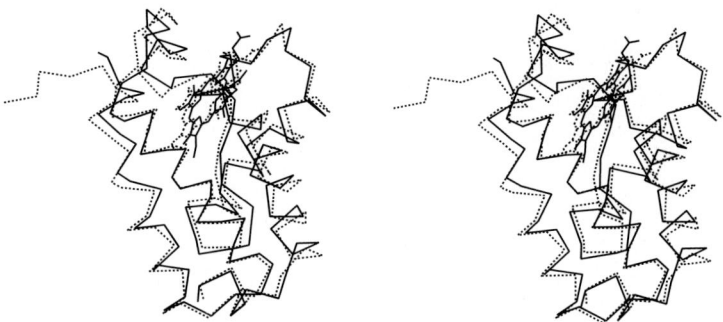

Fig. 3.2. Superposition of mainchains of *P. caudatum* globin (1dlw) (*solid lines*) and *C. eugametos* globin (1dly) (*broken lines*)

Fig. 3.3. Superposition of mainchains of *M. tuberculosis* globin (1idr) (*solid lines*) and *C. eugametos* globin (1dly) (*broken lines*)

eugametos globin (1dly) are rather similar. Figure 3.3 compares the conformations of *M. tuberculosis* globin (1idr) and *C. eugametos* globin (1dly). These are also rather similar except for the chain termini and the region between the E and F helices.

Alignment of truncated and full-length globins is harder but possible. Figure 3.4 shows a structural alignment of sperm whale myoglobin (1mbo) and *P. caudatum* globin (1dlw). The sequence alignment appears in Fig. 3.5. The pairwise percentages of identical residues in this sequence alignment are shown in Table 3.3.

The three truncated globins are fairly close relatives; all are very distant relatives of sperm whale myoglobin (1mbo).

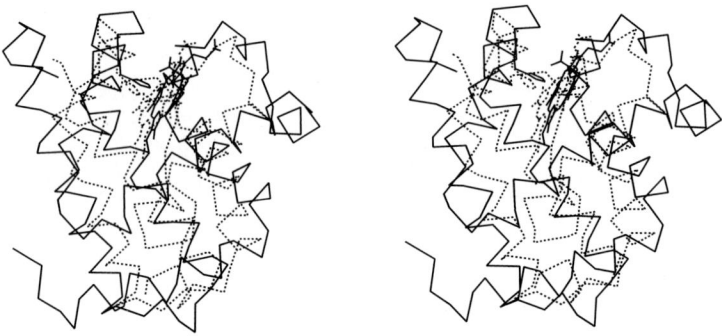

Fig. 3.4. Superposition of sperm whale myoglobin (1mbo) (*solid lines*) and *C. eugametos* globin (1dly) (*broken lines*)

```
                   10         20         30         40         50         60
                    |          |          |          |          |          |
         AAAAAAAAAAAAAAAAAA----BBBBBBBBBBBBBBBBBBBCCCCCCC-------DDDDDDD
    1mbo VLSEGEWQLVLHVWAKVEA---DVAGHGQDILIRLFKSHPETLEKFDRFKHLKTEAEMKA
    1dlw              SLFEQLGGQAAVQAVTAQFYANIQADA-TVATFFN-------------
    1dly              SLFAKLGGREAVEAAVDKFYNKIVADP-TVSTYFS-------------
    1idr    GLLSRLRKREPISIYDKIGGHEAIEVVVEDFYVRVLADD-QLSAFFS-------------

                   70         80         90        100        110        120
                    |          |          |          |          |          |
         EEEEEEEEEEEEEEEEEEEEE--------FFFFFFFFFF------GGGGGGGGGGGGGGGG
    1mbo SEDLKKHGVTVLTALGAILKKKGHHEAELKPLAQSHATKH--KIPIKYLEFISEAIIHVL
    1dlw GIDMPNQTNKTAAFLCAALGGPNAWTGR--NLKEVHA-NM--GVSNAQFTTVIGHLRSAL
    1dly NTDMKVQRSKQFAFLAYALGGASEWKGK--DMRTAHK-DLVPHLSDVHFQAVARHLSDTL
    1idr GTNMSRLKGKQVEFFAAALGGPEPYTGA--PMKQVHQ-GR--GITMHHFSLVAGHLADAL

                  130        140        150
                    |          |          |
         GGG-----HHHHHHHHHHHHHHHHHHHHHHHH
    1mbo HSRHPGDFGADAQGAMNKALELFRKDIAAKYKELGYQG
    1dlw TGAGV---AAALVEQTVAVAETVRGDVVTV
    1dly TELGV---PPEDITDAMVVASTRTEVLNMPQQ
    1idr TAAGV---PSETITEILGVIAPLAVDVTS
```

Fig. 3.5. Sequence alignment of truncated globins and sperm whale myoglobin. The *top line* shows the helices in sperm whale myoglobin

Table 3.3. Pairwise percentages of identical residues in the structure derived sequence alignment of sperm whale myoglobin (1mbo) with the three truncated globins

		1mbo	1dlw	1dly	1idr
Sperm whale myoglobin	1mbo	100	12	8	10
P. caudatum globin	1dlw		100	39	36
C. eugametos globin	1dly			100	40
M. tuberculosis globin	1idr				100

3.4 Helix Contacts

3.4.1 Geometry of Inter-Helix Contacts

The basis of the globin structure is the interactions of packed helices, and of the contacts of the protein with the haem group. Chothia and co-workers [7,8]

analysed the general principles of the packing of α-helices in globular proteins. They found that the structures of most helix–helix interfaces are described by a ridges-into-grooves model: Sidechains on the surfaces of helices form ridges protruding from the helix surface. Ridges may form from residues separated by four positions in the sequence (most common), three positions, or one position. In tertiary structures of proteins, helices tend to pack so that ridges on the surface of one helix pack into grooves of the other. Because ridges formed from different sets of residues have different angles with respect to the helix axis, different combinations of ridge–groove structures produce different inter-axial angles in packed helices. For example, the most common packing – $i \pm 4$ ridges from both helices – corresponds to an inter-axial angle of $\Omega \simeq -40°$. The ridges-into-grooves model explains the distribution of inter-axial angles observed in α-helical proteins.

The angle Ω is defined as follows: Determine the axis of each helix, and the line perpendicular to both axes. Project the axes onto the plane perpendicular to this line. Then Ω is the angle, within the plane, between these projections. By convention, $\Omega < 0$ if the near helix is rotated clockwise relative to the far helix [7]. Ω suffices to define the relative geometry of helices that form face-to-face packings. (If the interface is formed by the end of one or both helices, the angle between the helix axes is reported as the angle τ.)

Although many helix packings in globins follow the ridges-into-grooves model, globins are unusually rich in helix packings of non-standard structure. In the B/E and G/H contacts of full-length globins, ridges that contain a notch (arising from a sidechain smaller than its neighbours along the ridge) pack so that a ridge from one helix crosses over a ridge from the other, at the notch. (The observation of this unusual structural feature in the corresponding helix packings of phycocyanin was adduced as evidence for the evolutionary relationship between globins and phycocyanins [9].)

3.4.2 Pairs of Helices Making Contacts

Of the five major helix contacts in normal-length globins: A/H, B/E, B/G, F/H and G/H, all but the F/H contact appear in the truncated globins. (*P. caudatum* (1 dlw) contains an F/H contact.)

In addition to the five major contacts conserved in full-length globins, other contacts occur sporadically among them: A/E, C/G, B/D and E/H. These do not appear among truncated globins. The truncated globins contain an A/B contact, not observed in typical full-length globins. Table 3.4 shows the relative geometry of the helices packed in sperm whale myoglobin and the truncated globins.

In the analysis of full-length globins, it was observed that the pattern of residue–residue contacts tends to be conserved. That is, if two residues are in contact in the helix interface of one globin, the two homologous residues in

Table 3.4. Relative geometry of packed helices in sperm whale myoglobin and truncated globins

		A+A'			B			C			E			F			G			H+H'		
		r	Ω	τ	r	Ω	τ	r	Ω	τ	r	Ω	τ	r	Ω	τ	r	Ω	τ	r	Ω	τ
A'	1mbo																			8.4	82.8	
+	1dlw	6.3	−98.8	40.6																10.6	42.9	
A	1dly	5.9	−107.7	36.1																10.5	41.8	
	1idr	5.8	−107.0	36.5																11.4	33.7	16.8
B	1mbo							3.9	38.1	31.1												
	1dlw							5.4	61.8	32.7												
	1dly																					
	1idr							5.4	59.8	37.0												
C	1mbo				3.9	38.1	31.1															
	1dlw				5.4	61.8	32.7															
	1dly																					
	1idr				5.4	59.8	37.0															
E	1mbo	10.8	28.6																	7.5	−78.1	
	1dlw																			8.6	−50.6	
	1dly																			9.5	−48.4	
	1idr	4.9	2.5	32.9																9.2	−51.7	
F	1mbo																			9.3	118.3	
	1dlw																			7.3	115.1	31.7
	1dly																			7.0	117.4	
	1idr																			8.3	113.1	32.7
G	1mbo																			8.7	−52.8	
	1dlw																			8.2	−47.7	
	1dly																			8.8	−48.7	
	1idr																			8.6	−44.4	
H'	1mbo	8.4	82.8								7.5	−78.1		9.3	118.3		10.2	−18.3				
+	1dlw	10.6	42.9								8.6	−50.6		7.3	115.1	31.7	9.4	−25.3				
H	1dly	10.5	41.8								9.5	−48.4		7.0	117.4		10.4	−23.9				
	1idr	11.4	33.7	16.8							9.2	−51.7		8.3	113.1	32.7	9.8	−28.8				

another globin are also likely to be in contact, even if the sizes of the residues have changed substantially, and even if the changes are not complementary.[1]

Such conservation of the reticulation of the residues is among the most well-preserved structural feature of distantly related proteins. It forms the basis for the ability of the program DALI to recognise very distantly related proteins [10].

Of the 59 positions described in [2] as common positions involved in the packing of helices and in the interactions between the helices and the haem group in at least 7 of the 9 full-length globins then studied, 41 make contacts to the same helix or the haem in all three truncated globins, and 4 (E19, G9, H15 and H20) make contacts to the same helix or the haem in two of the three truncated globins. Of the other 14 positions, 7 are in the A helix, which is deleted in *P. caudatum* globin (1dlw) and *C. eugametos* globin (1dly); 1 is in the BC corner, which shows a 1-residue deletion in the truncated globins; 3 are involved in a shift of partners in the B/E contact; 1 is involved in a truncation of the C-terminus of the E helix and 2 are in the part of the F helix lost in the truncated globins.

3.4.3 Structures of Helix Interfaces in Truncated Globins, Compared to Those in Sperm Whale Myoglobin

Of the five common helix interfaces observed in full-length globins, the B/G and F/H contacts are $i \pm 4/i \pm 4$ packings, the A/H contact is formed from $i \pm 1$ ridges on the A helix packed against $i \pm 4$ ridges on the H helix and the B/E and G/H contacts have crossed-ridge structures [2].

3.4.4 The B/G Interface

The B/G interfaces in sperm whale myoglobin and those of the truncated globins share a similar structural pattern. In the B/G interface of sperm whale myoglobin (Fig. 3.6), a ridge on the surface of the G helix created by the sidechains of residues 106Phe-110Ala-114Val-118Arg packs into a groove between ridges on the surface of the B helix created by the sidechains of residues 24His-28Ile-32Leu and 31Arg-35Ser. This is a typical $i \pm 4/i \pm 4$ packing, and an inter-axial angle of 127.2°.

In the B/G interface of *P. caudatum* globin (1dlw) (Fig. 3.7), a ridge on the surface of the G helix created by the sidechains of residues 84His-88Ala-92Ala packs into a groove between ridges on the surface of the B helix created by the sidechains of residues 15Val-19Phe and 22Asn-26Asp. This interface has the same $i \pm 4/i \pm 4$ packing, and similar values of the inter-axial distance

---

[1] Complementary mutations preserve the sum of the volumes of the residues in contact, and are the exception rather than, as was once believed, the rule. Complementary mutations are a special case of correlated mutations; correlated mutations in general do not preserve the sum of the volumes of residues in contact.

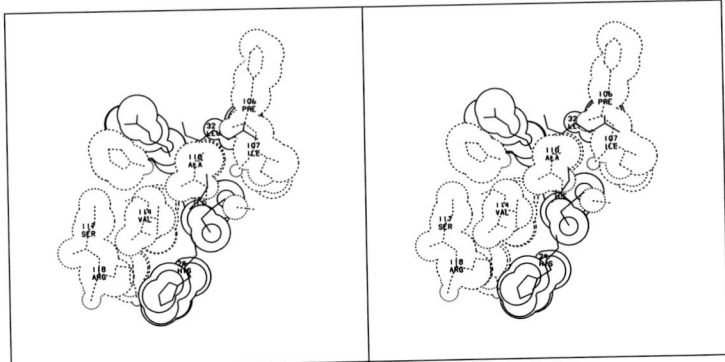

Fig. 3.6. B/G helix interface in sperm whale myoglobin (1mbo). B helix residues in *solid lines* and G helix residues in *broken lines*

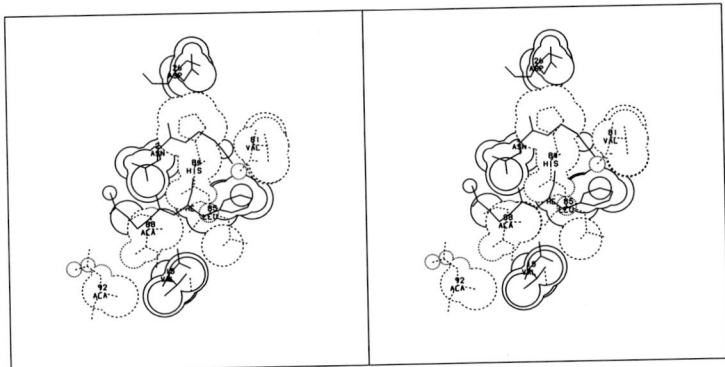

Fig. 3.7. B/G helix interface in *P. caudatum* globin (1dlw). B helix residues in *solid lines* and G helix residues in *broken lines*

and angle and only somewhat smaller buried surface area. It is consistent with the overall sequence alignment, in that residues 110–114–118 of sperm whale myoglobin (1mbo) correspond to 84–88–92 of *P. caudatum* globin (1dlw), and residues 28–32 of sperm whale myoglobin (1mbo) correspond to 19–23 of *P. caudatum* globin (1dlw).

3.4.5 The A/H Interface

In full-length globins, the A/H interface involves an $i \pm 1$ ridge from the A helix packing against an $i \pm 4$ ridge from the H helix. In the truncated globins, the shortness of the A helix makes the contact shorter, leaving a much smaller interface (Fig. 3.8). The surface area buried by the A/H contact in truncated globins is ∼60% of that buried in the A/H contact of sperm whale myoglobin. Indeed, it may be that the requirements of an A/H interaction limit the extent to which N-terminal truncation can occur. This was noted by Pesce et al. [3].

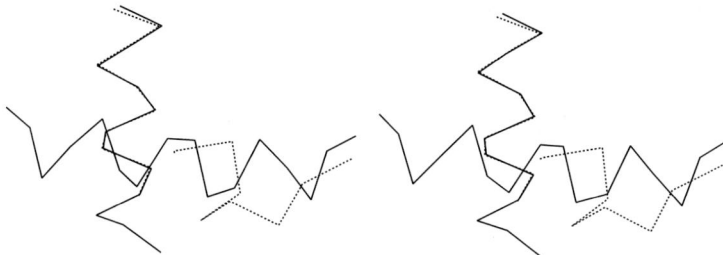

Fig. 3.8. The A and H helices of sperm whale myoglobin (1mbo) (*solid lines*) and *C. eugametos* globin (1dly) (*broken lines*). The structures are superposed on the H helix, the axis of which is vertical

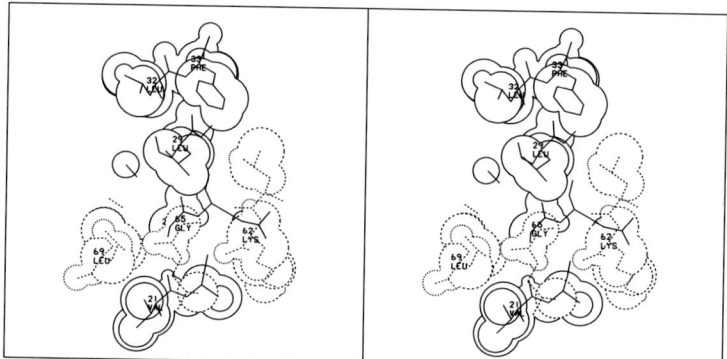

Fig. 3.9. Van der Waals slices through the B/E contact in sperm whale myoglobin (1mbo). In this and the next picture, the axis of the B helix is vertical and the B helix residues are shown in *solid lines*. The axis of the E helix is oblique and the E helix residues are shown in *broken lines*

(In 1966, at a CIBA Foundation discussion, Francis Crick asked about the structure of sperm whale myoglobin: '... it is very unclear to me why you cannot chop-off one or two helices; for example, why shouldn't you chop-off the first length of helix? If you did chop it off what would the molecule look like? Would it fall to pieces? Would it be just a bit unstable? Or would it be more or less the same? [11]'. Now we know.)

3.4.6 The B/E Interface

In full-length globins, the B/E contact has the unusual crossed-ridge structure (Fig. 3.9). However, the values of inter-axial radius and angle are similar to those expected for a standard $i \pm 4/i \pm 4$ contact. The B/E contacts in truncated globins have similar inter-axial distances and angles to the B/E contacts of full-length globins. However, surprisingly the B/E interfaces in truncated globins contain the standard $i \pm 4/i \pm 4$ packing, rather than the

Fig. 3.10. Van der Waals slices through the B/E contact in *C. eugametos* globin (1dly)

unusual crossed-ridge structure (Fig. 3.10). These are the first examples of B/E interfaces in globins with a regular $i \pm 4/i \pm 4$ packing.

The author, in a talk the title of which – 'Protein structures that might exist but don't' – is now seen to contain an erroneous assumption [12], noting the very similar inter-axial distances and angles in the F/H contact in lupin leghaemoglobin, which has a common $i \pm 4/i \pm 4$ packing, and the B/E contacts in several other globins, which have crossed-ridge structures (see Fig. 4 of [2]), proposed as a protein design challenge the engineering of a globin with an $i \pm 4/i \pm 4$ B/E interface. Although no protein engineer took up this challenge, the new structures show that this is indeed possible.

In sperm whale myoglobin, the ridge crossing occurs at the opposition of residues 25Gly and 65Gly (Fig. 3.9). Comparing the relative positions and orientations of the B/E contacts in sperm whale myoglobin and *C. eugametos* globin (Fig. 3.11) with the alignment of the sequences, note that the residue 25Gly in sperm whale myoglobin corresponds to 16Val in *C. eugametos* globin, and 65Gly in sperm whale myoglobin corresponds to 42Arg in *C. eugametos* globin.

3.5 Patterns of Residue–Residue Contacts at Helix Interfaces

We have mentioned the tendency of patterns of residue–residue contacts in proteins to be conserved. In globins, comparisons of the structures of helix interfaces showed that in each helix–helix interface, roughly half the inter-residue contacts are preserved, in the sense that if two positions are in contact in one globin, homologous residues at these positions are likely to be in contact in other globins. Typically half the contacting pairs, those at the centre of the contact interface, are common to all globins studied. Individual globins

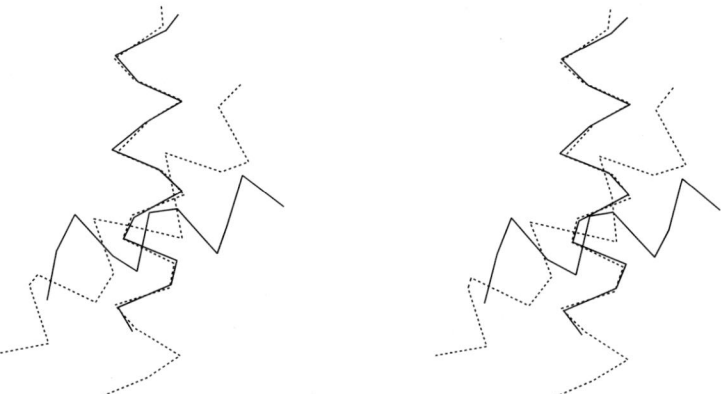

Fig. 3.11. Comparison of relative geometry of B and E helices in sperm whale myoglobin (1mbo) and *C. eugametos* globin (1dly). Sperm whale myoglobin is shown in *solid lines* and *C. eugametos* globin in *broken lines*. The molecules have been superposed on the B helix, the axis of which is vertical in the picture. The shift in relative position and orientation of the E helix with respect to the B helix is no greater than that described in homologous helix contacts that have *similar* packing patterns at their interfaces (see Fig. 6 of [2])

supplement these common contacting pairs with additional contacts that vary from molecule to molecule.

Figure 3.12 shows the inter-residue contacts at the B/G interface in the nine globins studied in previous work [2] and, superposed, the inter-residue contacts at the B/G interfaces in the three truncated globins. These results show that the pattern of contacts in the truncated globins is entirely normal. The sets of conserved contacts appear, and the additional peripheral contacts observed in truncated globins appear in full-length globins also.

In contrast, the inter-residue contacts at the B/E interface, which in truncated globins has a normal $i \pm 4/i \pm 4$ ridge/groove structure in contrast to the crossed-ridge structure of full-length globins, show a very different pattern of inter-residue contacts (see Fig. 3.13).

3.5.1 The G/H Interface

In full-length globins, the G/H contact has an unusual crossed-ridge structure, giving it a somewhat unusual inter-axial angle of $\Omega = -20°$. In truncated globins, the G/H interface is similar, both in inter-axial angle, and in the crossed-ridge structure. Unlike the B/E interface, the G/H interface shares the unusual structure with the full-length globins. It is not unreasonable to suggest that the unusual crossed-ridge structure is required to achieve the unusual inter-axial angle of the G/H interface, unlike the more common inter-axial angle achieved by the B/E interface, which is consistent with either the

70   A.M. Lesk

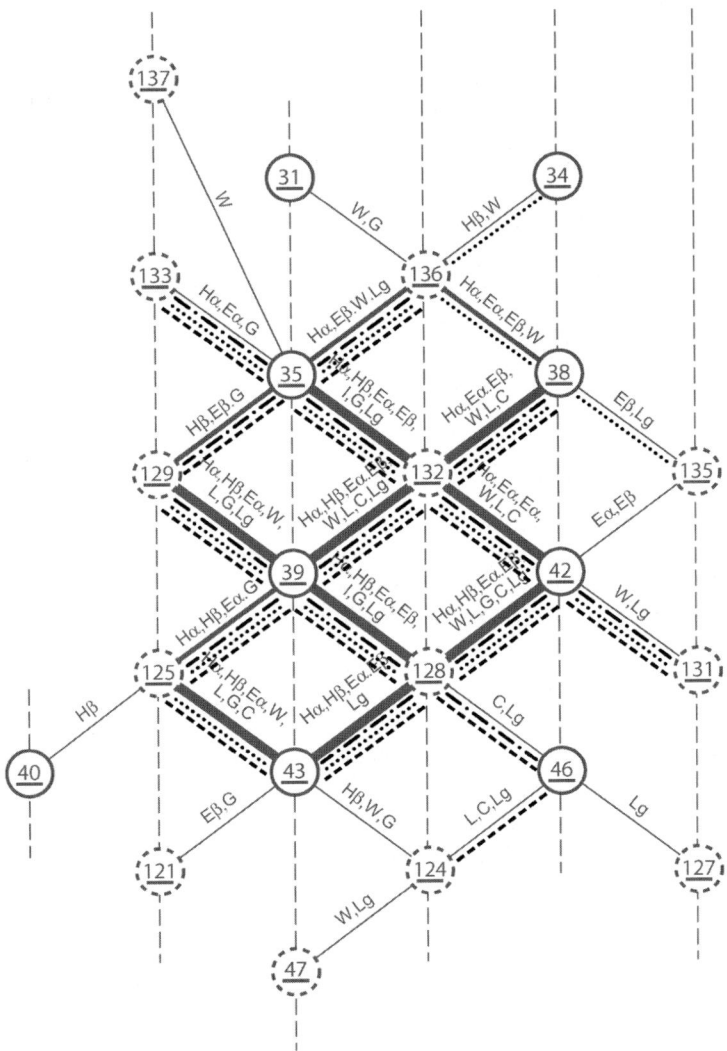

Fig. 3.12. The pattern of residue–residue contacts in the B/G interfaces of full-length (*light gray*) and truncated (*black*) globins. The *grey* background is a copy of Fig. 8a of [2] and shows the contact patterns in the full-length globins: Hα = human haemoglobin, α-chain; Hβ = human haemoglobin, β-chain; Eα = horse haemoglobin, α-chain; Eβ = horse haemoglobin, β-chain; W = sperm whale myoglobin; L = lamprey globin; G = glycera globin; C = *Chironomus* erythrocruorin; Lg = lupin leghaemoglobin. The *oblique lines* indicated the residues making contacts; the width of these lines in the original (*light grey*) figure reflects the number of full-length globins in which homologous contacts are observed. *Fully dark* are the residue–residue contacts in truncated globins: 1dly = ..., 1dlw = – – –, 1idr = – · – · –

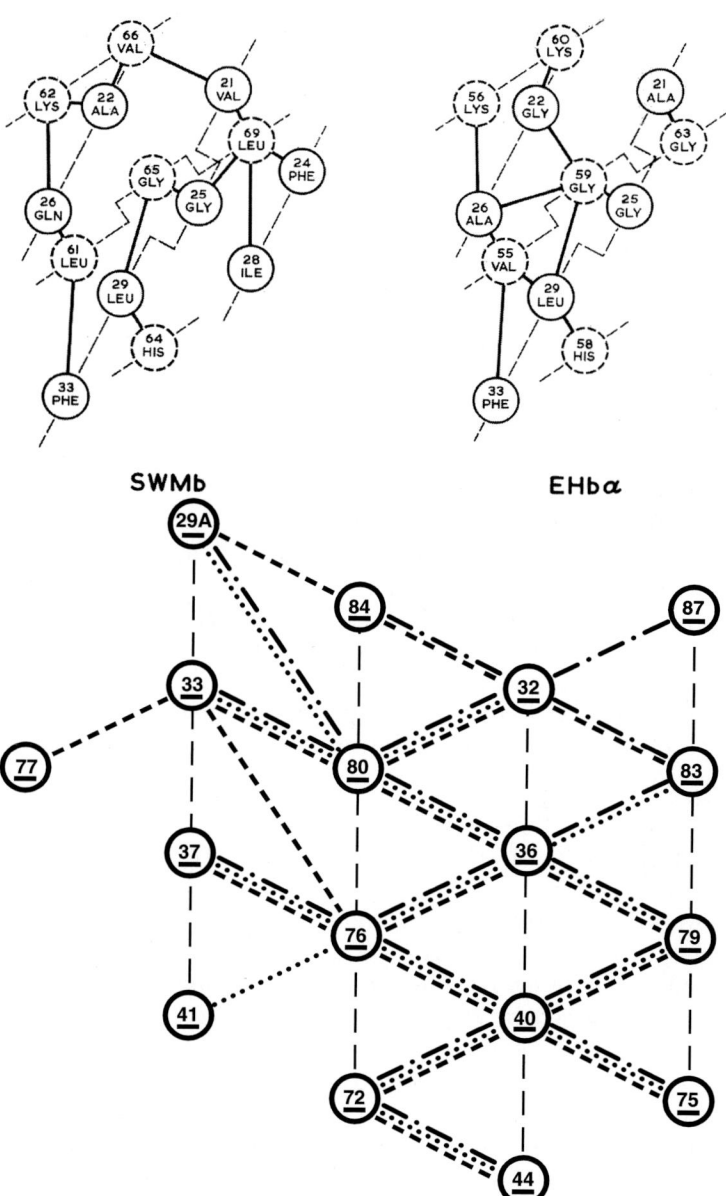

Fig. 3.13. The patterns of residue–residue contacts in the B/E interfaces of full-length and truncated globins. (*Top*) The crossed-ridge structure of the B/E contact in sperm whale myoglobin (Swmb) and Horse haemoglobin, α-chain. (This reproduces part of Fig. 9a of [2].) (*Bottom*) The normal $i \pm 4/i \pm 4$ ridge/groove structure in truncated globins: 1dly = ..., 1dlw = – – –, 1idr = – · – · –

common $i \pm 4/i \pm 4$ ridge–groove structure or with the unusual crossed-ridge structure.

3.6 Haem Contacts

With some significant differences, the haem-binding residues are conserved. In the previous investigation of full-length globins, it was found that residues at 15 positions make contact with the haem in seven or more of the nine globin structures studies (Table 6 of [2]). In 14 of these cases, the homologous residue makes contact with the haem in the three truncated globins.

3.7 The Tunnel

In full-length globins the O_2 binding site is blocked, and the molecules must partially unfold to permit ligand entry and exit. In contrast, in the truncated globins there are channels linking the ligand binding site to the surface of the molecule. In *M. tuberculosis* globin (1idr), the tunnel has one branch entering the molecule between the AB and GH inter-helical regions, and a second branch between the G and H helices [13].

What mutations were required to create the tunnel in truncated globins? Several structural features contribute to the blocking of the tunnel in sperm whale myoglobin. These include

1. A mutation from 94V in *M. tuberculosis* globin to 107I in sperm whale myoglobin; this position is in the middle of the G helix, and is part of the B/G contact.
2. A mutation from 15I in *M. tuberculosis* globin to 14W in sperm whale myoglobin, and also a change in position of helix A.
3. The contacts between residues 29V, 59V, 62F and 98L in *M. tuberculosis* globin are replaced by residues 28I, 72L, 69L and 111I in sperm whale myoglobin. These residues are homologous except for position 32F in *M. tuberculosis* globin that is really aligned to 28I in sperm whale myoglobin.
4. Residues 106G and 17D in *M. tuberculosis* globin are replaced by 119H and 16K in sperm whale myoglobin.

3.8 Conclusions

Study of distantly related proteins is a way to ask Nature to tell us what is essential for a protein folding pattern. The idea is that those features of the structure conserved throughout the family are essential to create the common topology, and those features that vary within the family are nonessential.

A difficulty with this approach is that one is always dealing with a finite set of structures. It may be possible to determine the common features of the structures available, but if at some future date (even 25 years later) new related structures appear, the extended family may have fewer common features than one had previously concluded. This is the problem which structural genomics projects are aimed at ameliorating.

The truncated globins are recognisable members of the family. They possess many but not all the common features derived from comparisons of full-length globins.

The conclusions of the earlier work were that the globin fold is created and stabilised by a common set of helix–helix and helix–haem contacts. Most but not all of the conserved positions in these internal interfaces are retained in the truncated globins.

The most striking structural differences between the full-length and truncated globins are the re-conformation of the N-terminal part of the F helix (keeping the proximal histidine in position to bind the iron but losing the F-helix/H-helix contact), the loss or re-conformation of the N-terminal part of the A helix, shortening of the CD region and the change in structure of the B/E helix contact, from a crossed-ridge structure in full-length globins to a standard $i \pm 4/i \pm 4$ ridge-groove packing in the truncated globins.

Do we know now what the minimal components of a globin are? It is known that a 108-residue peptide corresponding approximately to the central exon of horse myoglobin binds haem and must therefore retain some of the structure of the haem pocket [14]. It has been suggested that a small fragment of the globin fold is homologous to part of colicin A [15]. The story may not yet be over.

References

1. M.F. Perutz, J.C. Kendrew, H.C. Watson, J. Mol. Biol. **13**, 669 (1965)
2. A.M. Lesk, C. Chothia, J. Mol. Biol. **136**, 225 (1980)
3. A. Pesce, M. Couture, S. Dewilde, M. Guertin, K. Yamauchi, P. Ascenzi, L. Moens, M. Bolognesi, EMBO J. **19**, 2424 (2000)
4. M. Milani, A. Pesce, Y. Ouellet, M. Ascenzi, M. Guertin, M. Bolognesi, EMBO J. **20**, 3902 (2001)
5. J.B. Wittenberg, M. Bolognesi, B.A. Wittenberg, M. Guertin, J. Biol. Chem. **277**, 871 (2002)
6. H.M Berman, J. Westbrook, Z. Feng, G. Gilliland, T.N. Bhat, H. Weissig, I.N. Shindyalov, P.E. Bourne, Nucleic Acids Res. **28**, 235 (2000)
7. C. Chothia, M. Levitt, D. Richardson, Proc. Natl. Acad. Sci. USA **74**, 4130 (1977)
8. C. Chothia, M. Levitt, D. Richardson, J. Mol. Biol. **145**, 215 (1980)
9. A. Pastore, A.M. Lesk, Protein Struct Funct Genet **8**, 133 (1990)
10. L. Holm, C. Sander, J. Mol. Biol. **233**, 123 (1993)

11. F.H.C. Crick, in *Ciba Foundation Symposium: Principles of Biomolecular Organisation*, ed. by G.E.W. Wolstenholme, M. O'Connor (J. and A. Churchill Ltd., London, 1966), p. 98
12. A.M. Lesk, FASEB J. **11**, Abstract 89 (1997)
13. M. Milani, A. Pesce, Y. Ouellet, M. Ascenzi, M. Guertin, M. Bolognesi, EMBO J. **20**, 3902 (2001)
14. G. De Sanctis, G. Falcioni, B. Giardina, F. Ascoli, M. Brunori, J. Mol. Biol. **188**, 73 (1986)
15. L. Holm, C. Sander, Nature **361**, 309 (1993)

4

The Structurally Constrained Neutral Model of Protein Evolution

U. Bastolla, M. Porto, H.E. Roman and M. Vendruscolo

The observation that protein sequences accumulate substitutions in time at an almost regular rate [1] created a great interest in molecular evolution, suggesting that substitutions in protein sequences can be used as an effective 'molecular clock' for estimating the time elapsed from the last common ancestor among genes [1–5]. This approach opened a new avenue for reconstructing the tree of life by analyzing the sequences of orthologous genes, whose evolutionary tree coincides with the tree of the species containing them. The practical importance of the study of molecular evolution became therefore evident as a way to reconstruct natural histories.

In addition, the molecular clock hypothesis sparked a lively debate about the mechanisms of molecular evolution. Kimura [6, 7] and King and Jukes [8] proposed that most substitutions in protein sequences are fixed in evolving populations not because they offer a selective advantage but, rather, because they are effectively neutral and therefore invisible to natural selection. The 'neutral theory' could account for the regular rate in time of the accumulation of amino acid substitutions. It failed, however, to predict correctly other features of the evolutionary process, among which the variance of the number of substitutions [9].

One is now starting to understand the reasons for this apparent limitations of neutral theories, thanks to the recent progress in structural biology. This progress has begun to make possible the use of structural information in evolutionary studies, starting with the pioneering works of the Vienna group on the RNA model [10–12] (see also the chapter by Schuster and Stadler in this book), whereas the study of molecular evolution was initially almost entirely based on the analysis of macromolecular sequences [3, 4, 7]. It appears that a paradigm shift is taking place in the field of molecular evolution, from coding symbols (sequence) to coded meaning (structure and function). This book investigates this new approach at several levels of biological organization.

In this chapter, we review some results that were obtained through approaches in which the structural stability of the native state of proteins is taken explicitly into account as a constraint on the evolutionary process [13–30], and in particular through the Structurally Constrained Neutral (SCN) model of protein evolution [31, 32].

We will also show that several results of SCN simulations can be rationalized and rederived analytically by considering a vectorial representation of protein sequences and structures. In this approach, protein sequences are represented as hydrophobicity profiles HPs [33] and protein structures are represented through the principal eigenvector (PE) of the contact matrix [34–37]. As we have shown that the optimal HP and the structural profile are strongly correlated [38], an 'optimal' HP can be derived, i.e. the profile best compatible with a given protein structure. In simulations of SCN evolution, sequence vectors move around this optimal one. This scheme provides us with a framework that can be used to predict, by analytical calculations, site-specific conservation due to structural constraints and site-specific amino acid distributions [39, 40].

4.1 Aspects of Population Genetics

First of all, we need to state some terminology. A mutation is a microscopic event in which the sequence of a gene is altered in a single individual. At the population level, a substitution is a macroscopic event in which the representative, or wild-type, gene changes as a result of the fixation of a mutant gene.[1] Natural selection mediates this transition from the microscopic to the macroscopic level. In physical sciences, a similar role is played by statistical mechanics, which explains macroscopic phenomena in terms of the behaviour of their microscopic components. One of the aims of this chapter is to explore this analogy further.

Three main factors influence the fixation of a mutant allele in a population: the size of the population, M; the selective effect of the mutation, measured through its fitness relative to the wild-type, s; and the rate at which mutations occur, measured in mutations per gene and generation, μ.

4.1.1 Population Size and Mutation Rate

In most of this chapter, we will consider the limit of very small mutation rates, $M\mu \ll 1$, as it is customary in classical population genetics. For $M\mu \ll 1$, the time scale for the appearance of a new mutant $(1/\mu)$ is much larger than the time scale for fixation of a neutral allele, which spans on the average M generations. This limit implies that the population is fairly homogeneous genetically, and at any generation there is at most one mutant arising. This has been termed the 'blind-ant' regime [41] because the population can only test a very small neighbourhood in genotype space at any time. The opposite regime, $M\mu \gg 1$, is assumed to hold in the 'quasispecies' model [42, 43], which

---

[1] Fixation of a mutation takes place when all individuals in the population are descendent of one individual bearing that allele.

considers infinite population sizes (concerning this regime, see the chapters by Jain and Krug and by Lázaro in this book).

To justify the choice of the blind-ant regime, we note that the mutation rate in mammalian genomes was estimated to be 5×10^{-9} per nucleotide per year [3], which, for a species with generation time of two years and a protein of 600 nucleotides (i.e. 200 amino acids) yields $\mu = 6 \times 10^{-6}$. An even smaller value of μ would have resulted by considering that many mutations are synonymous. For a population of effective size $M = 10^5$ (already a quite large estimate)[2] one obtains $M\mu = 0.6$. Although this value is not so small, numerical studies reveal that the results valid in the blind-ant regime continue to be valid qualitatively for $M\mu$ of order one (see Sect. 4.1.5).

It has been argued that the opposite regime of large $M\mu$ is valid for RNA viruses (see the chapter by Lázaro in this book), which have very high mutation rates, of the order of one nucleotide per genome per year [44], corresponding to $\mu \approx 10^{-1}$. Their effective population size is, however, quite reduced because of the bottlenecks that the population suffers when transferred from one host to the other (in these cases, the effective population size essentially coincides with the population at the bottleneck [4]).

4.1.2 Natural Selection

The other relevant parameter for the evolutionary dynamics is the difference in fitness between competing alleles. Since reproduction is inherently stochastic, there is a chance that the less fit allele is fixed even starting as a single individual. Different stochastic models of the reproductive process give qualitatively similar results. We illustrate them through the Moran's birth and death process [45]. According to this model, the probability that a mutant allele B with fitness $F(B)$, arising as a single individual in a haploid[3] population of size M, substitutes the wild-type A with fitness $F(A)$, is given by

$$P_{\text{fix}}(A \to B) = \frac{1 - e^{f(B)-f(A)}}{1 - e^{M[f(B)-f(A)]}}, \tag{4.1}$$

where $f(x) = \log[F(x)]$ with $x = A, B$. We will define in the following $s = f(B) - f(A)$. Notice that if $|Ms|$ is small there is a significant probability that even deleterious mutations ($s < 0$) are eventually fixed in the population.

Berg et al. [46] and Sella and Hirsh [47] have recently noticed that the above formula has an interesting analogy with the stochastic processes used to simulate statistical mechanical systems, since it satisfies the condition of

---

[2] The effective population size is the effective number of breeding adults in a population after adjusting for diverse factors, including reproductive dynamics. The effective population size is usually much less than the actual number of living or reproducing individuals [7].

[3] Haploid organisms carry one single copy of each chromosome, in difference to diploid organism carrying two copies of each chromosome.

detailed balance, $\pi(A) P(A \to B) = \pi(B) P(B \to A)$, with respect to a stationary distribution $\pi(A)$ that is analogous to a Boltzmann distribution in statistical physics (see the chapter by Lässig in this book). If the mutation process satisfies detailed balance with respect to a stationary distribution $\pi_{\mathrm{mut}}(A)$, as it is assumed in many models of molecular evolution [4], then the stationary distribution of the substitution process is

$$\pi(A) = \frac{1}{Z} \pi_{\mathrm{mut}}(A) \, e^{Mf(A)}. \qquad (4.2)$$

This equation is formally identical to a Boltzmann distribution in statistical physics if one identifies the logarithmic fitness $f(A)$ as the energy and the population size M as the inverse temperature (Z is a normalization constant). Smaller populations evolve at higher temperature, in the sense that the evolution is more dominated by stochastic events, and their mean fitness is lower than for corresponding larger populations.

The above result is valid for the small mutation rate regime. It is interesting that a formal analogy between evolving systems and statistical mechanical systems can be derived also for the quasi-species regime, where the infinite population limit is considered. In this case, the mutation rate μ, considered to be vanishingly small in the previous approach, plays the role of the temperature [48, 49]. For a treatment of this subject (see Chap. 14 by Jain and Krug).

4.1.3 Mutant Spectrum

We now go back to classical population genetics. It is customary to divide mutations into four classes, depending on their fitness effect (for a deeper discussion of this topic, see Chap. 13).

1. *Strongly deleterious mutations*: $Ms \ll -1$. These mutations decrease significantly the fitness of the individuals carrying them and they are soon removed from the population through purifying selection.
2. *Nearly neutral mutations*: $-\log(M) \leq Ms \leq \log(M)$. The fitness effect of these mutations is of the same order of importance as are reproductive fluctuations, and their fate is determined both by selection and by random drift [50, 51] (see also the chapter by Ohta in this book). Deleterious mutations in this range have a non-vanishing probability to lead to substitutions. The detailed balance condition, satisfied by several models of the substitution process, including the one presented above, implies that the frequency of mildly deleterious and mildly advantageous substitutions must be equal on average [47], as also previously noted by several authors, which is in contrast with the emphasis of some studies on mildly deleterious substitutions. The advantageous compensatory substitutions play an important role in the dynamics of viral populations, as discussed in the chapter by Lázaro in this book. For small $|Ms|$, the average time required for fixation of these substitutions is of the order of the population size M.

3. *Neutral mutations*: They have negligibly small effects on the fitness, $Ms \approx 0$ and can spread in the population through random genetic drift. The probability of fixation of a neutral mutation is $1/M$, and the expected time for fixation is of order M.
4. *Advantageous mutations*: $Ms \gg 1$. These mutations are efficiently fixed in the population through natural selection with probability close to one, and the time for fixation increases only logarithmically with the population size as $\log(M)/s$.

This classification is useful for distinguishing between different evolutionary scenarios, as advantageous, neutral and nearly neutral mutations can lead to substitutions. In the early years of population genetics, the emphasis was placed on the positive selection of advantageous mutations as the dominant force acting on the substitution process [52]. However, the accumulation of protein sequences eventually changed this view. To explain the very high amount of heterozygosity found in natural populations, as well as the molecular clock hypothesis, at the end of the 1960s Kimura [6] and King and Jukes [8] proposed that most substitutions are selectively neutral. This hypothesis, provocative and controversial at that time, lead to a simple mathematical model of the substitution process that will be discussed in Sect. 4.1.4. The neutral model is now considered by many as the null model of molecular evolution, and distinguishing positive selection from a neutral background is the subject of a vast area of evolutionary sequence analysis [53,54]. Subsequently, Ohta and Kimura [50] introduced the concept of nearly neutral substitutions, and Ohta [51] proposed that most substitutions belong to this class.

As more specifically discussed in the chapter by Ohta in this book, there are testable differences between neutral and nearly neutral substitutions, in particular: (a) The rate of nearly neutral substitutions, especially non-synonymous ones, is expected to decrease with population size.[4] This dependence can explain the discrepancies observed between various mammalian groups in the substitution rates per generation [55]. (b) The presence of nearly neutral substitutions implies that compensatory substitutions must be positively selected. This might explain the surprisingly high level of positive selection detected recently [54] using the McDonald and Kreitman test [53]. (c) In nearly neutral, but not in neutral, evolution, macromolecular properties are expected to be less optimized in smaller populations. Studies of endosymbiotic bacteria, which have small effective populations because of the bottleneck in the transmission from one host to its offsprings, have predicted that r-RNA molecules coded in the genomes of endosymbiotic bacteria have lower thermodynamic stability [56] and that their proteins are less stable with respect to misfolding [57]. These findings are consistent with the high expression of chaperones, which are proteins that assist the folding of other proteins, observed in endosymbiotic

---

[4] In principle, also the neutral substitution rate should decrease with the population size since the condition for a mutation to be neutral is $Ms \approx 0$. This effect, however, is usually neglected in mathematical models.

bacteria [58], and that can favour fitness recovering in a bacterial population subject to strong bottlenecks [59] (see Chap. 7).

4.1.4 Neutral Substitutions

The neutral theory of Kimura is based on the assumption that the fitness effect of a mutation with respect to the wild-type, s, has a bimodal distribution, with the most likely effects corresponding either to strongly disadvantageous ($Ms \ll -1$) or to neutral mutations ($Ms \approx 0$). Advantageous mutations are not considered because they are expected to be rare, at least for proteins that maintain the same function and evolve in the rather stable cellular environment [60]. The neutral theory therefore applies to families of orthologous proteins, whose evolutionary tree coincides with the species tree, and whose function and structure is expected to be conserved in evolution. On the other hand, paralogous proteins, which diversified after an event of gene duplication specializing into different functions (as for instance myoglobin and the two hemoglobin chains), undergo several positively selected substitutions in the process of developing a new function, as it is witnessed by the acceleration of the substitution rate after gene duplication [3]. Nearly neutral mutations are not considered for the sake of mathematical simplicity. From the point of view of the neutralist–selectionist controversy that was discussed for several decades in the molecular evolution literature, nearly neutral substitutions were often considered on the same ground as strictly neutral one, despite the differences discussed in the previous section.

In Kimura's model, neutral mutations undergo a diffusion process that in the population genetics literature receives the name of 'random genetic drift'. The rate at which neutral mutations occur in individual genes is μx, where μ is the mutation rate and x is the probability that a mutation is neutral. This probability is considered to be independent of population size M, even though, strictly speaking, the condition that a mutation is neutral is $s \ll 1/M$. The connection between the population size and the substitution rate lays at the heart of the nearly neutral theory and distinguishes it from the original neutral theory.

The number of neutral mutations arising in one generation is therefore $M\mu x$ and, since the probability that one of them substitutes the wild-type is $1/M$ (all the M genes have the same selective value), the neutral substitution rate per generation is given by

$$\frac{\mathrm{E}[S_t]}{t} = \mu x \qquad (4.3)$$

and it is independent of M. Here, S_t is the number of accepted neutral mutations in a time interval t. This provides a sort of molecular clock, in agreement with the earliest empirical observations [7], but in worse agreement with the so-called generation time effect (see the chapter by Ohta in this book).

Another assumption, which we call the 'homogeneity hypothesis', is that the neutral mutation rate $x(\mathbf{A})$,[5] which in principle may be different for all sequences $\mathbf{A}$, is constant throughout evolution, $x(\mathbf{A}) \equiv x$. As shown later, this hypothesis implies that the number of neutral substitutions has a Poissonian distribution in the low mutation limit $M\mu \ll 1$. The population, as we mentioned above, is fairly homogeneous in this limit and there is at most one mutant arising at each generation. The number of mutations taking place in time t in an individual lineage is a Poissonian variable with mean value μt. For a population, the number of mutations is the sum of M Poissonian variables, and it is still Poissonian with mean $M\mu t$. The probability that one of these mutants become fixed is the product of the probability that the mutation is neutral, x, times $1/M$. Since at every generation there is at most one mutant, the probability of n out of m mutants becoming fixed is $\binom{m}{n}(x/M)^m(1-x/M)^{m-n}$. Therefore, the probability that there are n neutral substitutions within a time interval t is given by

$$P\{S_t = n\} = \sum_{m=n}^{\infty} e^{-M\mu t}\frac{(M\mu t)^m}{m!}\binom{m}{n}\left(\frac{x}{M}\right)^n\left(1-\frac{x}{M}\right)^{m-n}$$
$$= e^{-\mu x t}\frac{(\mu x t)^n}{n!}. \qquad (4.4)$$

As one can see, the result is a Poissonian variable with average value $\mu x t$. The homogeneity hypothesis seems at first sight very plausible since the neutral fraction x results from the average over a large number of sites in a gene. If the evolving sites are uncorrelated, the law of large numbers implies that the fluctuations of x vanish. However, as we shall see later, stability constraints introduces global correlations between the sites of protein coding genes, so that the homogeneity hypothesis is violated in models that take into account such stability constraints.

4.1.5 Beyond the Small $M\mu$ Regime: Neutral Networks

In the next sections, we shall consider the small $M\mu$ limit (the blind-ant regime). In this regime, the substitution process can be represented through the evolution of a single wild-type sequence. It should be emphasized that this set-up does not correspond to a one-individual population, but rather to a large population with a small mutation rate $\mu \ll 1/M$, so that most individuals have the same genotype. The population maintains the wild-type genotype until one of the possible neutral mutations is fixed. One time step in this set-up corresponds to the typical time for the fixation of a neutral mutation, M.

---
[5] We adopt a notation in this chapter where bold-face mathematical symbols such as $\mathbf{A}$ indicate vectors (sequences) or matrices, whereas A_i indicates the i-th component of $\mathbf{A}$.

When the mutation rate is not small, however, the fate of a genotype depends not only on its fitness $F(\mathbf{A})$, as indicated in (4.2) but also on the fitness of its neighbours in sequence space that can be connected to it through point mutations. An important quantity in this regime is the mutation load, i.e. the fraction $\mu(1 - x(\mathbf{A}))$ of offsprings of individuals with genotype $\mathbf{A}$ that undergo lethal mutations. If the homogeneity hypothesis does not hold and $x(\mathbf{A})$ fluctuates in sequence space, the population dynamics may favour genotypes with large neutrality fraction $x(\mathbf{A})$ and hence small mutation load. The parameter that controls whether this is the case is the product $M\mu$. As discussed earlier, a population with very small $M\mu$ can be represented through a single effective sequence evolving in the blind-ant regime. In the opposite limit of very large $M\mu$ (the quasi-species regime [42]), the distribution of the population in sequence space can be obtained analytically for a neutral model in which all viable sequences have the same fitness $F(\mathbf{A})$.

The result can be cast into a simple form [41]: Define the neutral connectivity matrix $x(\mathbf{A}, \mathbf{A}')$ to be 1 if $\mathbf{A}$ and $\mathbf{A}'$ are two viable sequences that can be connected through one point mutation and 0 otherwise. This matrix describes a neutral network of viable sequences interconnected through point mutations [10]. The stationary distribution of the fraction of individuals with genotype $\mathbf{A}$, $\rho(\mathbf{A})$, has to satisfy the stationarity condition $\rho(\mathbf{A}) = \sum_{\mathbf{A}'} \rho(\mathbf{A}') \, x(\mathbf{A}', \mathbf{A})$ and therefore it is proportional to the component of the PE of the neutral connectivity matrix for genotype $\mathbf{A}$. This component constitutes a sort of effective neutral connectivity of sequence $\mathbf{A}$ and it is positively correlated with the fraction of neutral neighbours $x(\mathbf{A})$ (see Sect. 4.4.1). Therefore, sequences with large $x(\mathbf{A})$ are more populated, and the mutation load is reduced.

Van Nimwegen et al. [41] simulated population dynamics on a neutral network $x(\mathbf{A}, \mathbf{A}')$, obtained from the predicted folding properties of a small RNA molecule. They found that the blind-ant regime is a good approximation up to $M\mu \approx 10$ and the large $M\mu$ regime is approached at $M\mu \approx 200$. Similar results were obtained by Wilke [61] using the neutral network obtained through the predicted folding thermodynamic properties of a model protein. We argue that the value of $M\mu$ at which the cross-over of the two regimes takes place depends on the correlation length of $x(\mathbf{A})$ in sequence space, ℓ_x. In fact, in neutral evolution the population occupies a region in sequence space around the wild-type with radius of order $M\mu$ mutations [45]. If this radius is smaller than ℓ_x, then all values of $x(\mathbf{A})$ in the population are fairly similar and the small differences in the mutation load can not be fixed in the population.

For animal and plant populations, characterized by small mutation rate and effective population sizes of tens of thousands of individuals, $M\mu$ is of order one and one would expect that the blind-ant regime is still a good approximation to the neutral dynamics. On the contrary, viral populations have large $M\mu$, compatible with the cross-over region towards the quasi-species regime.

We end this section with a summarizing comparison between the two limiting regimes of population genetics. Population genetics models can be simplified in two opposite regimes: very small (blind-ant regime) and very large (quasi-species regime) $M\mu$. In both cases, a formal analogy with statistical mechanical systems can be established. For $M\mu \ll 1$, when the population is fairly homogeneous, the negative of the logarithmic fitness plays the role of the energy function and the inverse of the population size plays the role of temperature. For $M\mu \gg 1$, when the population is very spread in sequence space, a combination of the negative of the logarithmic fitness with a mutation term plays the role of the energy and the mutation rate plays the role of temperature [48, 49] (see the chapter by Jain and Krug in this book). As the simulations by van Nimwegen et al. [41] and by Wilke [61] show in this case, even when mutant alleles are completely neutral under the point of view of the fitness, they may not be neutral under the point of view of mutation resistance. In the following, only the small $M\mu$ regime will be examined, since this is the relevant regime for many biological populations, most notably higher eukaryotes.

4.2 Structural Aspects of Molecular Evolution

4.2.1 Neutral Theory and Protein Folding Thermodynamics

The thermodynamic stability of the native state is a strong constraint on molecular evolution, and a consequence of the more general requirement of maintaining the biological function [62]. The native state of a protein must be stable with respect to both unfolding and misfolding [63]. However, the stability against unfolding and stability against misfolding are anticorrelated [57, 64]. Therefore, natural selection cannot achieve simultaneously the optimal value for both stability requirements and has to trade off between them.

Natural selection eliminates mutations that reduce folding stability and favors the fixation of more stable proteins. Nevertheless, natural proteins are only marginally stable against unfolding [65], and it is not difficult to engineer protein mutants to improve their stability. Moreover, a large number of mutations do not alter significantly the measured thermodynamic stability or the function of the protein. In the framework of the neutral theory of molecular evolution [6], these results can be interpreted, assuming that changes increasing folding stability are selectively neutral above some specific thresholds. According to this hypothesis, the threshold values are most frequently realized in protein evolution, because they correspond to an overwhelming portion of sequence space. This framework provides a possible explanation for the relatively low stability of native states of proteins [22] and for the fact that the observed amino acid occurrences are very close to the ones predicted from nucleotide occurrence frequencies [66, 67].

4.2.2 Structural Conservation and Functional Changes in Protein Evolution

It has since long been established that protein structures evolve much more slowly than protein sequences [68,69]. Methods of protein structure prediction on the basis of sequence homology are therefore quite successful [70]. Algorithms for comparing protein structures typically reveal distant evolutionary relationships between proteins having low sequence similarity [68]. Although these observations can be attributed to both sequence divergence and structure convergence, careful analysis of specific cases and more accurate methods for detecting sequence homology [71] suggest that sequence divergence beyond the limits of detectable homology is rather common (see e.g. [72] and the chapter by Dokholyan and Shakhnovich in this book). This prevalence of structural conservation has made it possible to create databases in which protein structures are classified into distinct structural groups with the same overall architecture (folds) [68,73,74]. For example, proteins classified in the same fold in the FSSP database [68] show a distribution of sequence identity comparable to that of random pairs of sequences [69]. Nevertheless, other indicators of structural changes often show a regular behaviour. For instance, within a given fold, the root mean square deviation between homologous proteins increases as sequences diverge [75].

Protein function, instead, is not as much conserved as the underlying structure, making its prediction rather difficult [76]. New functions are often created through gene duplication followed by differential regulation and recruitment of one of the copies to a new function [3]. In the transition to a new function, proteins accumulate substitutions, which may be fixed through positive selection, in a process that usually does not change significantly the overall fold.

Despite these general rules, several examples of proteins with detectable homology and yet different folds have been provided [77]. In these cases, the evolutionary changes are usually mediated through large scale mutations, such as insertion or deletions of entire secondary structure elements and circular permutations. As a consequence, the concept of protein fold has been reconsidered, and it has been suggested that insertions or deletions of secondary structure elements can provide a mechanism to connect many known folds [78]. Significant similarities between folds previously classified as distinct, possibly pointing at distant evolutionary relationships, were identified by Orengo and colleagues through an algorithm of protein structure comparison at the level of secondary structure [79] (see also the chapter by Ranea et al. in this book). In the majority of cases, however, point mutations and insertions or deletions of single residues do not seem to have produced evolutionary transitions to different protein folds. Therefore, in particular in the evolution of proteins that retain their function, the concept of protein fold can still be considered useful.

4.2.3 Models of Molecular Evolution with Structural Conservation

Structural stability was first considered in models describing the molecular evolution of RNA structures [10]. Schuster and co-workers described neutral networks in sequence space, associated to specific macromolecular structures (see also the chapter by Schuster and Stadler in this book).

In this view, structurally constrained molecular evolution proceeds along neutral networks, whose properties have a large impact on the evolutionary process. Schuster et al. showed that, in the case of some common RNA secondary structures, the neutral networks are dense in sequence space, and that networks of different common structures can be connected through a small number of point mutations [10]. These results suggest a view of RNA structural evolution as adaptation through neutrality, in which evolution proceeds along a neutral network until a crossing point to a fitter structure is found [11, 12].

Inspired by these studies, several authors introduced models of protein evolution with structural conservation. In this section, we shortly review some of these models. These models differ in the way the molecular structure is represented and the requirement of thermodynamic stability of the target structure is implemented. In the case of RNA, efficient algorithms can determine, approximately but reliably, the secondary structure of minimal energy for a given sequence [80]. Equivalent algorithms do not exist for protein tertiary structures. Therefore, several groups represented protein structures as self-avoiding walks on the simple cubic or square lattice, studying them by means of Monte Carlo simulations. The idea behind this approach is that qualitative properties of the evolution of lattice models can be transferred to real proteins. Other groups also adopted simplified off-lattice representations of protein structures, which were studied through effective energy functions, analogous to those used for lattice models. The two approaches usually yield qualitatively similar results. One should also distinguish between the approaches that impose only the requirement that the target structure has minimal energy, from those that further require that the energy landscape is well correlated. In the latter, all structures that are very different from the native one are energetically separated by a large energy gap from it, therefore favouring stability against misfolding.

Bornberg-Bauer and co-workers [13, 14] studied lattice polymers by imposing the condition that, for sequences in the neutral network, the energy of the target structure should be lower than that of all alternative structures, thus following closely the original RNA model. They studied the structures on a two-dimensional lattice and represented the sequences by a two-letter (hydrophobic-polar) code. Such a simplified protein model is amenable to exact enumeration of both conformations and sequences, and enabled Bornberg-Bauer and co-workers to establish that in the case of lattice proteins, neutral networks are disconnected in sequence space. They also discovered that these neutral networks are centred around the so-called *prototype sequence*, which is the sequence of maximal stability for a given structure, both mutationally and thermodynamically. Furthermore, these studies indicated that protein structures can be changed through point mutations, analogously to what was previously found for the RNA model.

Babadje et al. [15] adopted simplified representations of real protein structures, evaluating how well test sequences fit the target structure through a measure (the Z-score [81]) of the energy difference with respect to a set of alternative structures. They found that protein sequences can diverge almost as much as random pairs of sequences despite maintaining a high compatibility with the original structure.

Shakhnovich and Gutin [16] proposed an evolutionary model in which selection for fast folding is imposed in the framework of a lattice model, but without requiring the conservation of a particular structure. Later, Dokholyan and Shakhnovich [17] extended this approach considering sequences of fixed composition for which the target structure was required to have low energy. Evolution was modelled as a Monte Carlo process in sequence space, and large entropy barriers were found to separate clusters in sequence space. Mirny and Shakhnovich [18] analysed amino acid conservation in five of the most populated protein folds, identifying structural features correlated with conservation.

Dokholyan and Shakhnovich [19] modelled the process of gene duplication followed by structural divergence, showing that it can account for some of the statistical features of observed protein folds, most notably the almost power law distribution of the number of proteins per fold, and in addition that the model provides useful predictions concerning protein function (see also the chapter of Dokholyan and Shakhnovich in this book).

Goldstein and colleagues [20, 21] used lattice polymers to study a fitness landscape where the fitness of protein structures is given by their foldability, a concept borrowed from the spin-glass model of protein folding. They found that foldability can vary broadly, where structures with similar and large foldabilities are clustered together in structure space. When the selective pressure is increased, evolutionary trajectories become increasingly confined to 'neutral networks', where the sequence can be significantly changed while a constant structure is maintained. In a subsequent work, Taverna and Goldstein [22] showed that the marginal stability of proteins is a direct consequence of the hypothesis that changes in stability are neutral above some threshold and also of the high dimensionality of the sequence space.

Bussemaker et al. [23] obtained the interesting prediction that, in the lattice model they studied, the stability of small proteins is rather insensitive to random mutations. Tiana et al. [24] performed an exhaustive study of the effects that single mutations have on the stability of the native structure of a lattice protein, simulating the folding dynamics through a Monte Carlo approach. They classified protein sites into three types according to their robustness to mutations: 'green' sites, where mutations do not produce any relevant effect on stability (typically at the surface of the structure), 'yellow' sites for which the structure is slightly modified and 'red' sites (typically at the core of the structure) where mutations have a disruptive effect.

Parisi and Echave [25, 26] studied the impact of structural conservation on protein evolution, in a similar spirit to the SCN model that will be described in next section; the main difference is that they did not impose conditions on the stability of alternative structures. They simulated site-specific amino acid transition matrices, which were used in the calculation of the likelihood of families of protein sequences given their phylogenetic tree. In this way, they showed that the use of structural information can improve notably the likelihood of evolutionary models, and their ability to distinguish between different phylogenies.

Xia and Levitt [27, 28] used a two-dimensional lattice model and performed an exhaustive enumeration of the space of all sequences and the space of all structures. They found that, when evolution is dominated by mutation, the preference of the prototype sequence is not strong enough to offset the huge size of sequence space, so that most native sequences are located near the boundary of the fitness region and are marginally compatible with the native structure, in agreement with the results by Taverna and Goldstein [22]. On the other hand, when evolution is dominated by recombination events, the evolutionary preference for the prototype sequence is strong enough so that most native sequences are located near the centre of sequence–structure compatibility.

Aita et al. [29] identified amino acid sequences that fold into a target structure, imposing that the energy of the target must be much lower than that of alternative structures. They found that the neutral networks of different structures are separated by 5–30 mutations in sequence space, with separation increasing with the required threshold stability. Bloom et al. [30] studied the impact of random mutations on the stability of a wild-type structure, and found that the probability that a protein retains its structure declines exponentially with the number of mutations.

4.3 The SCN Model of Evolution

The SCN model is based on the observation that evolution conserves protein structure much more than protein sequence (see e.g. [68, 69]). It assumes that all mutations that maintain protein stability above a predefined threshold are selectively neutral, and all other mutations are strongly deleterious, thus resulting in a neutral model. These assumptions are consistent with the observation that many mutations do not significantly modify the activity of a protein and its thermodynamic stability, while mutations that improve substantially protein functionality are rare [60].

4.3.1 Representation of Protein Structures

In the SCN model, the structure of a protein of N residues is represented through an $N \times N$ contact matrix $\mathbf{C}$. This matrix is defined as $C_{ij} = 1$ if sites i and j are in contact, and $C_{ij} = 0$ otherwise. Two sites are considered in contact if any two of their heavy atoms are closer than a given cut-off distance, which we take as 4.5 Å. The effective free energy associated to a sequence of amino acids $\mathbf{A}$ in the configuration $\mathbf{C}$ is, in this type of approach, assumed to have the form of a sum of pairwise contact interactions,

$$E(\mathbf{A}, \mathbf{C}) = \sum_{i<j} C_{ij} U(A_i, A_j), \qquad (4.5)$$

where A_i labels one of the 20 amino acid types and $\mathbf{U}$ is a 20×20 symmetric interaction matrix, so that $U(a,b)$ is the interaction energy between amino acids a and b when in contact. A useful choice for the latter is the matrix derived in [82] in such a way to assign high thermodynamic stability to the native states of a large set of monomeric proteins [83].

Three remarks need to be made here: (a) The effective energy parameters take implicitly into account the effect of the solvent and they depend on temperature, thus they express free energies rather than energies. (b) The effective energy of a structure is defined with respect to a completely extended reference structure where no contacts are formed and which sets the zero of the energy scale. (c) The chain entropy sN is not included into the effective energy, as it is constant for constant chain length N.

4.3.2 Stability Against Unfolding

The stability of the native state against unfolding can be estimated from the negative of the native contact energy, $-E(\mathbf{A}, \mathbf{C}^*)$, neglecting changes of conformational entropy with the protein sequence. In the SCN model, we impose that $-E(\mathbf{A}, \mathbf{C}^*)$ is larger than a positive threshold $-E_{\text{thr}}$ for sequences $\mathbf{A}$ belonging to the neutral network.

As an alternative measure of stability, one can also use the Z-score of the native energy, $Z(\mathbf{A}, \mathbf{C}^*)$ [81,84], which gives the difference between the energy of sequence $\mathbf{A}$ in configuration $\mathbf{C}^*$ and its average energy in a set of alternative configurations, $\{\mathbf{C}\}$, in units of the standard deviation of the energy

$$Z(\mathbf{A}, \mathbf{C}^*) = \frac{E(\mathbf{A}, \mathbf{C}^*) - \langle E(\mathbf{A}, \mathbf{C}) \rangle_{\{\mathbf{C}\}}}{\sqrt{\langle E(\mathbf{A}, \mathbf{C})^2 \rangle_{\{\mathbf{C}\}} - \langle E(\mathbf{A}, \mathbf{C}) \rangle^2_{\{\mathbf{C}\}}}}. \qquad (4.6)$$

When a sequence $\mathbf{A}$ folds into a structure $\mathbf{C}^*$, the corresponding Z-score is negative and very large in absolute value. This measure is, however, better suited for estimating the stability against misfolding (see Sect. 4.3.3).

4.3.3 Stability Against Misfolding

For a given sequence $\mathbf{A}$, the energy landscape is defined to be well correlated if all configurations of low energy are very similar to the configuration of minimal effective energy, $\mathbf{C}^*$. Structure similarity is measured by the overlap $q(\mathbf{C}, \mathbf{C}^*)$, which counts the number of contacts that two structures have in common. This number is normalized by the maximal number of contacts, so that q ranges between zero and one. In a well-correlated energy landscape, the inequality

$$\frac{E(\mathbf{A}, \mathbf{C}) - E(\mathbf{A}, \mathbf{C}^*)}{|E(\mathbf{A}, \mathbf{C}^*)|} \geq \alpha(\mathbf{A})\left(1 - q(\mathbf{C}, \mathbf{C}^*)\right), \tag{4.7}$$

with a large $\alpha(\mathbf{A})$ holds. This inequality indicates that the energy gap between the ground state $\mathbf{C}^*$ of sequence $\mathbf{A}$ and any alternative structure $\mathbf{C}$, measured in units of the ground state energy, is larger than a quantity $\alpha(\mathbf{A})$ times the structural distance $1 - q(\mathbf{C}, \mathbf{C}^*)$. The dimensionless quantity $\alpha(\mathbf{A})$, which is the largest quantity for which the above inequality holds, can be used to evaluate the folding properties of sequence $\mathbf{A}$. For random sequences, the lowest energy configurations are structurally different and have similar energy, hence $\alpha(\mathbf{A})$ is close to zero. In this case, the energy landscape is rugged, the folding kinetics is very slow, and the thermodynamic stability is low. In contrast, computer simulations of well-designed sequences have shown that, when $\alpha(\mathbf{A})$ is large, the folding kinetics is fast and the stability with respect to changes in the energy parameters as well as mutations in the sequence is very high [16, 31]. In the SCN model, we impose that $\alpha(\mathbf{A})$ is larger than a positive threshold α_{thr} for sequences $\mathbf{A}$ belonging to the neutral network.

Further, it is assumed that the ground state structure $\mathbf{C}^*$ coincides with the target structure defining the neutral network. Indeed, in all the simulations performed using the SCN model, it was never found a sequence whose ground state structure was different than the target one and simultaneously had a sufficiently large energy gap. Therefore, imposing a well-correlated energy landscape through a condition on the normalized energy gap makes it very difficult to change the native structure into a new structure, which is also stable against misfolding. This result agrees qualitatively with the simulations of Aita et al. [29]. It illustrates the difference between RNA and proteins, since it is in contrast with the findings of Schuster et al., who showed that the neutral networks of two different RNA secondary structures can be separated by just one point mutation [10].

4.3.4 Calculation of $\alpha(\mathbf{A})$

Candidate structures for a protein sequence were generated from all possible alignments of the sequence with structures in the PDB. This procedure is called *gapless threading*. To speed up the computation, we considered a non-redundant subset of the PDB in which proteins with homologous sequences are excluded [85]. About 10^6 alternative structures were obtained for proteins of 100 amino acids, with this number decreasing for longer proteins. The energy function correctly assigns the lowest energy to the native structure for most proteins of known structure, and it generates a well-correlated energy landscape in which structures very different from the native have high energy gaps, so that $\alpha(\mathbf{A})$ is large.

Most of the computer time of these simulations is spent in the calculation of $\alpha(\mathbf{A}')$ for all possible point mutants of the actual sequence $\mathbf{A}$. To speed up this calculation, we note that $\alpha(\mathbf{A})$ is obtained from the configuration $\mathbf{C}$ with the highest destabilizing power, i.e. the highest value of the energy gap divided by the structural distance from the native configuration. This structure changes through evolution, but it is expected that the set of high scoring structures remains the same for neighbouring sequences. Therefore, for each actual sequence, we store a sufficiently large number of configurations with the highest destabilizing powers (typically 50, see [86]), and we compute their destabilizing power in the mutated sequences $\mathbf{A}'$. This procedure may slightly overestimate $\alpha(\mathbf{A}')$, since not all configurations are used, but the fraction of sequences for which $\alpha(\mathbf{A}')$ crosses the acceptance threshold is below 0.1% [86].

One drawback of the computation of $\alpha(\mathbf{A})$ based on gapless threading is that the number of alternative structures generated in this way decreases with the length of the sequence, N. Therefore, the actual value of $\alpha(\mathbf{A})$ is overestimated for longer sequences. This is not a significant problem when, as here, one is interested in comparing values of $\alpha(\mathbf{A})$ for different sequences of the same length. Nevertheless, it can be convenient, in particular for long chains, to evaluate $\alpha(\mathbf{A})$ using a different method [87]. This method estimates the minimal energy for non-native structures through a theoretical prediction based on the random energy model (REM) [88,89],

$$E_{\text{REM}}(\mathbf{A}) \approx N_c \langle U \rangle_\mathbf{A} - \sigma_{U,\mathbf{A}} \sqrt{2N_c \log(m_N)}, \qquad (4.8)$$

where N_c is the number of native contacts, $\langle U \rangle_\mathbf{A}$ and $\sigma_{U,\mathbf{A}}$ are the mean and standard deviation of the interaction energy for all possible contacts, native and non-native ones, within sequence $\mathbf{A}$, and m_N is the number of independent contact matrices for a protein of length N, satisfying physical constraints of hard core repulsion, hydrogen bonding and compactness. The minimal energy estimated in this way, $E_{\text{REM}}(\mathbf{A})$, is in very good agreement with the minimal non-native energy found by threading, $E_{\min}(\mathbf{A}) \approx (1.003 \pm 0.009) E_{\text{REM}}(\mathbf{A}) - (0.0016 \pm 0.0012)$, when m_N is set equal to the number of structures generated through threading, with a correlation coefficient $r = 0.96$ [87]. Using this estimate, one can evaluate the normalized energy gap as

$$\alpha'(\mathbf{A}) = \frac{E(\mathbf{A}, \mathbf{C}^*) - N_c \langle U \rangle_\mathbf{A} + \sigma_{U,\mathbf{A}} \sqrt{2N_c \log(m_N)}}{E(\mathbf{A}, \mathbf{C}^*)(1 - q_0)}. \qquad (4.9)$$

The number of alternative structures m_N is expected to increase exponentially with chain length as

$$\log(m_N) \approx AN + B. \tag{4.10}$$

The parameters have been approximately estimated as $A \approx 0.1$ and $B \approx 4$ in such a way that the minimal energy coincides with the one evaluated through threading for short chains ($N < 50$) and the estimated minimal effective energy is higher than the native energy for most proteins in the PDB [87]. Finally, one sets $q_0 = 0.1$ as the typical overlap between unrelated structures, disregarding the length dependence of this quantity.

4.3.5 Sampling the Neutral Networks

The neutral network of a given protein structure is defined as the set of sequences $\mathbf{A}$ for which the stability against both unfolding and misfolding, measured through $E(\mathbf{A}, \mathbf{C}^*)$ and $\alpha(\mathbf{A}, \mathbf{C}^*)$, respectively, exceed predetermined thresholds, chosen as 98.5% of the values of those parameters for the wild-type sequence in the PDB. The threshold chosen enforces conservation of the thermodynamic stability of the native structure $\mathbf{C}^*$. We verified that the qualitative behaviour of the model does not change in the range between 95% and 100% of the values for PDB sequences.

The SCN algorithm [31,32] explores the neutral network of a given protein starting from its PDB sequence $\mathbf{A}_1 = \mathbf{A}_{\mathrm{PDB}}$ and iterating the following procedure: At iteration n, (a) the number $X(\mathbf{A}_n)$ of viable neighbours of sequence $\mathbf{A}_n$ is computed, and (b) the sequence $\mathbf{A}_{n+1}$ is extracted randomly among all the viable neighbours of $\mathbf{A}_n$. In this way, we generate a stochastic process that explores the neutral network. This process looses rather quickly the memory of the initial sequence. The total number of viable point mutations, $X(\mathbf{A})$, expresses the local connectivity of the neutral network. This number is normalized by the total number of attempted mutations, X_{tot},[6] thus obtaining the fraction of neutral neighbours, $x(\mathbf{A}) = X(\mathbf{A})/X_{\mathrm{tot}} \in (0,1]$.

4.3.6 Fluctuations and Correlations in the Evolutionary Process

In contrast with the homogeneity assumption of Kimura's neutral model, the SCN model shows that stability constraints produce a broad distribution of the fraction of neutral neighbours $x(\mathbf{A})$. This distribution $P(x)$ is shown in Fig. 4.1

---

[6] We impose conservation of the starting cysteine residues in the sequence, and do not allow that other residues mutate into cysteine. These requirements are imposed because a mutation that changes the number of cysteine residues by one would leave the protein with a very reactive unpaired cysteine that would most likely affect its functionality and would be therefore rejected with very high probability. The maximum number of attempted mutations is therefore $X_{\mathrm{tot}} = 18(N - N_{\mathrm{cys}})$, where N is the number of residues and N_{cys} is the number of cysteine residues in the starting sequence.

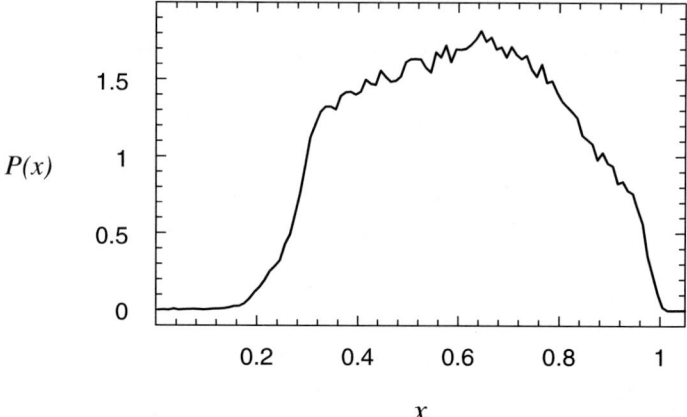

Fig. 4.1. Probability distribution $P(x)$ of the fraction x of neutral neighbours for myoglobin, as obtained by the SCN model (adapted from [90])

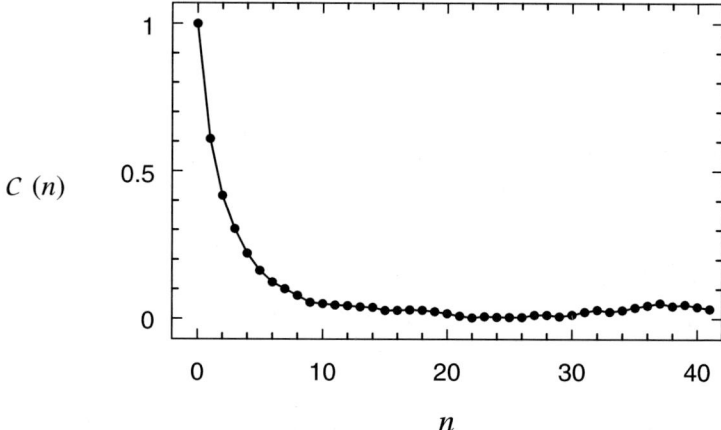

Fig. 4.2. Auto-correlation function $\mathcal{C}(n) \equiv \mathcal{C}(x(\mathbf{A}_k), x(\mathbf{A}_{k+n}))$ of neutral connectivities for sequences separated by n substitutions for myoglobin, as obtained by the SCN model (adapted from [86])

for the neutral network of myoglobin (PDB id. 1a6g). Other proteins yield qualitatively the same results. Besides this distribution being very broad, the fraction of neutral neighbours is strongly auto-correlated along a trajectory. In Fig. 4.2, we show the auto-correlation function $\mathcal{C}(x(\mathbf{A}_k), x(\mathbf{A}_{k+n}))$ of $x(\mathbf{A}_k)$, defined as

$$\mathcal{C}(x(\mathbf{A}_k), x(\mathbf{A}_{k+n})) = \frac{\frac{1}{m-n}\sum_{k=1}^{m-n} x(\mathbf{A}_k)\, x(\mathbf{A}_{k+n}) - \overline{x}^2}{\sigma_x^2}, \quad (4.11)$$

where the mean value $\bar{x} = (1/m) \sum_{k=1}^{m} x(\mathbf{A}_k)$ and the variance $\sigma_x^2 = \overline{x^2} - \bar{x}^2$ are calculated over the whole trajectory. Our results show that the auto-correlation decays exponentially as

$$\mathcal{C}(x(\mathbf{A}_k), x(\mathbf{A}_{k+n})) \approx \exp(-n/\ell_x), \qquad (4.12)$$

with ℓ_x of the order of three substitutions [90] and, as we shall see, it has important consequences on the statistics of the substitution process.

Broad fluctuations and strong auto-correlations of the neutral connectivity are a general feature of the SCN model, and distinguish it from the standard neutral model by Kimura. They have a rather simple explanation. Defining $x_i(\mathbf{A})$ as the fraction of neutral neighbours when mutation occurs at site i, one has $x(\mathbf{A}) = \sum_{i=1}^{N} x_i(\mathbf{A})/N$. If the fraction of neutral neighbours at different sites are not correlated, their mean $x(\mathbf{A})$ is expected to have fluctuations vanishing as $1/\sqrt{N}$. The broad distribution of $x(\mathbf{A})$ that we found indicates that this is not the case. In fact, there are significant positive correlations between almost all pairs of variables $x_i(\mathbf{A})$ and $x_j(\mathbf{A})$ [90]. These correlations are induced by the fact that the $x_i(\mathbf{A})$ at each site are significantly correlated with some global variable, for instance, the mean fraction of neutral neighbours $x(\mathbf{A})$. This is shown in Fig. 4.3 for the case of myoglobin, defining

$$\mathcal{C}_i = \frac{1}{m} \sum_{k=1}^{m} \frac{\left(x_i(\mathbf{A}_k) - \overline{x_i}\right)\left(x(\mathbf{A}_k) - \bar{x}\right)}{\sigma_{x_i} \sigma_x}. \qquad (4.13)$$

The figure shows that all the correlations $\mathcal{C}_i$ are positive and significant (they were computed from order of 10^6 sequences, with significance threshold of order 10^{-3}), and moreover, they are positively correlated with the robustness of site i to mutation, measured by $\overline{x_i}$ [90].

Therefore, sequences with large $x(\mathbf{A})$ are more rebust to mutation at all sites. As also found by Bornberg-Bauer for prototype sequences [13], and as we will discuss in next section, these more robust sequences have higher thermodynamic stability, so that mutations applied to them produce more often other stable sequences. Figure 4.3 goes one step further, and shows that there are some sites with small $\overline{x_i}$ that are less tolerant to mutations both in general and in mutationally robust sequences (the correlation between $x_i(\mathbf{A})$ and $x(\mathbf{A})$ is minimal for these sequences). The structural determinants of strong structurally constrained sites will be investigated in the next section.

4.3.7 Substitution Process

Amino acid substitutions within the SCN model are controlled by two independent events: Random mutations, described by a Poissonian process, and an acceptance process, which consists in testing whether the sequence is viable. The acceptance probability for a mutation that takes place in a protein of

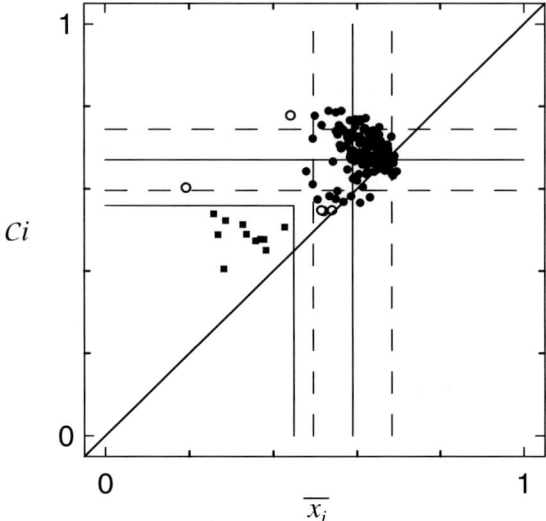

Fig. 4.3. Comparison between cross-correlations and conservations for myoglobin. The fraction of neutral mutations at site i, $\overline{x_i}$, is shown on the abscissa, and the correlation between x_i and the overall neutral connectivity x, C_i, is shown on the ordinate. The *dashed horizontal and vertical lines* indicate one standard deviation from the mean (*full horizontal and vertical lines*). Additionally, *horizontal and vertical lines* at the threshold of 1.5 standard deviations below the mean are also shown. The sites above the threshold for both quantities are shown as *full circles*, the sites below the threshold for both quantities are shown as *full squares*, whereas the sites that are above the first threshold but below the second, or vice versa, are shown as *open circles* (adapted from [86])

sequence $\mathbf{A}$ is given by the neutral connectivity $x(\mathbf{A})$. As a result of the broad distribution of this variable, the resulting substitution process is not Poissonian. For a given evolutionary trajectory (i.e. for a given sequence of neutral connectivities $\{x(\mathbf{A}_1), x(\mathbf{A}_2), \ldots\}$) one can compute the probability that the number S_t of accepted mutations in a time interval t equals n. This probability is the product of the Poissonian probability that k mutations take place in the time interval t, times the conditional probability that n of these are accepted,

$$P\{S_t = n\} = \sum_{m=n}^{\infty} e^{-\mu t}\frac{(\mu t)^m}{m!} P_{\text{acc}}(n|m), \qquad (4.14)$$

where the conditional acceptance probability of n mutations out of m is given by

$$P_{\text{acc}}(n|m) = \left(\prod_{i=1}^{n} x(\mathbf{A}_i)\right) \sum_{\{m_j\}} \prod_{j=1}^{n+1} [1 - x(\mathbf{A}_j)]^{m_j}. \qquad (4.15)$$

Here, the $\{m_j\}$ are all integer numbers between zero and $m - n$ satisfying $n + \sum_{j=1}^{n+1} m_j = m$. The probability that a mutation is accepted is thus $x(\mathbf{A}_1)$, as long as the protein sequence is $\mathbf{A}_1$, $x(\mathbf{A}_2)$ as long as the sequence is $\mathbf{A}_2$ and so on.

If all sequences have the same fraction of neutral neighbours $x(\mathbf{A}) = x$, (4.14) coincides with (4.4), and the number of substitutions in a branch of length t, S_t, is a Poissonian variable with mean $\mu t x$ and the substitution rate equals μx, as in Kimura's model. If the variance of the neutral connectivity is not zero, the moments of the substitution distribution can be computed in the long-time limit using the central limit theorem. Define τ_i as the time interval between the i-th and $i+1$-th substitutions. The τ_i are independent variables with exponential distribution and expectation values $\mathrm{E}[\tau_i] = \overline{1/\mu x}$, $\mathrm{E}[\tau_i^2] = \overline{2/\mu x^2}$. If S_t is large, we can apply the central limit theorem to the mean value $\sum_{i=1}^{S_t} \tau_i / S_t$, finding

$$\sum_{i=1}^{S_t} \tau_i \approx S_t \frac{1}{\mu x} \left[1 + \frac{zB}{\sqrt{S_t}} + \frac{1}{2} \frac{z^2 B^2}{S_t} \right] \approx t, \qquad (4.16)$$

where z is a normalized Gaussian variable, and

$$B^2 = \frac{\mathrm{E}[\tau_i^2]}{\mathrm{E}^2[\tau_i]} - 1 = \left(1 - e^{-1/\ell_x}\right)^{-1} \left(\frac{\overline{1/x^2}}{\overline{1/x}^2} - 1\right) + \frac{\overline{1/x^2}}{\overline{1/x}^2}. \qquad (4.17)$$

The normalized variance B^2 is larger than one because (a) the distribution of x is broad, so that $\mathrm{E}[1/x^2] > \mathrm{E}^2[1/x]$ and (b) trajectories are correlated (the term $[1 - \exp(-1/\ell_x)]^{-1}$ tends to one if the correlation length ℓ_x tends to zero). The first two moments of S_t can be calculated as

$$\mathrm{E}[S_t] \approx \frac{\mu t}{\overline{1/x}} \qquad (4.18)$$

$$R(t) \equiv \frac{\mathrm{E}[S_t^2] - \mathrm{E}^2[S_t]}{\mathrm{E}[S_t]} \approx B^2 \left(1 - \frac{3B^2 \overline{1/x}}{4t}\right). \qquad (4.19)$$

The normalized variance $R(t)$ is called the 'dispersion index'. Notice that if the substitution process is Poissonian one has $R \equiv 1$. The asymptotic value of the dispersion index for large time is $R(t \to \infty) = B^2$, which is larger than one due to the broad fluctuations and time correlations of x. Therefore, the substitution process is overdispersed. For small t, when the process probes only one sequence, the substitution process is expected to behave as a Poissonian process with $R(t \to 0) = 1$.

We compared the above predictions to the expectation values calculated from the probability defined in (4.14). The values of the neutral connectivities were obtained from the evolutionary trajectories $\{x(\mathbf{A}_1), x(\mathbf{A}_2), \ldots\}$ simulated with the SCN model (details of the calculation are given in [90]). Averages along an evolutionary trajectory are indicated with angular brackets $\langle \cdot \rangle$,

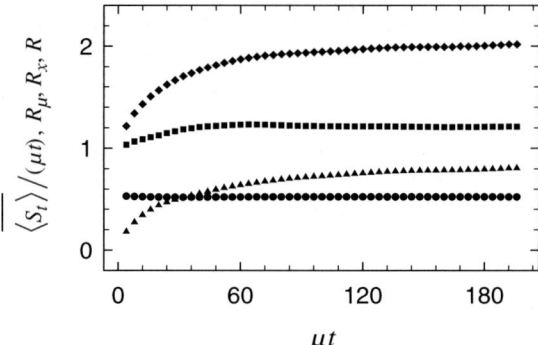

Fig. 4.4. Statistical properties of the substitution process of myoglobin, showing the average number of substitutions $\overline{\langle S_t \rangle}$ divided by μt (*circles*), the normalized mutation variance $R_\mu(t)$ (*squares*), the normalized trajectory variance $R_x(t)$ (*triangles*) and the normalized total variance $R(t) = R_\mu(t) + R_x(t)$ (*diamonds*)

whereas averages over evolutionary trajectories are indicated with an overline $\overline{\cdot}$. The mean and the normalized variance of the number of substitutions are shown in Fig. 4.4 for the case of myoglobin. In the plot, we distinguish the normalized mutation variance

$$R_\mu(t) = \frac{1}{\overline{\langle S_t \rangle}} \left(\overline{\langle S_t^2 \rangle} - \overline{\langle S_t \rangle^2} \right), \tag{4.20}$$

the normalized trajectory variance

$$R_x(t) = \frac{1}{\overline{\langle S_t \rangle}} \left(\overline{\langle S_t \rangle^2} - \overline{\langle S_t \rangle}^2 \right), \tag{4.21}$$

and the normalized total variance (the dispersion index) $R(t) = R_\mu(t) + R_x(t)$. Notice that if $x(\mathbf{A}) = x$, one obtains $R_\mu(t) \equiv 1$ as for all Poissonian processes, and the normalized trajectory variance $R_x(t) \equiv 0$. From the plot, it is also clear that most of the overdispersion comes from $R_x(t)$, i.e. from the variance between different evolutionary trajectories, which can generate rather different substitution rates.

The quantitative agreement of the dispersion index $R(t)$ of the SCN process with the prediction (4.19) is quite good as far as the long-time limit is concerned, but the temporal dependence is not well captured by this first-order approximation. The dispersion index of the SCN process is compatible with empirically obtained dispersion indices, which are usually in the range 1.5–5 [9, 50, 91]. Hence, these observed dispersion indices may be to a large extent due to the correlations present in the evolutionary process both in space and in time [90]. This result provides a mechanistic explanation of the fluctuating neutral space model proposed by Takahata to account for the observed statistics of the substitution process [92].

Here, we should notice a difference between the results shown in Fig. 4.4 and those presented in [86]. In [86], we reported that the average substitution rate $\overline{\langle S_t \rangle}/t$ decreases in time in the SCN model, tending to the asymptotic value $\mu \overline{1/x}$. Recently Ho et al. [93], analyzing protein sequences, observed an apparent decrease of the substitution rate through time that would match qualitatively the SCN prediction. However, in obtaining the results presented in [86], we sampled the initial sequences of the evolutionary trajectories with equal probability. This procedure is not entirely consistent, since the time spent at sequence $\mathbf{A}$ is proportional on the average to $1/x(\mathbf{A})$, so that the process spends more time in sequences with small neutral connectivity x. Taking this into account, we have sampled the initial sequence $\mathbf{A}_1$ with probability proportional to $1/x(\mathbf{A}_1)$. The initial rate is therefore $\int_0^1 P(x)\,(1/x)\,(\mu x)\,\mathrm{d}x / \int_0^1 P(x)\,(1/x)\,\mathrm{d}x$, which is equal to the final rate $\mu \overline{1/x}$, so that the rate is now constant in time. Figure 4.4 refers to this new sampling protocol. This does not modify significantly the normalized variances presented in Fig. 10 of [86], which was obtained with homogeneous sampling. Therefore, the results of [86] cannot explain the empirical observations of non-constant rate by Ho et al. [93].

4.4 Site-Specific Amino Acid Distributions

The reconstruction of phylogenetic trees from sequence alignments requires the use of a model of protein evolution [4,94] (see also the chapters by Xia and by Liò et al. in this book). In this context, the effects of both the mutational and the selection processes on protein folding and function must be taken into account. It is well known for instance that the local environment of a protein site within the native structure influences the probability of acceptance of a mutation at that site [95]. Nevertheless, such a view, which is based on structural biology, has a relatively limited impact on studies of phylogenetic reconstruction, where the corresponding models usually rely on substitution matrices that do not consider the structural specificity of different sites. The most used substitution matrices, such as JTT [96], are obtained by extrapolating substitution patterns observed for closely related sequences, and they have low performances when distant homologs are concerned [97].

To account for selection at the protein level, it is necessary to consider site-specific amino acid distributions within a protein family [98]. The use of site-specific substitution matrices improves substantially maximum likelihood methods for reconstructing phylogenetic trees [99–103]. In the studies mentioned above, site-specific constraints are obtained either through simulations of a protein evolution model or by fitting the corresponding parameters within a maximum likelihood framework. As we will discuss in the following, it is possible to deduce from the SCN model an analytical expression for site-specific amino acid distributions with no adjustable parameters. The resulting distributions are in very good agreement with model simulations and with site-specific amino acid distributions obtained from the PDB [39, 40].

Sites in the same protein evolve in a correlated way, because they undergo global stability constraints. However, Maximum Likelihood approaches become almost computationally unfeasible unless one assumes that sites evolve independently. Here, we will define a mean-field protein evolution model with independent sites that reproduces with great accuracy the results of the SCN model with global stability constraints. The price to pay for this simplification is that we shall consider an effective selection process that depends on the mutation process. At the mean-field level, mutation and selection, that are independent processes in the Darwinian framework, become effectively entangled.

4.4.1 Vectorial Representation of Protein Sequences

The interaction matrix $\mathbf{U}$ in (4.5) can be written in its spectral form as $U(a,b) = \sum_{\alpha=1}^{20} \epsilon_\alpha u^{(\alpha)}(a) u^{(\alpha)}(b)$, where ϵ_α are the eigenvalues, ranked by their absolute value, and $\mathbf{u}^{(\alpha)}$ are the corresponding eigenvectors. The main contribution to the interaction energy is given by $\epsilon_1 u^{(1)}(a) u^{(1)}(b)$, which has a correlation coefficient 0.81 with the elements $U(a,b)$ and a negative eigenvalue ϵ_1. It is well known that hydrophobic interactions constitute the most significant contribution to pairwise interactions in proteins, the components of the main eigenvector are strongly correlated with experimental hydropathy scales [104, 105]. By considering only this main component, one can define an effective energy function, $H(\mathbf{A}, \mathbf{C})$, which provides a good approximation to the energy, (4.5), as

$$\frac{H(\mathbf{A}, \mathbf{C})}{k_\mathrm{B} T} \equiv \epsilon_1 \sum_{i<j} C_{ij}\, h(A_i)\, h(A_j). \tag{4.22}$$

The vector $\mathbf{h}(\mathbf{A}) \equiv \mathbf{u}^{(1)}(\mathbf{A})$ is denoted as the Hydrophobicity Profile (HP) of sequence $\mathbf{A}$ [38]. This is an N-dimensional vector whose i-th component is given by $h(A_i) \equiv u^{(1)}(A_i)$. The 20 parameters $h(a) \equiv u^{(1)}(a)$, obtained from the PE of the interaction matrix, are called *interactivity* parameters, and are reported in Table 4.1.

Table 4.1. Interactivity scale used in this chapter and presented in [38]

A	R	N	D	C	Q	E	G	H	I
0.1366	0.0363	−0.0345	−0.1233	0.2745	0.0325	−0.0484	−0.0464	0.0549	0.4172
L	K	M	F	P	S	T	W	Y	V
0.4251	−0.0101	0.1747	0.4076	0.0019	−0.0432	0.0589	0.2362	0.3167	0.4083

4.4.2 Vectorial Representation of Protein Folds

A convenient vectorial representation of protein structures may be derived from the PE of the contact matrix $\mathbf{C}$, which we denote as $\mathbf{c}$. The latter maximizes the quadratic form $\sum_{ij} C_{ij} c_i c_j$ with the condition $\sum_i c_i^2 = 1$. In this sense, c_i can be interpreted as the effective connectivity at site i, since sites with large c_i are in contact with as many as possible sites j with large c_j. All the components of $\mathbf{c}$ have the same sign, which, by convention, is taken as positive. Moreover, if the contact matrix represents a single connected graph, as is the case for single-domain globular proteins, the information contained in the PE is in most cases sufficient to reconstruct the whole contact matrix [37], and consequently the full three-dimensional structure [106].

4.4.3 Relation Between Sequence and Structure

The constraint of thermodynamic stability predicts that there should be a correlation between the vectorial representations of protein sequences and structures.

For a given protein fold, we define the optimal HP, denoted as $\mathbf{h}_{\text{opt}}$, as the vector that minimizes the approximate effective free energy, (4.22), under the constraints that its mean hydrophobicity, $\langle h \rangle = N^{-1} \sum_i h_{\text{opt}}(A_i)$, and its mean square value, $\langle h^2 \rangle = N^{-1} \sum_i h_{\text{opt}}^2(A_i)$, are kept fixed.[7] These constraints imply that the mean and standard deviation of non-native interactions is also kept fixed, so that the normalized energy gap, (4.9), is also kept large. From the property of the PE that it maximizes $\sum_{ij} C_{ij} c_i c_j$ with the condition $\sum_i c_i^2 = 1$, it is clear that $\mathbf{h}_{\text{opt}}$ is strongly correlated with $\mathbf{c}$ [38]. In this formulation, $\langle h \rangle$ and $\langle h^2 \rangle$ are parameters not determined by the native structure, and they should guarantee a large normalized energy gap (in fact, in the approximation given by (4.22), the mean and mean square contact interactions that enter into the calculation of $\alpha(\mathbf{A})$ by (4.8) are $\langle U \rangle = \epsilon \langle h \rangle^2$ and $\langle U^2 \rangle = \epsilon^2 \langle h^2 \rangle^2$).

The optimal HP represents an analytical solution to the problem of sequence design for the effective energy function (4.22), and thus an approximate solution for the energy function, (4.5). In the SCN evolutionary model, mutations are accepted whenever the effective free energy and the normalized energy gap overcome predefined thresholds. Thus, the optimal HP is not expected to be ever realized during evolution, since they correspond to a negligible volume in the neutral network. However, thermodynamically stable sequences compatible with the given fold are expected to have HP values not too different from the optimal one. This is indeed observed in simulations of the SCN model. The mean correlation coefficient between the PE of the fold and the HP of the sequences generated through SCN simulations is typically 0.45, which is significant. The HP averaged over all sequences compatible with

---

[7] Here, we denote by angular brackets the average over all positions in a given protein sequence or structure.

a given fold, $[\mathbf{h}]_{\text{evol}}$, correlates much more strongly with the PE of that fold (and hence with the optimal HP), with a correlation coefficient larger than 0.95 for all of the studied folds [38]. These results show that one can recover the optimal HP through an evolutionary average of the HPs compatible with the protein fold.

Protein families represented in the FSSP [68] and in the PFAM [107] databases show qualitatively similar results. The correlation between the PE of the fold and the HP of individual sequences compares well with what was found in SCN simulations. The average HP over aligned sequences from the same protein family correlates more strongly with the PE than individual HPs: The average correlation coefficient is 0.58 for FSSP families and 0.57 for PFAM families [38]. This correlation is however much weaker than the analogous one for SCN protein families, which is 0.96. There are several explanations for this weaker correlation. First, this can be due to functional conservation, which plays an important role in protein evolution and is not represented in the SCN model. Part of the discrepancy can be also attributed to the approximate character of the effective energy function used to test the thermodynamic stability. Furthermore, real protein families are much smaller than SCN families, for which we generated of the order of 10^6 sequences. To test for such an effect, the average HP has been also computed using only few hundreds of SCN sequences, i.e. of the same order of magnitude as in FSSP or PFAM families. As a result, the correlation between the average HP and the PE was found to be reduced to values comparable to those observed for the FSSP and the PFAM sequence databases [38].

4.4.4 The PE as a Structural Determinant of Evolutionary Conservation

As showed by Bornberg-Bauer [13], thermodynamic stability and mutational stability are correlated. Sequences that are more stable can also bear a larger number of mutations. Bornberg-Bauer called the sequence of maximal mutational stability the *prototype sequence* of a fold, and showed that it has also maximal thermodynamic stability. In our model, the optimally stable sequence can be predicted analytically to have a HP that correlates very strongly with the PE. Sequences close to the optimal one, in the sense that they have a large correlation coefficient $r(\mathbf{h}(\mathbf{A}), \mathbf{c})$, are therefore expected to bear a large number of mutations and to have larger neutral connectivity $x(\mathbf{A})$. We verified this prediction using the SCN families. Although there is a significant correlation between the two quantities, the scattering of the data is very large. Thus in Fig. 4.5, we plot $x(\mathbf{A})$ averaged over protein sequences that have $r(\mathbf{h}(\mathbf{A}), \mathbf{c})$ in the same bin of width 0.02, as obtained for mesophilic rubredoxin (PDB id. 1iro). Sequences close to the optimal one have a very large fraction of neutral neighbours, as expected.

Thus, the relation $r(\mathbf{h}(\mathbf{A}), \mathbf{c})$ between PE and HP explains a significant part of the sequence variation of the overall mutational stability $x(\mathbf{A})$. As we

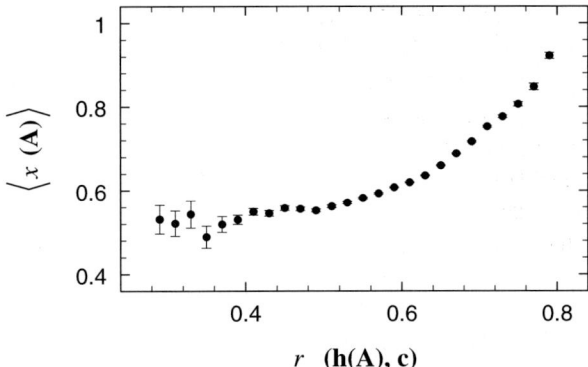

Fig. 4.5. Mean fraction of neutral neighbours $\langle x(\mathbf{A}) \rangle$ as a function of the correlation coefficient $r(\mathbf{h}(\mathbf{A}), \mathbf{c})$ between the vectorial representations of sequence $\mathbf{h}(\mathbf{A})$ and structure $\mathbf{c}$ for mesophilic rubredoxin (error bars indicate the standard deviation of the mean)

will see in the next section, the PE explains also a large part of the site-specific variation of mutational stability, with sites having PE components that are smaller or larger than the mean being more conserved through evolution.

4.4.5 Site-Dependent Amino Acid Distributions

The SCN model of protein evolution generates trajectories in sequence space for which the resulting HP fluctuates around the optimal HP, the latter being strongly correlated with the PE of the protein fold's contact matrix [38]. This remarkable feature can be used to compute analytically the site-specific distribution of amino acid occurrences $\pi_i(a)$, where i indicates a protein site and a one of the 20 amino acid types [39].

To derive an analytical expression for $\pi_i(a)$, the correlation coefficient between the PE $\mathbf{c}$ of the native contact matrix and the evolutionary average of the hydrophobicity vector, $[\mathbf{h}]_{\text{evol}}$, is assumed to be exactly 1, yielding that the two vectors are linearly related as

$$[h_i]_{\text{evol}} \equiv \sum_{\{a\}} \pi_i(a) h(a) = A(c_i/\langle c \rangle - 1) + B, \quad (4.23)$$

where the sum over $\{a\}$ is over all amino acids, and

$$A = \sqrt{\frac{\langle [h]^2_{\text{evol}} \rangle - \langle [h]_{\text{evol}} \rangle^2}{(\langle c^2 \rangle - \langle c \rangle^2)/\langle c \rangle^2}} \quad \text{and} \quad B = \langle [h]_{\text{evol}} \rangle. \quad (4.24)$$

In the above equations, two kinds of averages have been introduced: The angular brackets, denoting the average over the N sites of the protein,

$\langle f \rangle = N^{-1} \sum_i f_i$, where the corresponding standard deviation is denoted as $\sigma_f^2 = \langle f^2 \rangle - \langle f \rangle^2$, and the square brackets, denoting site-specific evolutionary averages, $[f_i]_{\text{evol}} = \sum_{\{a\}} \pi_i(a) f(a)$.

Equations (4.23) and (4.24) represent the conditions that the stationary distributions $\pi_i(a)$ have to fulfil in order to guarantee a perfect correlation between PE and the average HP. Assuming that these conditions are the only requirement that the $\pi_i(a)$ have to meet, we require that the $\pi_i(a)$ are the distributions of maximum entropy having the given average values. It is well known that the solution of this optimization problem are Boltzmann-like (exponential) distributions, characterized by an effective 'temperature' $|\beta_i|^{-1}$ that, in this context, varies from site to site and measures the tolerance of site i to accept mutations over very long evolutionary times,

$$\pi_i(a) = \frac{\exp[-\beta_i h(a)]}{\sum_{\{a'\}} \exp[-\beta_i h(a')]}, \qquad (4.25)$$

with the constraint, (4.23),

$$\sum_{\{a\}} \exp[-\beta_i h(a)] [h(a) - A(c_i/\langle c \rangle - 1) - B] = 0. \qquad (4.26)$$

Equation (4.26) states an analytical relation between the 'Boltzmann parameter' β_i and the PE component c_i, given the two evolutionary parameters A and B. This equation indicates that β_i equals zero if $c_i/\langle c \rangle = 1 + A^{-1}(\sum_{\{a\}} h(a)/20 - \langle [h] \rangle_{\text{evol}})$, and that β_i becomes negative for larger c_i and positive for smaller c_i. The relationship between β_i and c_i is expected to be almost linear in the range $|c_i/\langle c \rangle - 1| \ll 1$. In addition, β_i tends to minus infinity when the average hydrophobicity at site i, $[h_i]_{\text{evol}}$, tends to the maximally allowed value, and to plus infinity when the average hydrophobicity at site i tends to the minimum allowed value.

Equation (4.26) has a simple qualitative interpretation. Positions with large eigenvector component c_i are buried in the core of the protein structure and are therefore with high probability occupied by hydrophobic amino acids (positive $h(a)$), having a large and negative β_i. Conversely, surface sites with small c_i are more likely occupied by polar amino acids (negative $h(a)$), having large and positive β_i. Intermediate sites are the most tolerant to mutations, having a small $|\beta_i|$ corresponding to high substitution temperature.

The distributions derived here refer to very long evolutionary times, when memory of the starting sequence has been lost. We recall the three assumptions that have been made for deriving the site-specific distributions: (a) The first assumption is that selection on folding stability can be represented effectively as a maximal correlation between the HP of sequences compatible with a given fold and the optimal HP of that fold, the latter nearly coinciding with the PE. This assumption follows directly from an approximation of the effective free energy function with its principal (hydrophobic) component, (4.22). (b) The second assumption is that the average of the HP of selected sequences over very long evolutionary times has a correlation coefficient of unity with the PE, i.e. all other energetic contributions average out. (c) The third assumption is that this correlation is the only relevant property of the site-specific amino acid distributions, indicating that these distributions are the distributions of maximum entropy whose site-specific averages have correlation one with the PE, thus fulfilling the stability requirement. From these three assumptions, the Boltzmann form of the amino acid distributions follows in a straightforward manner. To compute the site-specific Boltzmann parameters, however, one still needs to determine the positional mean and standard deviation of the site-specific HPs. These quantities depend on the mutation process and the selection parameters. They were computed directly from the data in such a way that the analytical prediction does not contain any free parameter.

The agreement between the predicted site-specific amino acid distributions and those observed in SCN simulations is very good [39], showing that this analytical approach reproduces quantitatively the statistics of the much more complex SCN process.

Boltzmann distributions have a long history in studies of protein structure and evolution. Structural properties of native protein structures, as for instance amino acid contacts, have been assumed to be Boltzmann-distributed [108], and Boltzmann statistics for structural elements was predicted in stable folds of globular proteins [109]. Our work points out to a complementary explanation for such distributions.

Shakhnovich and Gutin [110] proposed a model of sequence design through Monte Carlo optimization, which produced a Boltzmann distribution in sequence space. A mean-field approximation of this model [17, 111] results in site-specific amino acid distributions of the form

$$\pi_i(a) \propto \exp[-\beta \, \phi_i(a)], \qquad (4.27)$$

formally similar to (4.25). There are, however, three important differences between the present formulation and (4.27). First, (4.27) was derived as a mean-field approximation to a Boltzmann distribution for entire sequences, whereas (4.25) was derived from the relationship between average hydrophobicity at a given site and the PE component. Second, in (4.27), the Boltzmann parameter β is the same for all sites, whereas β_i, obtained here from the PE, changes along the protein structure. Third and most important, to compute (4.27), aligned families of natural proteins were used in [17, 111], whereas the present computation only requires the PE and two empirical values, the average and the standard deviation of the HP.

In [100, 101], Goldstein and co-workers assumed that the site-specific distributions of physico-chemical amino acid properties have a Boltzmann form. From this assumption they derived a protein evolution model to be used in phylogenetic reconstruction within a maximum likelihood framework. Since the properties that were used in these studies are hydrophobicity and amino acid size, the proposed distributions are a general case of those discussed here.

However, differently from [100, 101], here we classify sites according to the PE component, which is a structural indicator strongly correlated with conservation, and we compute the Boltzmann parameters analytically, whereas in [100, 101] they are fitted using a maximum likelihood framework.

4.4.6 Sequence Conservation and Structure Designability

We have shown that there is a direct relationship between a structural indicator, the PE, and site-specific measures of long-term evolutionary conservation that imposes limits to divergent evolutionary changes. This relationship also provides a link between the topology of a fold and its designability.

One convenient measure of the amino acid conservation at a given site is given by the rigidity, defined in terms of $\pi_i(a)$ as

$$R_i \equiv \sum_{\{a\}} [\pi_i(a)]^2 = \frac{\sum_{\{a\}} \exp[-2\beta_i\, h(a)]}{\left\{\sum_{\{a\}} \exp[-\beta_i\, h(a)]\right\}^2}. \quad (4.28)$$

The value $R_i = 1$ means that the same amino acid is present at site i in all sequences, leading to complete conservation and $\beta_i^{-1} = 0$. In general, the rigidity decreases with increasing temperature $|\beta_i|^{-1}$. One can use (4.26) and (4.28) to compute the rigidity directly from the PE.

A standard information-theoretic measure of site-specific sequence conservation is given by the entropy of the amino acid distribution

$$S_i \equiv -\sum_{\{a\}} \pi_i(a) \log [\pi_i(a)] = \log [Z(\beta_i)] + \beta_i\, [h_i]_{\text{evol}}, \quad (4.29)$$

where $Z(\beta_i) \equiv \sum_{\{a\}} \exp[-\beta_i\, h(a)]$. The entropy attains its maximum value, $S_i = \log(20)$, at $\beta_i = 0$, and it decreases with increasing $|\beta_i|$. Predictions of the entropy based on a different approach, (4.27), using aligned protein families have been obtained in [17, 111].

An important property of the entropy is that its exponential, $\exp(S_i)$, provides an estimate of the average number of amino acid types acceptable at site i over very long evolutionary times. Assuming that the amino acid distributions at different sites are independent from each other, the exponential of the sum of all site-specific entropies, $\exp(\sum_i S_i)$, gives an estimate of the region of the sequence space compatible with a given fold. The size of this region represents the designability of the fold. Although the independence assumption is a clear oversimplification, the estimate of designability that can be obtained should be a valuable approximation, and the present approach allows to connect it explicitly to a topological feature of the protein native structure [112, 113].

Kinjo and Nishikawa [114] have recently pointed out the existence of a strong relationship between hydrophobicity and the main eigenvector of substitution matrices derived from protein alignments with various values of the sequence similarities of the aligned proteins. They considered the eigenvector corresponding to the largest eigenvalue (in absolute value) of the substitution matrices. For high sequence similarities (above 35%), this eigenvector indicates the propensity of the amino acid to mutate over short evolutionary times (mutability). For low sequence identities (below 35%), corresponding to long evolutionary times, this eigenvector is very strongly correlated with hydrophobicity. This correlation is easily understood in the light of the results presented here. In fact, Kinjo and Nishikawa used Henikoffs' method [115] for deriving substitution matrices from observed frequencies of aligned amino acids at sites with various PE values. In the present notation, these substitution matrices can be indicated as $M(a,b) \approx \log\left[\langle\pi_i(a)\,\pi_i(b)\rangle/\langle\pi_i(a)\rangle\langle\pi_i(b)\rangle\right]$, where the angular brackets denote positional average. In other words, these substitution matrices measure the tendency of two residue types a and b to co-occur at the same sites. The relationship between large time substitution matrices and hydrophobicity gives therefore independent support to the results discussed here.

4.4.7 Site-Specific Amino Acid Distributions in the PDB

We tested how the predicted site-specific distributions compare to those obtained from a representative subset of the PDB [39]. For this comparison, we considered a non-redundant subset of single-domain globular proteins in the PDB, with a sequence identity below 25% [85]. Globular folds were selected by imposing that the fraction of contacts per residue was larger than a length-dependent threshold, $N_c/N > 3.5 + 7.8 N^{-1/3}$. This functional form represents the scaling of the number of contacts in globular proteins as a function of chain length (the factor $N^{-1/3}$ comes from the surface to volume ratio), and the two parameters are chosen so as to eliminate outliers with respect to the general trend, which represents mainly non-globular structures. Single-domain folds were selected by imposing that the normalized variance of the PE components is smaller than a threshold, $(1 - N\langle c\rangle^2)/(N\langle c\rangle^2) < 1.5$. In fact, multi-domain proteins have PE components that are large inside their main domains and small outside them (the PE components would be exactly zero outside the main domains if the domains would not share contacts). Therefore, multi-domain proteins are characterized by a larger normalized variance of PE components with respect to single-domain ones. It has been verified that the threshold of 1.5 is able to eliminate most of the known multi-domain proteins and very few of the known single-domain proteins.

In [39], we selected 404 sequences of less than 200 amino acids, and classified sites according to the value of $c_i/\langle c\rangle$ into bins, where $\langle c\rangle$ denotes the average over a single structure. For each bin, the observed distributions $\pi_{c_i/\langle c\rangle}(a)$ were fitted with an exponential function of the hydrophobicity parameters, $\pi_{c_i/\langle c\rangle}(a) \propto \exp[-\beta_{c_i/\langle c\rangle}\,h(a)]$. As in the case of the SCN simulations, the interactivity scale derived from the effective free energy function, (4.22), was used. The exponential fit was sufficiently good, and yielded the observed

Boltzmann parameters $\beta_{c_i/\langle c \rangle}$ as a function of the normalized PE components, $c_i/\langle c \rangle$.

Next, one can calculate the predicted Boltzmann parameters $\beta_{c_i/\langle c \rangle}$ from the relation

$$c_i/\langle c \rangle = 1 + \tilde{A}^{-1} \left[\frac{\sum_{\{a\}} h(a) \exp[-\beta_{c_i/\langle c \rangle} h(a)]}{\sum_{\{a\}} \exp[-\beta_{c_i/\langle c \rangle} h(a)]} - \tilde{B} \right], \qquad (4.30)$$

where $\tilde{A}$ and $\tilde{B}$ are defined as the analogous terms in (4.24), and the averages indicated by the square brackets in (4.24) now denote, instead of the evolutionary averages over a protein family, the average over all sites with $c_i/\langle c \rangle$ in the same bin, even belonging to different structures, whereas angular brackets in (4.24) now denote the average over all values of $c_i/\langle c \rangle$ weighted with the number of sites in the bins.

The observed Boltzmann parameters are compared in Fig. 4.6 to the predictions of (4.30). The agreement is remarkable, as the predictions do not involve any adjustable parameter, since $\tilde{A}$ and $\tilde{B}$ are calculated from the PDB data [39].

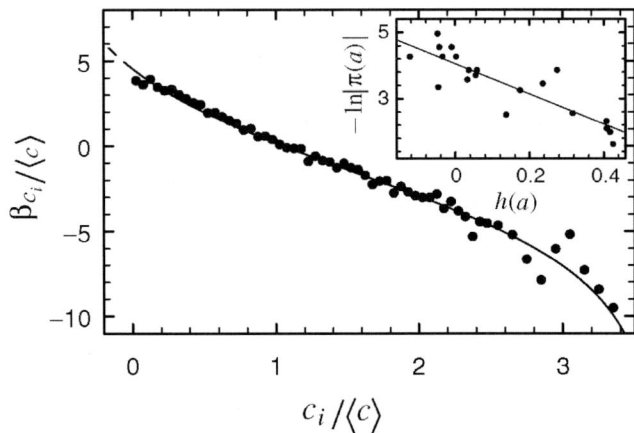

Fig. 4.6. 'Boltzmann parameter' $\beta_{c_i/\langle c \rangle}$ as a function of the normalized PE component $c_i/\langle c \rangle$ (*symbols*) obtained by analysing a subset of 404 non-redundant single-domain globular structures. The *continuous line* shows the analytical prediction, (4.30), obtained using the mean hydrophobicity $\langle [h]_{\text{PDB}} \rangle = 0.128$ and the variance $\langle [h]^2_{\text{PDB}} \rangle - \langle [h]_{\text{PDB}} \rangle^2 = 0.009$ as obtained from this set. The *dashed part* of the curve indicates the forbidden area $c_i < 0$. The *inset* exemplifies the numerically obtained $-\ln[\pi(a)]$ vs. hydrophobicity $h(a)$ of amino acid a (*symbols*), as obtained for $2.45 \leq c_i/\langle c \rangle < 2.5$, yielding via a linear fit (shown as *line*) a value of $\beta = -4.53$ for this bin (adapted from [39])

4.4.8 Mean-Field Model of Mutation plus Selection

Despite the good agreement with observations, the predicted distributions do not take into account the mutation process acting at the DNA level, but consider that all mutations from one amino acid to another are equiprobable. To incorporate the DNA level into the SCN scheme, we represent protein evolution at site i as an effective stochastic process with transition matrix

$$T(a,b) = P_\mu(a,b) \, P_{\text{fix},i}(a,b) \qquad (4.31)$$

for a substitution from a to $b \neq a$. The first factor represents the mutation process, and it is the same at all positions, and the second one represents the site-specific neutral fixation of mutations that conserve thermodynamic stability. Results from the SCN model show that, for what concerns the stationary distribution, the fixation term can be written as

$$P_{\text{fix},i}(a,b) = \min\{1, \exp(-\beta_i \, [h(b) - h(a)])\}, \qquad (4.32)$$

where the Boltzmann parameter β_i takes the value that fulfils (4.23). The larger the absolute value of β_i is, the larger is the fraction of mutations that are eliminated by negative selection for protein stability and the larger is the mutational load.

The stationary distribution of the complete transition matrix has now the form $\pi(a, \beta) \propto w_\beta(a) \exp[-\beta \, h(a)]$, where $w_\beta(a)$ satisfies the equations

$$0 = \sum_{\{b\}, b \neq a} \min\{\exp[-\beta \, h(b)], \exp[-\beta \, h(a)]\}$$
$$\times [w_\beta(a) \, P_\mu(a,b) - w_\beta(b) \, P_\mu(b,a)], \qquad (4.33)$$

for all final amino acid states b. If the mutation matrix satisfies the detailed balance equation, $w(a) \, P_\mu(a,b) = w(b) \, P_\mu(b,a)$, which is called 'reversibility' in the molecular evolution literature, then the stationary distribution of the mutation plus fixation process becomes

$$\pi_i(a) = \frac{w(a) \exp[-\beta_i \, h(a)])}{\sum_{\{a'\}} w(a') \exp[-\beta_i \, h(a')]}, \qquad (4.34)$$

where $w(a)$ is the stationary distribution of the mutation process, which is also the stationary distribution of the protein evolution process at sites where β_i equals zero (no mutations are rejected).

Within this more general context, the case $w(a) \equiv 1$, which corresponds to $P_\mu(a,b) = 1/20$, is the mutational model that was adopted in the previous subsection. Despite its simplicity, it provides already a surprisingly good prediction of the observed amino acid frequencies. If we adopt a reversible mutational model at the nucleotide level, we find

$$w(a) \propto \sum_{\text{codons}(a)} f(n_1) \, f(n_2) \, f(n_3), \qquad (4.35)$$

where $f(n)$ is the stationary frequency of the four nucleotides A, T, G and C. Using uniform nucleotide frequencies, $f(n) \equiv 1/4$ (or, in other words, $w(a)$ proportional to the number of codons) improves the prediction by 40% when measuring the similarity using the Jensen-Shannon (JS) divergence [40], without introducing any free parameter. By fitting the nucleotide frequencies, we can further improve significantly the prediction by 30% with only three free parameters [40]. The optimal nucleotide frequencies are $f(T) = 0.24$, $f(A) = 0.31$, $f(C) = 0.19$ and $f(G) = 0.26$. Notice that the optimal nucleotide frequencies violate Sueoka's parity 2 rule $f(A) = f(T)$ and $f(C) = f(G)$ [116], hinting at an asymmetric distribution of coding sequences on the two DNA strands [117].

Note that the site-specific mean hydrophobicity now depends on the parameters of the mutation process, so that

$$[h_i] \equiv \frac{\sum_{\{a\}} h(a) w(a) \exp[-\beta_i h(a)]}{\sum_{\{a\}} w(a) \exp[-\beta_i h(a)]} = A \left(c_i/\langle c \rangle - 1 \right) + B. \quad (4.36)$$

Therefore, the selection parameters β_i, defined implicitly by the above equation, also depend on the parameters of the mutation process. This looks at first sight in contradiction with the Darwinian paradigm according to which selection and mutation are independent forces. However, the contradiction is only apparent, as shown by the fact that the predicted distributions agree very well with simulations of the SCN model with mutations at the DNA level [117], for which the Darwinian paradigm holds. In the SCN model protein sites evolve in a correlated way as a result of global stability constraints. The effective model presented here is a mean-field model in which sites evolve independently, which constitutes a considerable simplification, in particular with respect to the task of evaluating likelihoods. The price to pay is that the selection parameter has to be computed self-consistently as the result of the mean hydrophobic environment created by other residues, in which the mutation process enters. For instance, when mutations favour the T nucleotide, that in second codon positions mostly codes for hydrophobic amino acids, the β parameter vanishes at hydrophobic positions with large $c_i/\langle c \rangle$, whereas, with the opposite mutation pattern, the β parameter vanishes at hydrophilic positions with small $c_i/\langle c \rangle$. Therefore, mutation and selection, although independent processes at a mechanistic level, become effectively entangled in the mean-field model. Accordingly, the mutation load, i.e. the fraction of mutants eliminated by negative selection, depends on the mutation bias [117], and so do the properties of protein folding thermodynamics: When the bias favours hydrophobic mutations, the balance between stability with respect to unfolding and stability with respect to misfolding shifts towards the former [57, 117]. In this way, the mutation process has a deep influence on the properties of proteins.

4.5 Conclusions

We have described how the conservation of protein structures influences the statistical properties of the evolutionary process by reviewing results that were obtained by using the SCN model. We have given particular emphasis to the effects of structure conservation on the topology of the neutral networks in sequence space and on the correlations during evolutionary trajectories, including the mutual effects on connected structural sites. Additionally, we have explained how the site-specific distributions of amino acids can be derived from the SCN model are consistent with those obtained from an analytically solvable mean-field model.

As illustrated by the results that we discussed, the inclusion of structure conservation in evolutionary models represents a powerful source of insight into the rules that determine molecular evolution. With the advent of structural genomics initiatives and the constant advances in computer technology, the range of applications of this approach is expected to expand considerably in the future.

Acknowledgements

UB would like to thank Peter Grassberger for interesting discussions on the SCN model, and Maya Paczusky for interesting discussions, for the hospitality offered at the Perimeter Institute (Waterloo, Canada) where part of this chapter was written, and for pointing out an inconsistency in the previous treatment of the substitution process.

References

1. E. Zuckerkandl, L. Pauling, in *Horizons in Biochemistry*, ed. by M. Kasha, B. Pullman (Academic Press, New York, 1962), pp. 189–225
2. E. Margoliash, Proc. Natl. Acad. Sci. USA **50**, 672 (1963)
3. D. Graur, W.H. Li, *Fundamentals of Molecular Evolution* (Sinauer, Sunderland, 2000)
4. M. Nei, S. Kumar, *Molecular Evolution and Phylogenetics* (Oxford University Press, 2000)
5. L. Bromham, D. Penny, Nat. Rev. Genet. **4**, 216 (2003)
6. M. Kimura, Nature **217**, 624 (1968)
7. M. Kimura, *The Neutral Theory of Molecular Evolution* (Cambridge University Press, 1983)
8. J.-L. King, T.H. Jukes, Science **164**, 788 (1969)
9. J.H. Gillespie, *The Causes of Molecular Evolution* (Oxford University Press, Oxford, 1991)
10. P. Schuster, W. Fontana, P.F. Stadler, I.L. Hofacker, Proc. R. Soc. London B **255**, 279 (1994)
11. M.A. Huynen, P.F. Stadler, W. Fontana, Proc. Natl. Acad. Sci. USA **93**, 397 (1996)

12. W. Fontana, P. Schuster, Science **280**, 1451 (1998)
13. E. Bornberg-Bauer, Biophys. J. **73**, 2393 (1997)
14. E. Bornberg-Bauer, H.S. Chan, Proc. Natl. Acad. Sci. USA **96**, 10689 (1999)
15. A. Babajide, I.L. Hofacker, M.J. Sippl, P.F. Stadler, Folding Des. **2**, 261 (1997)
16. A.M. Gutin, V.I. Abkevich, E.I. Shakhnovich, Proc. Natl. Acad. Sci. USA **92**, 1282 (1995)
17. N.V. Dokholyan, E.I. Shakhnovich, J. Mol. Biol. **312**, 289 (2001)
18. L.A. Mirny, E.I. Shakhnovich, J. Mol. Biol. **291**, 177 (1999)
19. N.V. Dokholyan, B. Shakhnovich, E.I. Shakhnovich, Proc. Natl. Acad. Sci. USA **99**, 14132 (2002)
20. S. Govindarajan, R.A. Goldstein, Biopolymers **42**, 427 (1997)
21. S. Govindarajan, R.A. Goldstein, Procl. Natl. Acad. Sci. USA **95**, 5545 (1998)
22. D.M. Taverna, R.A. Goldstein, Proteins **46**, 105 (2002)
23. H.J. Bussemaker, D. Thirumalai, J.K. Bhattacharjee, Phys. Rev. Lett. **79**, 3530 (1997)
24. G. Tiana, R.A. Broglia, H.E. Roman, E. Vigezzi, E.I. Shakhnovich, J. Chem. Phys. **108**, 757 (1998)
25. G. Parisi, J. Echave, Mol. Biol. Evol. **18**, 750 (2001)
26. G. Parisi, J. Echave, Gene **345**, 45 (2005)
27. Y. Xia, M. Levitt, Proc. Natl. Acad. Sci. USA **99**, 10382 (2002)
28. Y. Xia, M. Levitt, Curr. Op. Struct. Biol. **14**, 202 (2004)
29. T. Aita, M. Ota, Y. Husimi, J. Theor. Biol. **221**, 599 (2003)
30. J.D. Bloom, J.J. Silberg, C.O. Wilke, D.A. Drummond, C. Adami, F.H. Arnold, Proc. Natl. Acad. Sci. USA **102**, 606 (2005)
31. U. Bastolla, H.E. Roman, M. Vendruscolo, J. Theor. Biol. **200**, 49 (1999)
32. U. Bastolla, M. Porto, H.E. Roman, M. Vendruscolo, J. Mol. Evol. **56**, 243 (2003)
33. R.M. Sweet, D. Eisenberg, J. Mol. Biol. **171**, 479 (1983)
34. L. Holm, C. Sander, Proteins **19**, 256 (1994)
35. N. Kannan, S. Vishveshwara, J. Mol. Biol. **292**, 441 (1999)
36. N. Kannan, S. Vishveshwara, Prot. Eng. **13**, 753 (2000)
37. M. Porto, U. Bastolla, H.E. Roman, M. Vendruscolo, Phys. Rev. Lett. **92**, 218101 (2004)
38. U. Bastolla, M. Porto, H.E. Roman, M. Vendruscolo, Proteins **58**, 22 (2005)
39. M. Porto, H.E. Roman, M. Vendruscolo, U. Bastolla, Mol. Biol. Evol. **22**, 630; Mol. Biol. Evol. **22**, 1156 (2005)
40. U. Bastolla, M. Porto, H.E. Roman, M. Vendruscolo, Gene **347**, 219 (2005)
41. E. van Nimwegen, J.P. Crutchfield, M. Huynen, Proc. Natl. Acad. Sci. USA **96**, 9716 (1999)
42. M. Eigen, Naturwissenschaften **58**, 465 (1971)
43. M. Eigen, J. Mc Caskill, P. Schuster, Adv. Chem. Phys. **75**, 149 (1989)
44. J.W. Drake, J.J. Hollandy, Proc. Natl. Acad. Sci. USA **96**, 13910 (1999)
45. R. Durrett, *Probability Models for DNA Sequence Evolution*, (Springer, Berlin Heidelberg New York 2002)
46. J. Berg, S. Willmann, M. Lässig, BMC Evol. Biol. **4**, 42 (2004)
47. G. Sella, A.E. Hirsh, Proc. Natl. Acad. Sci. USA **102**, 9541 (2005)
48. I. Leuthauser, J. Stat. Phys. **48**, 343 (1987)
49. P. Tarazona, Phys. Rev. A **45**, 6038 (1992)
50. T. Ohta, M. Kimura, J. Mol. Evol. **1**, 18 (1971)

51. T. Ohta, Nature **246**, 96 (1973)
52. R.A. Fisher, *The Genetic Theory of Natural Selection* (Dover, 1930)
53. J. McDonald, M. Kreitman, Nature **351**, 652 (1991)
54. N.G.C. Smith, A. Eyre-Walker, Nature **415**, 1022 (2002)
55. T. Ohta, J. Mol. Evol. **41**, 115 (1995)
56. D.J. Lambert, N.A. Moran, Proc. Natl. Acad. Sci. USA **95**, 4458 (1998)
57. U. Bastolla, A. Moya, E. Viguera, E. van Ham, J. Mol. Biol. **343**, 1451 (2004)
58. S. Aksoy, Insect Mol. Biol. **4**, 23 (1995)
59. M.A. Fares, M.X. Ruiz-Gonzalez, A. Moya, S.F. Elena, E. Barrio, Nature **417**, 398 (2002)
60. M.C. Orencia, J.S. Yoon, J.E. Ness, W.P. Stemmer, R.C. Stevens, Nat. Struct. Biol. **8**, 238 (2001)
61. C.O. Wilke, BMC Genet. **5**, 25 (2004)
62. A.R. Fersht, *Structure and Mechanism in Protein Science: A Guide to Enzyme Catalysis and Protein Folding* (W.H. Freeman, 1999)
63. C.M Dobson, Nature **426**, 884 (2003)
64. V.N. Uversky, Cell. Mol. Life Sci. **60**, 1852 (2003)
65. K.A. Bava, M.M. Gromiha, H. Uedaira, K. Kitajima, A. Sarai, Nucl. Ac. Res. **32**, D120 (2004)
66. N. Sueoka, Proc. Natl. Acad. Sci. USA **47**, 469 (1961)
67. J.R. Lobry, Gene **205**, 309 (1997)
68. L. Holm, C. Sander, Science **273**, 595 (1996)
69. B. Rost, Folding Des. **2**, S19 (1997)
70. D. Cozzetto, A. Di Matteo, A. Tramontano, FEBS J. **272**, 881 (2005)
71. S.F. Atschul, T.L. Madden, A.A. Schaffer, J. Zhang, Z. Zhang, W. Miller, D.J. Lipman, Nucl. Acids Res. **25**, 3389 (1997)
72. N. Nagano, C.A. Orengo, J.M. Thornton, J. Mol. Biol. **321**, 741 (2002)
73. A.G. Murzin, S.E. Brenner, T. Hubbard, C. Chothia, J. Mol. Biol. **247**, 536 (1995)
74. C.A. Orengo, A.D. Michie, S. Jones, D.T. Jones, M.B. Swindells, J.M. Thornton, Structure **5**, 1093 (1997)
75. C. Chothia, A.M. Lesk, EMBO J. **5**, 823 (1986)
76. D. Devos, A. Valencia, Proteins **41**, 98 (2000)
77. N.V. Grishin, J. Struct. Biol. **134**, 167 (2001)
78. W.R. Taylor, Nature **416**, 657 (2002)
79. A. Harrison, F. Pearl, R. Mott, J. Thornton, C. Orengo, J. Mol. Biol. **323**, 909 (2002)
80. M. Zuker, D. Sankoff, Bull. Math. Biol. **46**, 591 (1984)
81. J.U. Bowie, R. Lüthy, D. Eisenberg, Science **253**, 164 (1991)
82. U. Bastolla, M. Vendruscolo, E.W. Knapp, Proc. Natl. Acad. Sci. USA **97**, 3977 (2000)
83. U. Bastolla, J. Farwer, E.W. Knapp, M. Vendruscolo, Proteins **44**, 79 (2001)
84. R.A. Goldstein, Z.A. Luthey-Schulten, P.G. Wolynes, Proc. Natl. Acad. Sci. USA **89**, 4918 (1992)
85. U. Hobohm, C. Sander, Protein Sci. **3**, 522 (1994)
86. U. Bastolla, M. Porto, H.E. Roman, M. Vendruscolo, J. Mol. Evol. **57**, S103 (2003)
87. U. Bastolla, L. Demetrius, PEDS **18**, 405 (2005)
88. B. Derrida, Phys. Rev. B **24**, 2613 (1981)

89. E.I. Shakhnovich, A.M. Gutin, Biophys. Chem. **34**, 187 (1989)
90. U. Bastolla, M. Porto, H.E. Roman, M. Vendruscolo, Phys. Ref. Lett. **89**, 208101 (2002)
91. C.H. Langley, W.M. Fitch, J. Mol. Evol. **3**, 161 (1974)
92. N. Takahata, Genetics **116**, 169 (1987)
93. S.Y.W. Ho, M.J. Phillips, A. Cooper, A.J. Drummond, Mol. Biol. Evol. **22**, 1561 (2005)
94. J. Felsenstein, J. Mol. Evol. **17**, 368 (1981)
95. J. Overington, M.S. Johnson, A. Sali, T.L. Blundell, Proc. Roy. Soc. Lond. B **241**, 132 (1990)
96. D.T. Jones, W.R. Taylor, J.M. Thornton, Comp. Appl. Biosci. **8**, 275 (1992)
97. S. Henikoff, J.G. Henikoff, Proteins **17**, 49 (1993)
98. A.L. Halpern, W.J. Bruno, Mol. Biol. Evol. **15**, 910 (1998)
99. P. Liò, N. Goldman, Genome Res. **8**, 1233 (1998)
100. J.M. Koshi, R.A. Goldstein, Proteins **32**, 289 (1998)
101. J.M. Koshi, D.P. Mindell, R.A. Goldstein, Mol. Biol. Evol. **16**, 173 (1999)
102. J.L. Thorne, Curr. Opin. Genet. Dev. **10**, 602 (2000)
103. M.S. Fornasari, G. Parisi, J. Echave, Mol. Biol. Evol. **19**, 352 (2002)
104. G. Casari, M.J. Sippl, J. Mol. Biol. **224**, 725 (1992)
105. H. Li, C. Tang, N.S. Wingreen, Phys. Rev. Lett. **79**, 765 (1997)
106. M. Vendruscolo, E. Kussell, E. Domany, Folding Des. **2**, 295 (1997)
107. A. Bateman, E. Birney, R. Durbin, S.R. Eddy, K.L. Howe, E.L.L. Sonnhammer, Nucl. Ac. Res. **28**, 263 (2000)
108. S. Miyazawa, R.L. Jernigan, Macromolecules **18**, 534 (1985)
109. A.V. Finkelstein, A.Ya. Badretdinov, A.M. Gutin, Proteins **23**, 142 (1995)
110. E.I. Shakhnovich, A.M. Gutin, Proc. Natl. Acad. Sci. USA **90**, 7195 (1993)
111. N.V. Dokholyan, L.A. Mirny, E.I. Shakhnovich, Physica A **314**, 600 (2002)
112. P. Koehl, M. Levitt, Proc. Natl. Acad. Sci. USA **99**, 1280 (2002)
113. J.L. England, E.I. Shakhnovich, Phys. Rev. Lett. **90**, 218101 (2003)
114. A.R. Kinjo, K. Nishikawa, Bioinformatics **20**, 2504 (2004)
115. S. Henikoff, J.G. Henikoff, Proc. Natl. Acad. Sci. USA **89**, 10915 (1992)
116. N. Sueoka, J. Mol. Evol. **40**, 318 (1995)
117. U. Bastolla, M. Porto, H.E. Roman, M. Vendruscolo, BMC Evol. Biol. **6**, 43 (2006)

5
Towards Unifying Protein Evolution Theory

N.V. Dokholyan and E.I. Shakhnovich

5.1 Two Views on Protein Evolution

There are many proposed models for protein domain evolution. The main contention [1], however, is between divergent [2] and convergent [3] models of evolution. Divergent evolution proposes that there was an ancestor to all domains that we see today and consequently all domains are historically related to each other. Convergent evolution claims that chance is largely responsible for the appearance of new kinds of domains. Convergent evolution explains the prevalence of certain types of structures by stating that those structures are more favorable and are often recycled for different functions. The distinction between convergent and divergent evolution is neither abstract nor purely academic. If divergent evolution is the primary model responsible for the appearance of new domains, then it is possible to reconstruct lineages of domains, follow structural change, and predict contextual functional ranges that occur due to well-defined evolutionary pressures. However, if convergent evolution is the main driving force behind the appearance of new structures and functions, in principle there should be no discernable structure–function relationship. An important step in understanding evolution is the quantification of dominant evolution scenario(s) as theoretical background for derivation of the structure–function correlation. Here, we describe recent advances in this direction.

5.2 Challenges in Functionally Annotating Structures

To outline the steps that are needed to create a theory of structure–function correlation, we would like to first outline the challenges in doing so. First, we have a problem of finding the "atomic unit" of functionality in proteins. This atomic unit is commonly taken to be a protein domain. This is so because domains can be both structurally sound and functional outside the protein that they are a part of. The most common ways of finding protein domains rely heavily on personal "intuition" [4] and are often points of vigorous debate [5,6]. However, such an atomic unit must be agreed upon if a theory of structure–function relation is to succeed. An atomic unit is integral to addressing the structure–function correlation because of the need to "compare" these units. If we pick incorrect atomic units, comparisons will be plagued by problems like "flow of structure" and overlapping sets of nonunique functions.

The determination of the function for a hypothetical protein is currently based on three strategies [7]. The first strategy is based on finding sequence similarity to known proteins. Even at low sequence similarities, there may be a set of conserved amino acids constituting an active site or a conserved hydrophobic core [8–11]. In the case of a conserved active site, similar amino acids may indicate the function of a hypothetical protein. The second strategy involves the search for protein surface cavities using sequence and structural similarities to protein with known function. As in the first strategy, the extent of the success of this methodology depends strongly on the conservation of local sequence and structural motifs. In addition, the second strategy relies on the knowledge of the protein structure. The driving assumption for these strategies is the possible similarity of the active sites between proteins sharing the same or similar function [12,13]. Also, there have been several mechanisms proposed to search for local functional motifs by comparison to libraries of three-dimensional structural templates [13–15] and the analysis of the physical properties of protein surfaces. Teichmann and Thornton [7, 16] described two examples of correct functional annotation of hypothetical proteins: the HdeA protein from *Escherichia coli* [17] and the protein corresponding to gene 226 from *Methanococcus janaschii* [18, 19]. Finally, the third strategy is based on the crystalographic studies of the bound cofactors in the native protein structure. The main limitation of this strategy is that it requires experimental reconstruction of the three-dimensional structures of protein–ligand complexes. The efforts to annotate function exclusively based on structure and sequence, as discussed above, are complicated by the fact that different sequences may fold into similar structures [4] but have different functions. A notable example of functional diversity inside a structurally homologous family is the case of the P-loop NTPases. The structures of RecA (2reb) [20] and adenylate kinase (2ak3) [21] proteins are similar. Both are alpha and beta proteins. Both contain P-loop topology. Both are placed in the same SCOP family [4]. Yet, their functions are quite distinct. RecA is a DNA repair protein, while the adenylate kinase is a transfer protein transferring phosphate groups from AMP to ADP.

The problem is further complicated when some clearly homologous genes, found through high scoring sequence comparison, fold into the same structure, but perform different functions depending on the genome they are expressed in. For example, two sequence homologues of an alpha/beta hydrolase share more than 75% sequence identity. Yet, the *E. coli* version has the function of a serine-activating enzyme, while the human homologue is a lysosomal carboxypeptidase [22].

Another challenge in understanding the structure–function relationship in protein domains is that it is difficult to define quantitatively the metric of the "function space." While structural relationships between protein domains are easily quantifiable, so that several good measures of it exist – Z-score [23–25] and RMSD, the distance in functional space is poorly defined. A way to address this crucial issue is by using a hierarchical description of protein functionality and probabilistic models that aim to quantify functional proximity based on dominating functional annotations.

Finally, the idea of understanding the protein function from individual characteristics is exacerbated by some puzzling findings where evolution cannot be easily traced. These findings were attributed most often to convergent evolution. For example, Ponting and Russel [26] describe the case of the Ser/His/Asp catalytic triad [27], which has been identified in five completely different protein folds. These folds apparently do not share any common ancestor based on extensive sequence similarity searches. Therefore, these folds are not considered homologous, yet their functions are the same and the catalytic triad is also strikingly similar.

5.3 The Importance of the Tree of Life

Along with creating methods for measuring sequence, structure, and metabolic pathway variability, we also have to create a way to describe the context (see [9]) for each sequence and structure. A hierarchical description of the context exists now and is referred to as the "tree of life" [28, 29]. The tree of life describes the most probable divergence of organisms (or most parsimonious way to describe their interrelationship based on multiple sequence alignment of a single ubiquitous protein). An example of a tree based on this principle of parsimony was constructed in the mid-1970s by Woese and collaborators who used SSU rRNA molecule as a molecular chronometer. On the basis of this information, most trees of life are now drawn [30–33]. SSU rRNA is used because it is widespread in organisms and its structure is highly conserved [33].

There is, however, considerable debate about the precision of the phylogeny predicted by a tree drawn purely on rRNAs alone. This dissention stems partly from the fact that other genes give believably different trees for the same set of organisms [34–42]. Further, apparently different trees can be produced by naturalistic methods of traditional phylogeny [43–45]. It has been noted that due to the existence of lateral transfer – a phenomenon where genes are transferred via a plasmid or vector or some other method other than divergence to another organism – nets should be used instead of trees to represent interrelationships between organisms [46]. Indeed, the building of an accurate tree of life or a net of life is very important, both because it is an interesting academic challenge and because it is a necessary part in understanding the context of the structure–function relationship elucidation of such aspects as functional pressure on domain evolution.

5.4 Building the PDUG

To consider sequence, structure, and function [47] information in a unified, systematic way, Shakhnovich et al. [48] defined both gene families and fold families (sequence and structure homologies) quantitatively using the Protein Domain Universe Graph (PDUG) [49]. The PDUG is a graph where nodes are sets of closely related sequences folding into well-defined domains [50, 51] and edges are connections between the nodes that are based on structural similarity (Fig. 5.1), while sequence identity between any pair of sequences belonging to different nodes is less than 25%. To build the PDUG, Dokholyan et al. [49] took all sequences from NRDB90 [52] and all structural domains from HSSP [51]. Shakhnovich et al. [48] further used BLAST [53] sequence homology to detect all sequences in NRDB90 with more than 25% sequence identity to each HSSP domain. That set of sequences was combined into a single gene family. Using cross-indexing between Swiss-Prot [54] and InterPro [55] all equilogs (different sequences with the same function) belonging to every gene family were identified. Those equilogs are further used to reconstruct the functional flexibility score (FFS; see [6]).

Using this PDUG formalism, it is possible to explore global correlations between sequence, structure, and function determinants. For example, we can define a gene family based on micro-evolutionary considerations: the PDUG represents on many evolutionary scales the variability accessible to a given

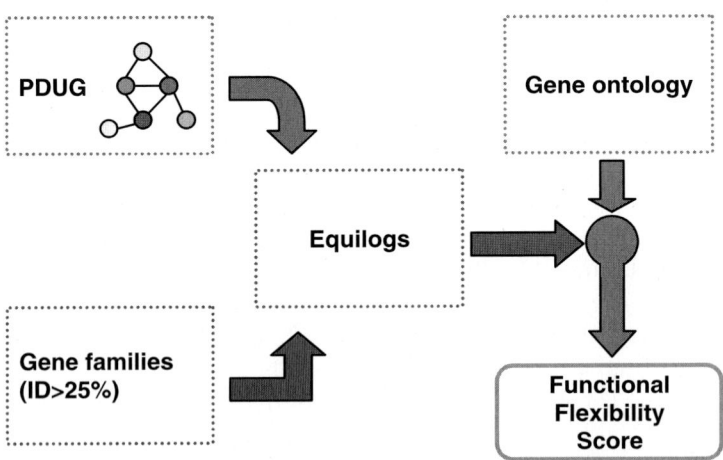

Fig. 5.1. A diagram of the scaled organization and intrinsic properties of the PDUG. The PDUG is built hierarchically: first, domains' structural similarities are compared to each other and from this information the structural graph is created. All the sequences from NRDB with more than 25% identity to the original sequence of each domain on the PDUG are collected into a gene family. All the equilogs (sequences with the same function) matching the gene family are collected and used to create a probabilistic GO tree from which the FFS is calculated using (5.1)

gene upon mutation, whether that variability is in sequence, function, or structure space. Unlike other definitions of gene families [50,56], the definition proposed by Shakhnovich et al. [48] is local, i.e., with respect to a particular gene. The gene family of a gene is, therefore, all the immediate sequence neighbors inside a PDUG node. Similarly, the fold family of a structure is all the structural neighbors of that domain on PDUG (Fig. 5.1). By defining the cutoff value for the sequence or structure comparison, it is possible to control the variability for that gene, thus implicitly controlling the allowed evolutionary divergence time on which structure–function determinants are calculated. This approach turns out to be invaluable in gleaning new insights into structure–function correlation, coevolution, and definition of important properties in the PDUG.

5.5 Properties of the PDUG: Power Laws on Very Different Evolutionary Scales

The properties of the PDUG have been recently addressed in several works [49,57,58]. The properties of the PDUG largest cluster were determined [49]. The size of the largest cluster in the PDUG and random control graph were determined and compared as a function of the structural similarity score $Z_{\min}$ [49] defined by FSSP [24]. The random control graphs were constructed by maintaining the same number of proteins and connections as in the actual PDUG, but reshuffling the connections between the nodes. Control random graphs represent an evolution process without any driving force, i.e., any node can be connected to another node by chance. Dokholyan et al. [49] found a pronounced transition of the size of the largest cluster in the PDUG at $Z_{\min} = Z_c \approx 9$. Random graphs featured a similar transition, but at a higher value of $Z_{\min} = Z_c \approx 11$. The distribution of cluster sizes depends significantly on whether $Z_{\min} > Z_c$ or $Z_{\min} < Z_c$ for both the PDUG and random graphs. It was also observed that the probability density $P(M)$ of cluster sizes M for both the PDUG and random graphs followed a power-law at their respective Z_c: $P(M) \propto M^{-2.5}$. The observed power-law behavior of $P(M)$ is simply a consequence of criticality at Z_c as it is featured prominently both for the PDUG and random graphs. The power-law probability density of cluster sizes is a generic percolation phenomenon that has been observed and explained in both percolation [59,60] and random graph theories [61].

To define more concretely the structural properties of the PDUG, Dokholyan et al. [49] computed the probability $P(k)$ of the number of edges per node k taken at $Z_{\min} = Z_c$ for individual clusters. It is known that $P(k)$ distinguishes random graphs from various graphs observed in science and technology [61]. In stark contrast with the equivalent random graph, the PDUG is scale-free with $P(k) \propto k^{-1.6}$ with a high degree of statistical significance (p-value less than 10^{-8}). The power law fit of $P(k)$ is most accurate at $Z \approx Z_c$, and noticeably deteriorates above and below Z_c. Similar calculations were also performed on the number of equilogs inside each PDUG node and the number of sequence members coding for individual domains [48]. These measures represent three different levels of evolutionary divergence and characterize the topology of the PDUG. Strikingly, it was discovered that not only do all these values similarly follow a power-law, but they also obey the same exponent $P(k) \propto k^{-1.6}$ This behavior turns out to be the fingerprint of divergent evolution and these data will help differentiating between the histories of divergent and convergent evolution, as well as building proper evolutionary models.

The power-law fit at $Z_{min} > Z_c$ quickly becomes meaningless as the range of values of connectivity k rapidly diminishes as greater Z_{min} lead to mostly disconnected domains. At $Z_{min} < Z_c$ the power law fit also becomes problematic in the whole range of k because at large values of k (50–100) $P(k)$ shows some nonmonotonic behavior that can be interpreted as a maximum at large k (the data are insufficient to conclude that with certainty). However, the remarkable property of a maximum $P(k)$ at $k = 0$ i.e., dominance of orphans remains manifest at all Z_{min} values. This is in striking contrast with random graph that is not scale-free at any value of Z_{min} and where $P(k)$ allows almost perfect Gaussian fit with maximum at higher values of k. This power-law behavior also turns out to be in contrast to that generated by convergent evolution models (see [8]). The criterion for selecting cutoff value based on transition in the giant component also turns out to be highly useful in determining proper functional clusters (see [10]).

It is worth noting that the exponent -1.6 in the connectivity distribution was recently corrected and is suggested to be closer to -1.0 [57]. The correction arises from the consideration of the exponential finite size effects, which significantly contribute to the power-law regime [62, 63]. In addition, recent examination of the origin of the scale-free properties of the PDUG suggested that the PDUG is not modular, i.e., it does not consist of modules with uniform properties. Instead, it was found the PDUG to be self-similar at all scales [57].

5.6 Functional Flexibility Score: Calculating Entropy in Function Space

Shakhnovich et al. [48] defined the function determinant of a gene family as entropy in function space. When this measure is calculated in the context of the PDUG, the Gene Ontology (GO) [24] is utilized to define the functional variability (functional flexibility score or FFS) of a set of genes. The FFS is a measure of the total amount of information needed to describe all the functionality of a gene family. Interestingly, the FFS statistically correlates with the logarithm of the total number of sequences in a gene family [48]. Contrary to merely counting the number of members in a gene family, the FFS appears to be a more robust (with respect to possible bias in the databases, as well as uneven sampling of phylogeny) measure of gene family diversity because it normalizes on the number equilogs, i.e., sequences that diverged far enough to represent functionally diverse proteins. The FFS is a characteristic of both the sequence topology of the PDUG as well as a functional determinant of neighborhoods.

To calculate functional entropy, all sequences were combined into a single set [48]. These sequences were then matched to InterPro [55] equilogs, proteins with different sequences but the same function. The complete GO tree was further reconstructed from the annotations of equilogs and the number of equilogs of the family, which is assigned a particular functional annotation normalized by total number of annotations at each level, is calculated (Fig. 5.1). Thus it is possible to calculate the average amount of information per level needed to fully describe the function of each gene family using the following equation [48]:

$$\text{FSS} = -\frac{1}{\text{Max}(L)} \sum_{l} \sum_{i \in \{\text{nodes on level } l\}} p_i \log(p_i). \qquad (5.1)$$

Here, $\text{Max}(L)$ is the maximal number of levels of annotation, the summation is taken over all levels l and over all nodes i filled by the gene family on the GO tree, and p_i is the fraction of the family that is annotated with function i.

5.7 Lattice Proteins and Its Random Subspaces: Structure Graphs

How important is the observation of power law in the organization of the PDUG? Is it a generic feature of proteins as compact polymers or a result of their evolutionary selection? To address this question, Deeds et al. [64] turned to a simple yet exact lattice model of compact 27-mers, whose fully compact conformations have been fully enumerated to yield 103,346 conformations. The structural comparison between all pairs of compact structures can be carried out in a similar way to the DALI method by calculating the number of native contacts that two conformations have in common. Then the lattice structure graph (LSG) is constructed in a similar way to the PDUG [64]. The evaluation of the node connectivity distribution for the complete lattice graph all 103,346 structures and various randomly selected (in a manner consistent with the convergent evolution scenario) subgraphs, as well as subgraphs of the 3,500 most designable lattice structures, yielded LSGs that feature Gaussian rather than power-law distribution of $p(k)$, in sharp contrast with the real PDUG. These results suggest the evolutionary origin of power-law degree distribution in the PDUG.

5.8 Divergence and Convergence Explored: What Power Laws Tell Us about Evolution

To advance toward a resolution of the debate distinguishing convergent and divergent evolution, Deeds et al. [65] explored the predictions of convergent models. The simplest such model assumes that nodes are discovered completely randomly – that is, the likelihood of adding a particular node has nothing to do with the number of structural neighbors it has. In this class of convergent models, as organisms evolve and speciate, nodes were added to each proteome randomly, producing organisms whose structural domains represent a random subset of the existing protein domains taken from all organisms. As with the entire PDUG, one can create a network from the structural similarity between the domains within a particular proteome, and under this type of convergent model, the resulting graph would be a random subgraph of the PDUG. Thus, unbiased convergence leads to the null hypothesis that proteome-specific subgraphs will be random subgraphs of the extant protein universe.

Given a template graph of N_0 nodes with a distribution of edges per node described by $p_{N_0}(k)$, the average distribution of edges per node in a subgraph of N nodes chosen completely randomly should follow:

$$p_N(k) = C \left[\sum_{s=k}^{\text{Max}k_{N_0}} \binom{s}{k} \left(\frac{N}{N_0}\right)^k \left(1 - \frac{N}{N_0}\right)^{s-k} \right] \tag{5.2}$$

where $\text{Max}k_{N_0}$ is the maximally connected node in the template graph and C is taken to normalize $p_N(k)$. This expression is similar in justification to those used in percolation theory to estimate the behavior of scale-free networks when a fraction of nodes are eliminated [66–68]. It is clear from the theory that the degree distributions of random subgraphs exhibit a power-law region that persists for a characteristic length at each subgraph size. Deviations to higher k's in subgraphs constitute an exponential tail to the distribution, indicating that random subgraphs are very unlikely to contain nodes with connectivity greater than this value. The maximum k in random subgraphs is thus a "fingerprint" of random subgraphs of a given size.

On the basis of the understanding of the behavior of random subgraphs, (5.2), Deeds et al. [65] turned to actual proteomes to determine whether they represented random subgraphs of the PDUG. The authors created the organismal subgraphs using homology to determine which of the domain structures in the PDUG occur in each organism. For each subset of nodes found in each proteome, Deeds et al. [65] calculated the degree distributions for these subgraphs.

In addition, Deeds et al. [65] performed a power-law regression on the degree distributions for the subgraphs of 59 fully sequenced bacterial proteomes, all of which were well fit by a power law. This raises the possibility that these subgraphs may indeed be random; however, the maximum k in the organismal subgraphs was considerably larger than that observed in arbitrary random subgraphs of the same size. The probability that organismal PDUGs are random subgraphs of the complete PDUG was estimated by comparison of their actual maximum k with distribution of maximum k in random subgraphs based on (5.2). Such probability was found to be extremely low in most organisms except few smallest ones where it reached about 0.01.

This finding suggests that the null hypothesis corresponding to a convergent evolution model is highly unlikely and, thus, that unbiased models have a low likelihood of explaining protein structural evolution. Even given the inability of convergent models to explain this behavior, it is not necessarily clear that this nonrandomness could be observed in a divergent model similar to those that have been proposed [49,69]. To explore this possibility, Deeds et al. [65] modified an earlier model described in detail by Dokholyan et al. [49] to include speciation (Fig. 5.2). Simulations of the speciation model that included generation of 3,500 model proteins and four organisms were performed, with speciation occurring after 1,000 and 2,000 steps, and then the degree distributions of the subgraphs in each organism were compared to that of the overall graph produced by the model. For ten realizations of the model, the model organism subgraphs exhibited power-law degree distributions that deviated to larger maximum k than expected at random. The p-values for the model

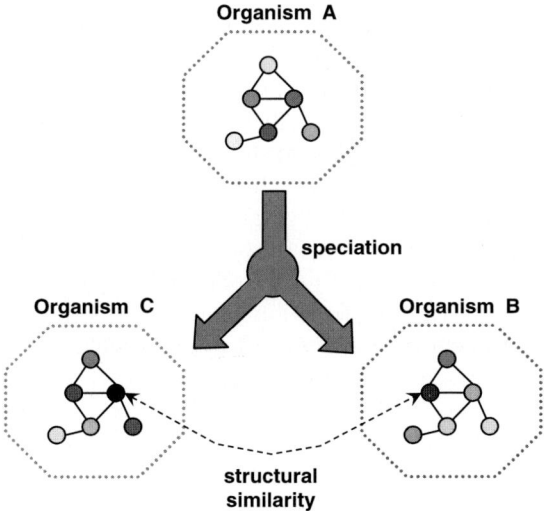

Fig. 5.2. A diagram of the speciation model. The PDUG evolves from all nodes belonging to one organism, and after a specified gene duplication event, an organism A undergoes speciation to create two organisms B and C with identical graphs. All the new nodes, however, are added to one organism or the other, given that the duplication events that give rise to new nodes will only occur within a single proteome. After the proteomes evolve independently for a number of steps, speciation occurs again, and so on

subgraphs indicated that the bulk of these graphs did not have a high probability of being random (although the probabilities were higher than observed for the actual organisms). Although this model is certainly not detailed enough to explain the evolution of real protein structures, it does demonstrate that a simple, dynamic process can produce model organism subgraphs that are similar to those observed in real organisms and that with careful modeling the debate between convergence and divergence can be resolved. Most importantly, these data show that convergent evolution is highly unlikely and that divergence is probably the dominating force in protein domain evolution.

5.9 Context Is Important

The above examples describing differences in function for the same protein between different genomes (see [2]) show that understanding the structure–function correlation involves buttressing structure information with "contextual" information. We see that functionality may vary for the same structure depending on the genome, or the metabolic pathway. This is intuitively understandable if we look at structure–function correlation from an evolutionary standpoint [7, 22, 26, 70, 71]. Domains that were put under different evolutionary pressures (which in our case translates directly into being in a different context) evolve different functions. Consequently, we think that to understand the relationship between structure and function, it is not enough only to enumerate the possibilities, but it is also necessary to understand the progression that has led to the state of protein domain universe as we observe it now.

In a recent work, Shakhnovich [72] has demonstrated that using phylogenetic information it is possible to dramatically and quantitatively improve functional annotation. He introduced, besides the structural similarity measure (Z-score) and the functional similarity (FFS), a measure of phylogenetic distance (P) and presented a 3-dimensional landscape in (Z, FFS, P) space. Dramatically, this landscape was found to be well-shaped. This finding points out that pronounced structure–function correlation is observed only for domains that are phylogenetically close, while for phylogenetically distant domains correlation between structure and function (i.e., between Z and FFS) essentially vanishes. This study shows clearly that "context" information, such as the origin and phylogenetic history of a protein domain, in some cases may influence the precise function of the gene more than the structure.

5.10 Not All Functions Are Created Equal and Neither Are Structures

With the PDUG and the newly developed techniques, it is possible to perform structure–function studies on a global scale capturing evolutionary relationships that are not easily revealed by anecdotal studies alone. An interesting problem to address is the coevolution of gene family size, functionality, and structure. For example, we can ask whether some functions require smaller gene families than others by computing correlations between particular functionality and the FFS. Since the FFS reflects the size of the gene family, this is equivalent to asking whether there is any bias in the kind of function performed by domains encoded by large gene families versus small ones. Previous research has shown that some functionalities may allow many analogous functions increasing the FFS of the family, while others tend to have stricter requirements. For example, some functions such as the kinase activity have varied specificities within a relatively small number of sequence mutations, [73] while others such as globins have much less functional flexibility despite, in some cases, substantial sequence divergence [74]. Another example is the eightfold (β/α) barrel structure, first observed in triose-phosphate isomerase, occurs ubiquitously in nature. It is nearly always an enzyme and most often involved in molecular or energy metabolism within the cell. This extreme example of the "one fold-many functions" paradigm illustrates the difficulty of assigning function through a structural genomics approach for some folds. Another example is the beta-propeller fold that appears as a very fascinating architecture based on four-stranded antiparallel and twisted beta-sheets, radially arranged around a central tunnel. Similar to the β/α-barrel (TIM-barrel) fold, the beta-propeller has a wide range of different functions. Some proteins containing beta-propeller domains have been implicated in the pathogenesis of a variety of diseases, such as cancer, Alzheimer's, Huntington's, Lou Gehrig's diseases, arthritis, familial hypercholesterolemia, retinitis pigmentosa, osteogenesis, hypertension, and microbial and viral infections. While some studies exist that suggest that gene families encoding enzymes and enzymatic folds are larger [75, 76], they do not provide us with an overall picture, if one exists, of general biases. Such biases may reveal important evolutionary pressures that determine the codependence of structure and function.

Shakhnovich et al. [48] computed the FFS of all domains on the PDUG and in turn assigned a functional category at the first level of GO (Fig. 5.1). It was found that as the FFS of the domain increases, the percentage of enzymes in the bin decreases, and consequently the percentage of domains with signal transduction activity increases. Other functions remain relatively constant, at least to the first approximation of these data. These results point to the tendency of larger gene families (tests were also done directly between gene family size and FFS with the same results) to encode domains with signal transduction activity and less diverse gene families to encode domains with enzymatic activity. These results are unexpected in the light of the studies mentioned above [75–77] and more research has to be done in this area to understand the biological reasons and evolutionary mechanisms that have enabled us to observe these trends. However, the obvious avenue of research from here would involve the determination of the interdependency between the allowed functional diversity for the gene family and the particular function.

An approach to begin addressing this question is to look at physical characteristics of the structure. From a physical perspective, the potential of a gene to obtain new function may depend on its ability to accept mutations without destroying the three-dimensional structure of a protein domain that it encodes. The PDUG enables us to begin testing such hypothesis. As mentioned before, we know from some studies that some folds such as TIM-Barrels [75] and beta-propellers [77] encode large sequence families that are functionally diverse. However, until now, no specific structural characteristic could be identified that corresponded to the allowed functional diversity. Using the PDUG and the FFS, it was possible to attribute functional diversity to structure, which is supported with previously reported results of different functional categories having different FFS [78]. For example, most $\alpha+\beta$ protein domains are involved in binding functions such as DNA binding, ATP grasp, and FAD binding activity.

These results could mean that transcription factors are not as selective as once thought and that noise in form of nondeleterious mutations to the fold may give rise to the necessary transcriptional noise in the expression of proteins. Of course, this finding sheds no light on the intrinsic qualities of all alpha proteins versus $\alpha + \beta$ proteins that enables the latter to support more sequences. It also says nothing about what biological mechanisms and pressures enabled separations between commonly used folds and orphans. Clearly coevolution of structure and function occurs.

5.11 Concluding Remarks

The presented overview provides strong evidence that methodologies based on the use of the PDUG work in a variety of applications. Such an approach makes it possible to address from a unique single prospective such seemingly disconnected issues as character of protein domain evolution, structure–function relationship, relation between structural and functional properties of proteins, and certain properties of genes that they encode. We are confident that the PDUG approach is likely to yield further important fundamental unifying insights into structural genomics, evolution, and structural biology.

References

1. N.V. Dokholyan, E.I. Shakhnovich in *Power Laws, Scale-free Networks and Genome Biology*, ed. by E.V. Koonin, Y.I. Wolf, G.P. Karev (Eurekah.com and Springer, Berlin Heidelberg New York, 2005)
2. M.Y. Galperin, D.R. Walker, E.V. Koonin, Genome Res. **8**, 779 (1998)
3. R.F. Doolittle, Trends Biochem. Sci. **19**, 15 (1994)
4. A.G. Murzin, S.E. Brenner, T. Hubbard, C. Chothia, J. Mol. Biol. **247**, 536 (1995)

5. A.S. Siddiqui, U. Dengler, G.J. Barton, Bioinformatics **17**, 200 (2001)
6. U. Dengler, A.S. Siddiqui, G.J. Barton, Proteins **42**, 332 (2001)
7. S.A. Teichmann, A.G. Murzin, C. Chothia, Curr. Opin. Struct. Biol. **11**, 354 (2001)
8. N.V. Dokholyan, E.I. Shakhnovich, J. Mol. Biol. **312**, 289 (2001)
9. N.V. Dokholyan, E.I. Shakhnovich, Proceedings of the International School of Physics "Enrico Fermi" **145**, 227 (2001)
10. U. Bastolla, M. Porto, H.E. Roman, M. Vendruscolo, Proteins **58**, 22 (2005)
11. U. Bastolla, M. Porto, H.E. Roman, M. Vendruscolo, Gene **347**, 219 (2005)
12. R.B. Russell, P.D. Sasieni, M.J. Sternberg, J. Mol. Biol. **282**, 903 (1998)
13. J.A. Irving, J.C. Whisstock, A.M. Lesk, Proteins **42**, 378 (2001)
14. A.C. Wallace, N. Borkakoti, J.M. Thornton, Protein Sci. **6**, 2308 (1997)
15. R.B. Russell, J. Mol. Biol. **279**, 1211 (1998)
16. S. Jones, J.M. Thornton, J. Mol. Biol. **272**, 133 (1997)
17. K.S. Gajiwala, S.K. Burley, J. Mol. Biol. **295**, 605 (2000)
18. K.Y. Hwang, J.H. Chung, S.H. Kim, Y.S. Han, Y. Cho, Nat. Struct. Biol. **6**, 691 (1999)
19. B. Stec, H. Yang, K.A. Johnson, L. Chen, M.F. Roberts, Nat. Struct. Biol. **7**, 1046 (2000)
20. R.M. Story, I.T. Weber, T.A. Steitz, Nature **355**, 318 (1992)
21. K. Diederichs, G.E. Schulz, J. Mol. Biol. **217**, 541 (1991)
22. A.E. Todd, C.A. Orengo, J.M. Thornton, J. Mol. Biol. **307**, 1113 (2001)
23. L. Holm, C. Sander, J. Mol. Biol. **233**, 123 (1993)
24. L. Holm, C. Sander, Nucleic Acids Res. **25**, 231 (1997)
25. L. Holm, C. Sander, Trends Biochem. Sci. **20**, 478 (1995)
26. C.P. Ponting, R.R. Russell, Annu. Rev. Biophys. Biomol. Struct. **31**, 45 (2002)
27. G. Dodson, A. Wlodawer, Trends Biochem. Sci. **23**, 347 (1998)
28. W.F. Doolittle, Science **284**, 2124 (1999)
29. A.J. Roger, O. Sandblom, W.F. Doolittle, H. Philippe, Mol. Biol. Evol. **16**, 218 (1999)
30. W.F. Doolittle, J.R. Brown, Proc. Natl. Acad. Sci. USA **91**, 6721 (1994)
31. N.R. Pace, Science **276**, 734 (1997)
32. C.R. Woese, Microbiol. Rev. **51**, 221 (1987)
33. C.R. Woese, O. Kandler, M.L. Wheelis, Proc. Natl. Acad. Sci. USA **87**, 4576 (1990)
34. K.S. Makarova et al., Genome Res. **9**, 608 (1999)
35. D. Tumbula et al., Genetics **152**, 1269 (1999)
36. J.A. Lake, R. Jain, M.C. Rivera, Science **283**, 2027 (1999)
37. R.F. Doolittle, Nature **392**, 339 (1998)
38. M.C. Rivera, R. Jain, J.E. Moore, J.A. Lake, Proc. Natl. Acad. Sci. USA **95**, 6239 (1998)
39. M. Ibba et al., Science **278**, 1119 (1997)
40. M. Ibba et al., Proc. Natl. Acad. Sci. USA **96**, 418 (1999)
41. T.D. Edlind et al., Mol. Phylogenet. Evol. **5**, 359 (1996)
42. J.R. Brown, W.F. Doolittle, Proc. Natl. Acad. Sci. USA **92**, 2441 (1995)
43. C.J. Bult et al., Science **273**, 1058 (1996)
44. R.P. Hirt et al., Proc. Natl. Acad. Sci. USA **96**, 580 (1999)
45. J.A. Lake, Proc. Natl. Acad. Sci. USA **91**, 1455 (1994)
46. E. Hilario, J.P. Gogarten, Biosystems **31**, 111 (1993)

47. B.E. Shakhnovich, N.V. Dokholyan, C. Delisi, E.I. Shakhnovich, J. Mol. Biol. **326**, 1 (2003)
48. B.E. Shakhnovich, E. Deeds, C. Delisi, E.I. Shakhnovich, Genome Res. **15**, 385 (2005)
49. N.V. Dokholyan, B. Shakhnovich, E.I. Shakhnovich, Proc. Natl. Acad. Sci. USA **99**, 14132 (2002)
50. C.A. Orengo, F.M. Pearl, J.M. Thornton, Methods Biochem. Anal. **44**, 249 (2003)
51. S. Dietmann et al., Nucleic Acids Res. **29**, 55 (2001)
52. L. Holm, C. Sander, Bioinformatics **14**, 423 (1998)
53. S.F. Altschul et al., Nucleic Acids Res. **25**, 3389 (1997)
54. B. Boeckmann et al., Nucleic Acids Res. **31**, 365 (2003)
55. R. Apweiler et al., Bioinformatics **16**, 1145 (2000)
56. E.L. Sonnhammer, S.R. Eddy, E. Birney, A. Bateman, R. Durbin, Nucleic Acids Res. **26**, 320 (1998)
57. N.V. Dokholyan, Gene **347**, 199 (2005)
58. G. Tiana, B.E. Shakhnovich, N.V. Dokholyan, E.I. Shakhnovich, Proc. Natl. Acad. Sci. USA **101**, 2846 (2004)
59. S. Havlin, D. Ben-Avraham, Adv. Phys. **36**, 695 (1987)
60. D. Stauffer, A. Aharony, *Introduction to Percolation Theory* (Taylor and Francis, Philadelphia, 1994)
61. B. Bollobas, *Random Graphs* (Academic, London, 1985)
62. N.V. Dokholyan et al., Physica A **266**, 55 (1999)
63. N.V. Dokholyan et al., J. Stat. Phys. **93**, 603 (1998)
64. E.J. Deeds, N.V. Dokholyan, E.I. Shakhnovich, Biophys. J. **85**, 2962 (2003)
65. E.J. Deeds, B. Shakhnovich, E.I. Shakhnovich, J. Mol. Biol. **336**, 695 (2004)
66. R. Albert, A.-L. Barabasi, Rev. Mod. Phys. **74**, 47 (2002)
67. R. Cohen, K. Erez, D. Ben-Avraham, S. Havlin, Phys. Rev. Lett. **85**, 4626 (2000)
68. R. Cohen, D. Ben-Avraham, S. Havlin, Phys. Rev. E **66**, 036113 (2002)
69. E.V. Koonin, Y.I. Wolf, G.P. Karev, Nature **420**, 218 (2002)
70. J.M. Thornton, C.A. Orengo, A.E. Todd, F.M. Pearl, J. Mol. Biol. **293**, 333 (1999)
71. A.G. Murzin, Curr. Opin. Struct. Biol. **8**, 380 (1998)
72. B.E. Shakhnovich, PLoS Computat. Biol. **1**, e9 (2005)
73. G. Manning, G.D. Plowman, T. Hunter, S. Sudarsanam, Trends Biochem. Sci. **27**, 514 (2002)
74. D. Bashford, C. Chothia, A.M. Lesk, J. Mol. Biol. **196**, 199 (1987)
75. N. Nagano, C.A. Orengo, J.M. Thornton, J. Mol. Biol. **321**, 741 (2002)
76. M.Y. Galperin, E.V. Koonin, Protein Sci. **6**, 2639 (1997)
77. V. Fulop, D.T. Jones, Curr. Opin. Struct. Biol. **9**, 715 (1999)
78. B.E. Shakhnovich, J.M. Harvey, J. Mol. Biol. **337**, 933 (2004)

Part II

Molecules: Genomes

A Twenty-First Century View of Evolution: Genome System Architecture, Repetitive DNA, and Natural Genetic Engineering

J.A. Shapiro

It is essential for nonbiologists to understand that evolutionary theory based on random mutation of autonomous genes is far from the last word on how genomes have changed in the course of biological evolution. The last 50 years of molecular genetics have produced an abundance of new discoveries and data that make it useful to revisit some basic concepts and assumptions in our thinking about genomes and evolution. Chief among these observations are the complex modularity of genome organization, the biological ubiquity of mobile and repetitive DNA sequences, and the fundamental importance of DNA rearrangements in the evolution of sequenced genomes. This review will take a broad overview of these developments and suggest some new ways of thinking about genomes as sophisticated informatic storage systems and about evolution as a systems engineering process.

6.1 Introduction: Cellular Computation and DNA as an Interactive Data Storage Medium

Cells are amazingly capable and sophisticated information processors. If we simply reflect on the incredible complexity and reliability of each cell division cycle, encompassing hundreds of millions of biochemical reactions and morphogenetic events, then we can only wonder at the control regimes that keep cellular just-in-time production facilities operating properly in the face of changing environments and damage. For example, a fast-growing *E. coli* cell replicates its DNA at a speed of almost 2,000 base-pairs per second with a final precision of better than one error per 10^9 nucleotides incorporated [1].

One of the keys to a twenty-first century vision of how genomes operate is to think about DNA as a data storage medium. Genomic storage operates over three different time scales that may be defined in terms of cellular and organismal generations:

1. *Many organismal generations.* Genetic storage in local DNA sequences and long range chromosome structure

2. *Multiple cell generations.* Epigenetic storage in covalent modifications and stable chromatin configurations
3. *Within a single cell cycle.* Computational storage in meta-stable nucleoprotein complexes.

These three time scales reflect the different ways that DNA interacts with the rest of the cell as it carries out computations and decision making. Cellular computations involve evaluation of multiple internal and external inputs. Inputs include the replication status of the genome, where the cell is in the cell cycle, what nutrients are available, what intercellular signaling molecules are present, and what other cells are touching the cell surface. Some situations require fast responses, such as a change in the nutritional environment or the detection of genome damage. Other situations result in longer term cellular differentiations, characterized by the formation of chromatin configurations (epigenetically stable complexes of DNA with protein and RNA heritable through multiple cell generations [2]). Certain conditions involve restructuring of genomic DNA molecules, either as part of the normal life cycle [3–6] or in response to a crisis situation [7–9]. When DNA restructuring occurs in DNA that is heritable through mating or vegetative reproduction, enduring evolutionary changes are the consequence.

6.2 Genome System Architecture and Repetitive DNA

Genomes contain many different kinds of functional information organized through the use of multiple codes, not just the triplet (three base) code for amino acids in proteins. In addition to coding sequences determining the primary structures of RNA and protein molecules (these may be considered as genomic data files), there is information for other essential processes:

1. Packaging DNA molecules within the nucleoid of prokaryotic cells or the nucleus of eukaryotic cells
2. DNA replication and transmission of genome copies to progeny cells
3. Repair of DNA damage
4. DNA restructuring.

Without effective genome packaging, replication, transmission, and repair, no cell-based life form could reproduce. Without DNA restructuring, no organism could evolve.

Our current understanding of how coding sequence expression (data file access) and other essential genome processes operate is based upon the revolutionary genetic studies of protein synthesis and genome reproduction carried out by Jacob and colleagues on the operon and replicon theories in the 1950s and 1960s [10,11]. These studies defined a new class of genome component that was unknown to classical genetics: signals written into the DNA that are recognized by other cellular molecules and which affect only the DNA carrying the signal. These cis-acting signals are fundamentally different from any classical definition of a gene. They serve to *format* coding sequences and genome architecture in the same way that generic bit strings format the encoded information in electronic data storage media and guide the computational hardware to the right data files and indicate the appropriate routines to apply. In an analogous fashion, cis-acting signals in the genome direct cellular hardware to form functional nucleoprotein complexes to carry out tasks such as transcription, replication, DNA distribution to daughter cells, and DNA rearrangements [12]. Since they are generic and work at many locations, cis-acting signals belong to the repetitive component of the genome [13]. By applying an informatic perspective, we can appreciate the functional relevance and interconnections of genome formatting features that have proved difficult to understand within the mechanistic conceptual framework of classical genetics.

Employing the informatic metaphor, it is possible to argue that each genome has a characteristic "system architecture," in much the same way that different computer systems do [13, 14]. The taxonomically specific system architecture includes elements such as

1. Transcription signals used to regulate expression of particular coding sequences as RNA copies
2. Signals for genome transmission (origins of DNA replication, centromeres for aligning and distributing chromosomes during cell division, and telomeres for completing replication at the ends of linear DNA molecules)
3. Signals for recombination and DNA rearrangement
4. Signals for compacting the genome with protein and RNA to form particular chromatin structures
5. Signals for attaching the genome to particular cellular or nuclear structures.

From the genome system architecture perspective, it is possible for two genomes in different species to have identical coding sequences but distinct sets of cis-acting signals and different genome system architectures. The result of different architectures would almost certainly be germ-line reproductive incompatibility and, quite probably, distinct patterns of coding sequence expression leading to phenotypic and ecological diversity. The major determinants of genome system architecture are the repetitive elements in the genome, such as the long head-to-tail tandemly arrayed repeats that flank the centromeres of most eukaryotic chromosomes [15], telomere repeats that permit the replication of chromosome ends [16], and dispersed repeats that contain many signals for transcription, chromatin organization, and nuclear localization [13, 17].

There is an extensive literature on the effects that repetitive DNA can exert on coding sequence expression. These effects include countless experiments with dispersed repeats that can migrate from one location to another in the genome (mobile genetic elements; [18–21]). There is also a growing number of studies of so-called "position effect" phenomena, where the expression of a particular genetic locus depends upon its location relative to dense concentrations of repetitive elements that are labeled "heterochromatic" because they can be seen in the microscope as chromosome regions that stain differently from the normal (or "euchromatic") regions of the chromosome [22, 23]. Of particular importance are trans position effects where repetitive element arrays affect the expression of genetic loci on different chromosomes [22]. From a mechanistic point of view, we now explain these dosage-dependent genome-wide effects as due to titration of a limited supply of heterochromatin-binding proteins [23]. From an organizational point of view, trans position effects tell us that the whole genome is a single integrated system, regulated both in cis and trans by networks employing DNA repeats, in which segments on different DNA molecules communicate with each other.

It has been evident for a long time that repetitive DNA is a more discriminating indicator of hereditary relationships than coding sequences. For example, as long as 25 years ago, it was possible to construct a phylogeny of primates by examining the distribution of restriction endonuclease cutting sites in tandem repeats of "alpha satellite" DNA at centromeres [24]. Moreover, each order of mammals can be distinguished by its content of highly repeated short interspersed nucleotide elements (SINEs) dispersed throughout the genome at between 10^4 and 10^6 copies per haploid genome (data tabulated in [25]). Plant species can also be distinguished by their centromeric repeats [13], and closely related "sibling" *Drosophila* species differ markedly in their content of both tandem satellite arrays and dispersed repeats [26, 27]. Indeed, we use repetitive microsatellite DNA for forensic DNA analysis to determine relationships between individuals [28]. In other words, the repetitive component of the genome is far more taxonomically specific than coding sequences. This conclusion is consistent with a key role for repetitive DNA in evolutionary diversification.

6.3 Genomes and Cellular Computation: *E. coli lac* **Operon**

In trying to understand how the genome serves as an information storage system, it is helpful to look at simple phenomena that have been subject to exhaustive detailed analysis for several decades. One of the most thoroughly understood genomic computation and decision-making systems is the classic case of the *E. coli lac* operon [10]. This system operates at short time scales. Readers unfamiliar with the *lac* operon can best inform themselves through a simple internet search that will turn up several basic descriptions of the system. For more detailed molecular or computational descriptions, they can consult reviews such as [29] or [30]. The *lac* system has been analyzed in logical circuit terms by Setty et al. [31], and it is useful to note in that paper how much more precise the natural system is than the synthetic constructs.

In regulating transcription of the *lac* operon, precise control results from a series of highly integrated molecular interactions that allow *E. coli* cells to distinguish between two sugars and execute the following nontrivial algorithm: "IF lactose is available AND IF glucose is not available AND IF the cell can synthesize beta-galactosidase and lactose permease, THEN transcribe *lacZYA* from the *lac* promoter." These molecular interactions are typical of the processes that occur in all cells to carry out information processing. They can be stated in the form of Boolean propositions, for the operations involving

Lac Operon Products
1. LacY + lactose (external) → lactose (internal)
2. LacZ + lactose → allolactose (minor product)
3. LacI + *lacO* → LacI-*lacO* (repressor bound, *lacP* inaccessible)
4. LacI + allolactose → LacI-allolactose (repressor unbound, *lacP* accessible)

Glucose Transport Components and Adenylate Cyclase
5. IIAGlc-P + glucose(external) → IIAGlc + glucose-6-P(internal)
6. IIAGlc-P + adenylate cyclase(inactive) → adenylate cyclase(active)
7. Adenylate cyclase(active) + ATP → cAMP + P~P

Transcription Factors
8. Crp + cAMP → Crp-cAMP
9. Crp-cAMP + *CRP* → Crp-cAMP-*CRP*
10. RNA Pol + *lacP* → unstable complex
11. RNA Pol + *lacP* + Crp-cAMP-*CRP* → stable transcription complex

Partial Computations
No lactose → *lacP* inaccessible (3)
Lactose + LacZ(basal) + LacY(basal) → *lacP* accessible (1, 2, 4)
Glucose → low IIAGlc-P → low cAMP
  → unstable transcription complex (5, 6, 7, 10)
No glucose → high IIAGlc-P → high cAMP
  → stable transcription complex (5, 6, 7, 8, 9, 11)

It is important to note that information processing involves metabolic enzymes in the bacterial cytoplasm and transport proteins in the bacterial membrane as well as DNA binding regulatory proteins and diffusible cytoplasmic molecules that serve as signals representing distinct environmental conditions. Thus, there is no single separate class of information processing molecules (i.e., no Cartesian dualism in the *E. coli* cell). Informatics is incorporated into functional cellular molecules (e.g. the glucose transporter), and a single node in the cellular computation network may have nonlinear microprocessing functions. In this regard, the components of the *lac* operon system are paradigmatic for all cellular control regimes.

Another paradigmatic feature of the *lac* operon system is the role of weak interactions, specific binding, and cooperativity to assemble and stabilize nucleoprotein complexes essential for carrying out the molecular computations. This is seen in the formation of the complex needed to block access of the *lac* promoter signal to the transcriptional apparatus (Boolean proposition

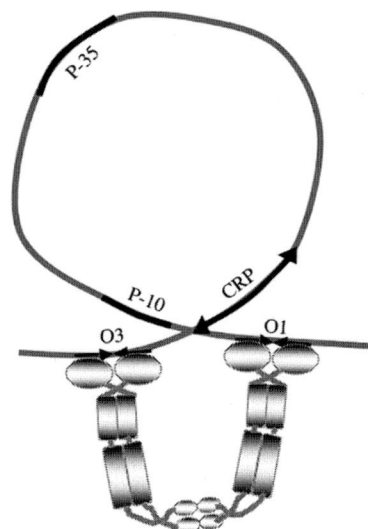

Fig. 6.1. DNA repression loop formed by interaction between LacI repressor molecules and two operators flanking the promoter region for initiating *lac* operon transcription

3 earlier). Graphically, this cooperativity can be seen in a cartoon of the repressor–DNA complex (Fig. 6.1). The looped repression complex depends on a series of cooperative protein–DNA binding events involving internal repeats within two different palindromic operator sequences and four separate DNA-binding domains of individual LacI repressor chains as well as protein–protein binding events between the four different copies of LacI. Any one of these interactions would be unstable, but together they form a complex that prevents *lac* operon transcription when inducer is not present.

The use of multiple cooperative interactions in cellular networks is a fundamental reason that genomes and protein structures are full of repetitive components. In turn, repetition in DNA and proteins means that specific logical operations arise through combinations of basic circuit elements (e.g., complex regulatory regions in DNA, intra- and intermolecular interactions between protein domains). In addition, the phenomenon known as "allostery" means that binding one ligand can alter the shape and behavior of a molecule, thus affecting its binding to other ligands. The allosteric properties of proteins and nucleic acids confer communication and processing capabilities on individual molecules and constitute the structural basis of how cellular network nodes act as complex microprocessors. Cooperative interactions between allosteric molecules thus endow cellular computing networks with combinatoric complexity as well as the Fuzzy Logic precision [32] that comes from the integration of multiple layers of approximation.

Working up from the details of a single basic cellular information-processing example leads us to a number of areas that are ripe for formal analysis and in silico simulation. It is important for the information scientists, mathematicians, and physicists who do this work to bear in mind that the principles underlying cellular analog computing may well be different from those that operate in electronic digital computers. For example, the direct participation of DNA in formation of nucleoprotein complexes suggests that it may not be useful to apply Turing's concepts of separate "machine" and "tape" [33] to cellular computations. Although such a fundamental difference does not invalidate the informatic metaphor, it does mean that we will have to be careful in applying existing computational models to cells.

Combinatorics, fuzzy logic models, and principles learned from linguistics and semiotics may all serve as key guides to a formal description of cellular information-processing networks. By the same token, bioinformatics that goes beyond the parsing of large databases has the potential to lead us to novel computing paradigms that may prove far more powerful than the Turing machine-based digital concepts we now use. To anyone who becomes familiar with the molecular details of cell biology, it is clear that no human contrivance operates with the degree of complexity, the efficiency, or the reliability of living cells. Thus, it is reasonable to argue that cells employ control regimes of great sophistication beyond our existing technologies.

6.4 New Principles of Evolution: The Lessons of Sequenced Genomes

Genome sequence analysis is one of our most important guides to disentangling how cellular systems operate and how function changes in the course of evolution. Here, we find abundant support for the general principles deducible from cases like *lac*. In particular, repetition, reuse, and combinatorics have proven to be fundamental in protein and whole genome evolution.

The insights from genome sequences have altered our thinking about the evolution of individual molecules, particularly proteins, as well as the evolution of overall genome structure. We have obtained abundant evidence that protein structures evolve by iterating, shuffling, and accumulating substructures called "domains," each comprising one of a finite sets of structural elements [34–37]. Figure 6.1 illustrates the different domains of the LacI repressor as shaded blocks. Moreover, genome sequences establish that protein families characterize individual taxa from bacteria through mammals. These families have their own phylogenetic structures and clearly evolved by iteration of existing coding sequences, not by de novo appearance of each family member [38, 39].

In addition, there are emerging patterns in genome structure that exist at hierarchical levels beyond the coding sequences (data files) determining individual protein and RNA products. We find that expression systems evolve by combining coding sequences, regulatory signals, and chromatin markers into higher order complexes that persist and diversify in the course of evolution [2]. At even higher levels of organization, we find that genomes contain extensive chromosome segments (syntenic regions) that may be duplicated at various locations within the genome [40, 41] and scrambled into new combinations during evolution [39].

6.5 Natural Genetic Engineering

All the preceding whole-genome sequence discoveries require mechanisms for rearranging DNA segments of all size classes as basic evolutionary processes. What do we know about the capacity of cells to carry out such natural genetic engineering? An important clue is the discovery that our own genomes are at least 43% composed of DNA segments that can move from one location to another [35]. Two classes of transposable or mobile genetic elements have been recognized from the work of Barbara McClintock and her molecular followers [18–21, 42, 43]: DNA transposons move exclusively at the level of DNA molecules, while retrotransposons and other retroelements move by means of an RNA intermediate that can be reverse-transcribed into genomic DNA [21, 44–46].

McClintock discovered mobile genetic elements in the first instance because they mediated chromosome rearrangements. Molecular analysis has confirmed that the same mechanisms that lead defined segments of DNA to move from one location to another (transpose) can also mediate both large- and small-scale rearrangements. There appears to be something of a molecular division of labor: DNA elements mediate rearrangements of large segments (Fig. 6.2), while retroelements mobilize smaller segments, generally not larger than several kilobases in length (Fig. 6.3).

The mechanisms underlying these rearrangements are just the kind of processes needed to explain the patterns of genome conservation and scrambling found by comparing whole genome sequences. The transposable elements (TEs) mediating genome rearrangements operate in natural populations [4, 48, 50–52]; they can execute key evolutionary processes in the laboratory, like exon shuffling [53]; and there is abundant documentation from sequence analysis that TE-based mechanisms have in fact been used in genome evolution [17, 49, 54–56].

An especially illuminating example of natural genetic engineering is the mammalian immune system. This system is important for several reasons. First, because of its medical implications, it has been the subject of intense investigation, and we know a great deal about how it operates and how it is regulated. Second, the evidence is quite convincing that the natural genetic engineering functions in our immune system arose from DNA transposons and cellular repair functions [5, 57, 58]. Third, and most important, the immune system is a system whose normal functions include ensuring rapid protein evolution of molecules for recognizing foreign invaders. In other words, our

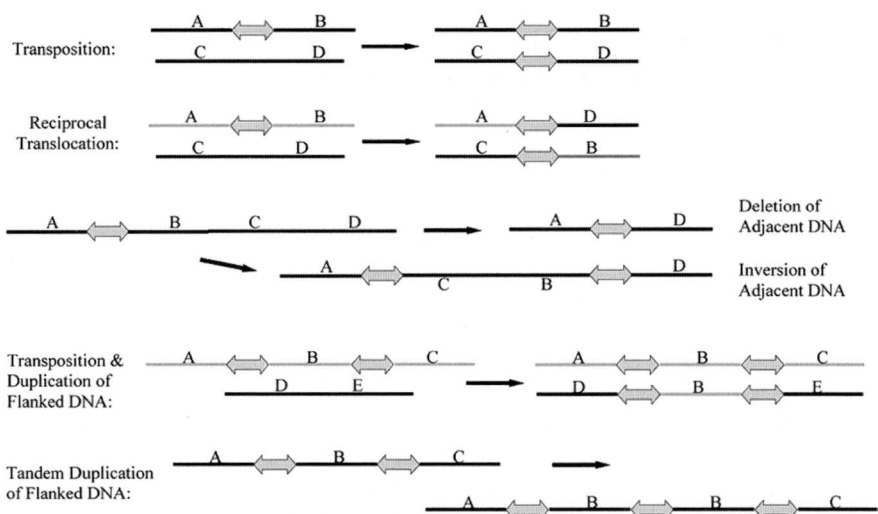

Fig. 6.2. Some of the rearrangements mediated by DNA transposons. Well-documented DNA rearrangements were carried out by the transposition systems of replicative or cut-and-paste DNA transposons (*double-headed arrows*). Often the structures of rearranged DNA generated by either mechanism are the same. These examples are based largely on the rearrangements described in [47], and by Engels (see, http://engels.genetics.wisc.edu/Pelements/HEI.html). The web site includes an animation of P factor-mediated duplication/deletion events. Note that DNA transposons have at least two ways of duplicating a sequence flanked by copies of the element: either a transposed duplication at a new genetic site or a tandem duplication at the original site. Both sorts of duplications are found in sequenced genomes, especially in loci encoding large paralogue families

immune systems have evolved to evolve. This example (or counter-example) of evolved evolvability demonstrates that there can be no theoretical obstacle to evolving capacities to improve the evolutionary process. This means that we should expect evolutionary processes to show the same degree of sophistication, complexity, and efficiency as other biological functions. The operation of nonrandom and adaptive mechanisms for restructuring DNA molecules in evolution should not be a surprise. Indeed, we should expect evolutionary processes to be specific and efficient because all existing organisms are survivors of repeated evolutionary competitions, and those organisms that are best at adapting their genomes to new conditions are most likely to have won those competitions.

The task of immune system cells (lymphocytes) is to produce a virtually infinite array of antigen-recognition proteins starting with a finite set of germline DNA coding elements. Accomplishing this task requires the lymphocytes to create genetic diversity in just the right *DNA* locations to encode unpredictable *protein* sequences at the appropriate protein locations. This localized diversification task is efficiently completed by a consecutive process

"RETROGENE" FORMATION:

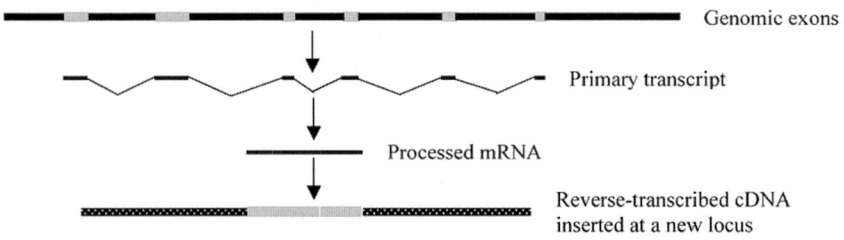

RETROTRANSDUCTION OF AN ADJACENT EXON:

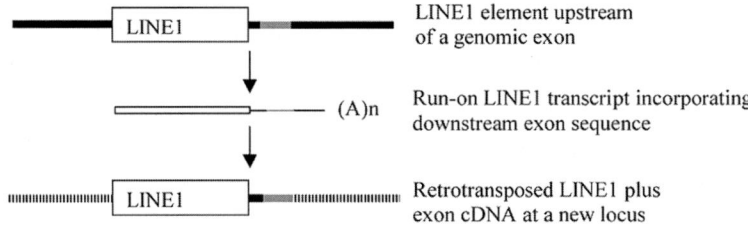

Fig. 6.3. Retrogene formation and retrotransduction of exons. The diagram summarizes how the reverse transcription and integration activities of LINE elements can create processed intron-free integrated cDNA copies of any cellular mRNA ("retrogenes") or can integrate DNA copies of exons located downstream of an active LINE element after read-through transcription ("retrotransduction"). See [49] for more details

of five highly regulated DNA rearrangements occurring at defined locations in the genome to generate an indeterminate set of DNA segments encoding the antigen-binding sites of immunoglobulin and T cell receptor molecules. The details of this process have been described in reviews and in the literature version of this Chap. [59]. For purposes of this volume, it is sufficient to summarize some basic features:

1. DNA rearrangements occur at specific sites in the genome because they are demarcated by repetitive DNA signals. Some of these signals are directly involved in DNA cleavage and rejoining, while other signals control transcription that is linked to DNA changes.
2. Although immune system rearrangements occur at well-demarcated locations, the actual sequences produced are highly flexible. This means that junctions of regions encoding antigen-binding domains can have many different restructured sequences. In some cases, the lymphocytes can even insert de novo synthesized DNA segments (i.e., with no germ-line template) into these junctions.

3. The DNA rearrangement process normally occurs only in cells destined to produce antibodies or T cell receptors and follows a highly determined series of rearrangements. Both intracellular signals, such as the ability to produce a particular immunoglobulin chain, and extracellular signals, such as lymphokine-directed transcription, regulate these DNA restructuring processes.

Altogether, the immune system provides one case where cells display a striking degree of control over DNA restructuring. Additional examples are found in other developmental DNA rearrangements [3, 4, 60]. Lymphocytes further illustrate the potential cells have to turn natural genetic engineering activities on and off in response to internal and external signals. They demonstrate how DNA rearrangements can be at one and the same time highly specific, directed by DNA sequences or transcriptional activity, and yet flexible, using untemplated nucleotides and variable internucleotide linkages to enhance combinatorial diversity. The mixture of specificity and flexibility enables the immune joining system to produce extraordinary protein diversity (on the order of 10^{12} combinations) while conserving antibody chain structures. By integrating specificity and flexibility, immune system engineering optimizes the chance to produce a functional antibody molecule with an indeterminate specificity.

It is highly significant that the degree of cellular control over natural genetic engineering exemplified by lymphocytes and other developmental systems is not an isolated or even unusual case. Experimentation with a number of different TE systems has shown that they can be activated temporarily by response to particular conditions. The conditions are quite varied, ranging from blockage of normal chromosome separation during early embryonic development [42] to osmotic and other physical stresses associated with protoplast regeneration [61] to oxidative starvation stress during "adaptive mutation" [62–66] to mating outside the normal breeding group causing "hybrid dysgenesis" [50, 51, 67, 68].

In the case of hybrid dysgenesis, it is especially worth noting that the DNA changes typically occur during the development of the germ line [69]. This has profound implications for the evolutionary success of massively restructured genomes. When change occurs during germ line development, cells that have undergone multiple DNA rearrangements have several offspring that produce multiple gametes. As a consequence, natural genetic engineering after hybrid dysgenesis can lead to the appearance of more than one individual progeny carrying similarly altered genomes. These progeny thus constitute a small interbreeding population that can transmit the massively restructured genome to future generations.

In addition to control over when and in what situations natural genetic engineering functions become active, there are a variety of examples where mobile elements display various degrees of targeting specificity in the genome (Table 6.1). In some cases, we know the molecular basis for targeting.

Table 6.1. Specificity of natural genetic engineering functions

Example	Observed specificity (mechanism)
Mating type cassette switching (*S. cerevisiae*)	Localized, directional gene conversion (HO endonuclease cleavage initiates homology-dependent recombination) [70]
Immune system V(D)J joining	Cleavage at specific recombination signal sequences (RSSs); flexible joining by nonhomologous end joining (NHEJ) functions (recognition of RSSs by RAG1+2 transposase) [5, 58]
Immune system somatic hypermutation	5' exons of immunoglobulin determinants (transcriptional specificity) [6]
Immune system class switching	Lymphokine-controlled choice of switch regions (promoter activation) [6]
Budding yeast (*S. cerevisaea*) retroviral-like elements Ty1-Ty4	Strong preference for insertion upstream of RNA polymerase III initiation sites (protein–protein interaction of integrase with RNA polymerase III factors) [71, 72]
Budding yeast retroviral-like element Ty1	Preference for insertion upstream of RNA polymerase II initiation sites rather than exons [73]
Budding yeast retroviral-like element Ty5	Strong preference for insertion in transcriptionally silenced regions of the yeast genome (protein–protein interaction of integrase with Sir4 silencing protein) [74–76]
Fission yeast (*S. pombe*) retroviral-like elements Tf1 & Tf2	Insertion almost exclusively in intergenic regions (>98% for Tf1); biased towards PolII promoter-proximal sites, 100–400 bp upstream of the translation start; preference for chromosome 3 [77–79]
Murine Leukemia Virus (MLV)	Preference for insertion upstream of transcription start sites in human genome [80, 81]
HIV	Preference for insertion into actively transcribed regions of human genome [81]
Drosophila P-factors	Preference for insertion into the 5' end of transcripts [82]
Drosophila P-factors	Targeting ("homing") to regions of transcription factor function by incorporation of cognate binding site; region-specific [83–86]
R1 and R2 LINE element retrotransposons	Insertion in arthropod ribosomal 28S coding sequences (sequence-specific endonuclease, reverse transcription) [87, 88]

Table continues on next page

Table 6.1. *Continued*

Example	Observed specificity (mechanism)
HeT-A and TART retrotransposons	Insertion at Drosophila telomeres [89]
Group I homing introns (DNA based)	Site-specific insertion into coding sequences in bacteria and eukaryotes (sequence-specific endonuclease) [90]
Group II homing introns (RNA based)	Site-specific insertion into coding sequences in bacteria and eukaryotes (RNA recognition of DNA sequence motifs, reverse transcription) [91, 92]

Connections between DNA rearrangement specificity, on the one hand, and transcriptional control or chromatin formatting functions, on the other, are particularly significant for the following reason. Most biologists recognize that signal transduction networks can direct transcriptional and chromatin formatting activities to particular regions or sites in the genome. Thus, connecting these activities to the operation of mobile elements establishes a readily understood molecular mechanistic basis for cellular control networks to target DNA rearrangements in response to internal and external signals. The targeting does not have to be (and in many cases is known not to be) rigidly deterministic. In this way, targeting may enhance the probability of adaptively useful genome changes without preventing the flexibility that is necessary to cope with unpredictable challenges.

6.6 Conclusions: A Twenty-First Century View of Evolution

On the basis of discoveries about genome system architecture and natural genetic engineering, it is now possible to formulate a series of basic concepts that lead to viewing evolution as something akin to a systems engineering process:

1. Genomes are formatted by repetitive elements and organized hierarchically for multiple information storage and transmission functions.
2. Major evolutionary steps occur by DNA rearrangements carried out by sophisticated cellular natural genetic engineering systems operating non-randomly.
3. Significant evolutionary changes can result from altering the repetitive elements formatting genome system architecture, not just from altering protein and RNA coding sequences.

4. Cellular regulation of natural genetic engineering activities makes evolutionary change responsive to biological inputs with respect to timing and location of DNA rearrangements.

These basic ideas about the role of cell-regulated natural genetic engineering of genome system architecture have implications for how we think about the evolutionary process, and previous articles have discussed some of these [12–14]. In the context of this volume, it is worthwhile to emphasize how natural genetic engineering (a) can increase the efficiency of searching for genome configurations that encode functional complex systems and (b) can favor the elaboration of hierarchic system architectures.

In the immune system, for example, natural genetic engineering takes existing functional coding modules and assembles them into new combinations. Since the rearranged DNA segments already have functionality, the potential of the newly assembled genomic structure for adaptive utility is greater than for a structure resulting from random changes. The same is true of other examples of natural genetic engineering. For example, insertion of a mobile element containing a package of integrated transcription and chromatin-formatting signals can place an existing coding region under novel controls. In this way, a working product can be expressed under conditions where it was previously absent (see [93]). The evidence is quite solid that this process has taken place during evolution [17,49,54]. Similarly, insertion of a DNA segment encoding a functional domain is more likely to add new capabilities to a protein than are random changes in sequence or addition of random polypeptide components. Domain addition is commonly used in laboratory engineering of proteins.

Acquisition of new DNA regulatory regions and protein domains are examples of engineering a new system by arranging known components in new combinations. The rearrangement process can always be followed, as it often is in human engineering, by fine-tuning or modification of individual components (microevolution). Here again, the immune system is instructive. A similar "rearrangement-followed-by-fine-tuning" sequence of events occurs in the immune system with targeted "somatic hypermutation" of the exons encoding antigen-binding domains of immunoglobulins [5,6].

The ability to regulate DNA rearrangements in time and location within the genome also adds significantly to the evolutionary efficiency of genome restructuring. By making sure that genomes in normally reproducing organisms are stable and that the genomes of cells under stress are mutable, networks activating natural genetic engineering functions provide hereditary variability when it is most needed [7]. Episodic activation of genome restructuring functions means that multiple changes can occur when complex rearrangements may be required to meet adaptive needs. Since they result from a common activation event, different genetic changes are not independent but coordinated. Episodic activation of genome restructuring functions further predicts that evolutionary change will be inherently intermittent and punctuated rather than continuous (cf. [94]).

Targeting of genetic change has potential advantages. It can limit change to regions where it is needed, as in restricting somatic hypermutation in B cells to exons for antigen binding domains, and can prevent damage to functional coding elements, as in the targeting of retrotransposon insertions to the upstream regions of genetic loci (Table 6.1). Restricting retrotransposon insertion sites thus enhances the potential for a constructive regulatory change. In yeast, selections for increased protein expression most commonly produce mutant strains carrying just such retrotransposon insertions [93]. Although virtually every mobile element displays some degree of target selectivity, there have been no careful studies of how selectivity may influence the ability of the element to make useful changes. Now that experimenters are learning how to target mobile elements (see Table 6.1 and [75, 91, 92, 95]), the time is ripe to investigate whether enhanced targeting alters their ability to generate adaptive changes.

In addition to increasing the efficiency of genome restructuring in response to challenge, the action of natural genetic engineering systems also imparts structural characteristics to genomes. Duplication and rearrangement of genomic segments can involve DNA sequences in the megabase range [56, 96]. So natural genetic engineering has the potential to facilitate the establishment and amplification of higher order genomic subsystems, as has clearly occurred in the evolution of homeodomain complexes [97]. This tendency to amplify progressively larger subsystems may help explain the hierarchic nature of genome coding [2].

Another result of change by natural genetic engineering is the tendency for genomes to accumulate dispersed copies of repeats. This tendency is usually explained by the "selfish DNA" hypothesis [98, 99]. However, the selfish DNA view does not take into consideration the well-documented functional information found in all classes of repetitive DNA elements [13]. Since dispersed repeats influence both coding sequence expression and physical organization of genomes, an alternative *functionalist* hypothesis must be entertained: namely, that repeat distribution reflects the establishment of a system architecture required for effectively integrated genome functioning.

6.7 Twenty-First Century Directions in Evolution Research

It appears from this discussion that a distinct twenty-first century view of evolution can stimulate research at the interface between experimental observation-based biology and mathematical analysis of complex systems. That was the objective of the present symposium. The ideas presented here are consistent with molecular genetics but are quite different from conventional evolutionary theory. Whether they prove to be predictive or not remains unknown until tests have been performed.

Experimental tests of evolutionary ideas can take two forms. One kind of test involves experimentation in the wet lab with evolving organisms. In most cases, these experiments will utilize microbes, such as bacteria and yeast, as the test organisms because microbes have the most easily manipulated genomes and because rare changes can be sought in large populations. Tests of evolutionary concepts need to ask questions that go beyond the standard mutation studies that examine changes at specific genetic loci. In particular, experimentalists should begin to examine how more complex multilocus systems arise in genomes. To see what mutational or natural genetic engineering processes are involved in system evolution, we need to devise effective selections that require the coordinated action of multiple genetic loci. Can these arise, for example, by the concerted insertion of DNA transposons in bacteria or retrotransposons in yeast, as suggested by the observations of Errede et al. [93] and Peaston et al. [100]?

A second form of evolutionary test requires simulated evolutionary processes, such as genetic programming [101]. Using simulated evolution models, computer scientists can determine whether the presence of mobile formatting repeats speeds up the evolutionary process, as suggested by their prominence in genomes. In addition, simulated evolution can examine how different evolutionary operators, modeled upon natural genetic engineering systems, influence the structure of the evolved programs. To achieve success, programs that accomplish real tasks, such as robotic control [102] will have to be subjected to evolutionary modification. Moreover, various programming languages or higher-level formalisms (e.g. object-oriented programming) will have to be used to find the most suitable evolutionary model, and devising operators modeled on genomic systems will require close collaboration between computer experts and molecular geneticists.

The preceding sketch of new experimental approaches in vivo and in silico should make it clear how alternative fundamental concepts of evolution open up new methods of scientific exploration. The integration of biology and information sciences in the twenty-first century will lead us to ask questions that could not have been imagined in the middle of the twentieth century.

References

1. T.A. Kunkel, J. Biol. Chem. **279**, 16895 (2004)
2. R. van Driel, P.F. Fransz, P.J. Verschure, J. Cell Sci. **116**, 4067 (2003)
3. S. Beermann, Chromosoma **60**, 297 (1977)
4. D.M. Prescott, Nat. Rev. Genet. **1**, 191 (2000)
5. C.H. Bassing, W. Swat, F.W. Alt, Cell **109**, S45 (2002)
6. K. Kinoshita, T. Honjo, Nat. Rev. Mol. Cell Biol. **2**, 493 (2001)
7. B. McClintock, Science **226**, 792 (1984)
8. J.A. Shapiro, Genetica **86**, 99 (1992)
9. J.A. Shapiro, Trends Genet. **13**, 98 (1997)
10. F. Jacob, J. Monod, J. Mol. Biol. **3**, 318 (1961)

11. F. Jacob, S. Brenner, F. Cuzin, Cold Spring Harbor Symposium Quantitative Biology **28**, 329 (1963)
12. J.A. Shapiro, in *From Epigenesis to Epigenetics: The Genome in Context*, ed. by L.V. Speybroeck, G.V. de Vijver, D. de Waele (Ann. N.Y. Acad. Sci. **981**, 111 (2002))
13. J.A. Shapiro, R.V. Sternberg, Biol. Revs. Camb. Philos. Soc. **80**, 227 (2005)
14. J.A. Shapiro, in Molecular Strategies for Biological Evolution, ed. by L. Caporale (Ann. N.Y. Acad. Sci. **870**, 23 (1999))
15. K.H. Choo, Dev. Cell **1**, 165 (2001)
16. E.H. Blackburn, Cell **106**, 661 (2001)
17. I.K. Jordan, I.B. Rogozin, G.V. Glazko, E.V. Koonin, Trends Genet. **19**, 68 (2003)
18. A.I. Bukhari, J.A. Shapiro, S.L. Adhya, *DNA Insertion Elements, Episomes and Plasmids* (Cold Spring Harbor Press, Cold Spring Harbor, NY, 1977)
19. J.A. Shapiro, *Mobile Genetic Elements* (Academic Press, New York, 1983)
20. D.E. Berg, M.M Howe (eds), *Mobile DNA* (ASM Press, Washington, D.C., 1989)
21. P.L. Deininger, J.V. Moran, M.A. Batzer, H.H. Kazazian, Curr. Opin. Genet. Develop. **13**, 651 (2003)
22. J.B. Spofford, in *The Genetics and Biology of Drosophila*, ed. by M. Ashburner, E. Novitski (Academic Press, New York, 1976) pp. 955
23. G. Schotta, A. Ebert, R. Dorn, G. Reuter, Semin. Cell Dev. Biol. **14**, 67 (2003)
24. L. Donehower, D. Gillespie, J. Mol. Biol. **134**, 805 (1979)
25. R.V. Sternberg, J.A. Shapiro, Cytogenet. Genome Res. **110**, 108 (2005)
26. A.P. Dowsett, Chromosoma **88**, 104 (1983)
27. A.K. Csink, S. Henikoff, Trends Genet. **14**, 200 (1998)
28. P. Bennett, Mol. Pathol. **53**, 177 (2000)
29. W.S. Reznikoff, Mol. Microbiol. **6**, 2419 (1992)
30. J.A. Shapiro, J. Biol. Phys. **28**, 745 (2002)
31. Y. Setty, A.E. Mayo, M.G. Surette, U. Alon, Proc. Natl. Acad. Sci. USA **100**, 7702 (2003)
32. L. Zadeh, in *Fuzzy Sets and Applications to Cognitive and Decision Making Processes*, ed. by L.A. Zadeh et al. (Academic Press, New York, 1975) pp 1
33. A.M. Turing, Mind (the Journal of the Mind Association) **59**, 433 (1950)
34. R.F. Doolittle, Ann. Rev. Biochem. **64**, 287 (1995)
35. International Human Genome Consortium, Nature **409**, 860 (2001)
36. L. Aravind, R. Mazumder, S. Vasudevan, E.V. Koonin, Curr. Opin. Struct. Biol. **12**, 392 (2002)
37. E.V. Koonin, Y.I. Wolf, G.P. Karev, Nature **420**, 218 (2002)
38. E.M. Zdobnov et al. Science **298**, 149 (2002)
39. Mouse Genome Sequencing Consortium, Nature **420**, 520 (2002)
40. E.E. Eichler, Trends in Genetics **17**, 661 (2001)
41. Arabidopsis Genome Initiative, Nature **408**, 796 (2000)
42. B. McClintock, *Discovery and Characterization of Transposable Elements: The Collected Papers of Barbara McClintock* (Garland, New York, 1987)
43. N.L. Craig, R. Craigie, M. Gellert, A.M. Lambowitz, *Mobile DNA II* (ASM Press, Washington, D.C., 2002)
44. J.M. Coffin, S.H. Hughes, H.E. Varmus, *Retroviruses* (Cold Spring Harbor Laboratory Press, Cold Spring Harbor, New York, 1997)

45. H.H. Kazazian, Jr., Science **289**, 1152 (2000)
46. H.H. Kazazian Jr., Science **303**, 1626 (2004)
47. J.A. Shapiro, Proc. Nat. Acad. Sci. USA **76**, 1933 (1979)
48. W.R. Engels, http://engels.genetics.wisc.edu/Pelements/HEI.html
49. J. Brosius, Gene **238**, 115 (1999)
50. J.C. Bregliano, M. Kidwell, in *Mobile Genetic Elements*, ed. by J.A. Shapiro (Academic Press, New York, 1983) pp. 363–410
51. W.R. Engels, *in Mobile DNA*, ed. by D.E. Berg, M.M Howe (ASM Press, Washington, D.C., 1989) pp. 437–484
52. D.N. Lerman, P. Michalak, A.B. Helin, B.R. Bettencourt, M.E. Feder, Mol. Biol. Evol. **20**, 135 (2003)
53. J.V. Moran, R.J. DeBerardinis, H.H. Kazazian Jr., Science **283**, 1530 (1999)
54. R.J. Britten, Gene **205**, 177 (1997)
55. A. Nekrutenko, W.-H. Li, Trends Genet. **17**, 619 (2001)
56. J.A. Bailey, G. Liu, E.E. Eichler, Am. J. Hum. Genet. **73**, 823 (2003)
57. A. Agrawal, Q.M. Eastman, D.G. Schatz, Nature **394**, 744 (1998)
58. M. Gellert, Ann. Rev. Biochem. **71**, 101 (2002)
59. J.A. Shapiro, Gene (Evolutionary Genomics) **345**, 91 (2005)
60. G.A. Wyngaard, T.R. Gregory, J. Exp. Zool. **291**, 310 (2001)
61. S.R. Wessler, Curr. Biol. **6**, 959 (1996)
62. J.A. Shapiro, Mol. Gen. Genet. **194**, 79 (1984)
63. B.G. Hall, Genetics **120**, 887 (1988)
64. G. Maenhaut-Michel, J.A. Shapiro, EMBO J. **13**, 5229 (1994)
65. G.J. McKenzie, R.S. Harris, P.L. Lee, S.M. Rosenberg, Proc. Natl. Acad. Sci. USA **97**, 6646 (2000)
66. H. Ilves, R. Horak, M. Kivisaar, J. Bacteriol. **183**, 5445 (2001)
67. R.J. O'Neill, M.J. O'Neill, J.A. Graves, Nature **393**, 68 (1998)
68. P.B. Vrana, J.A. Fossella, P.G. Matteson, M.J. O'Neill, S.M. Tilghman, Nature Genet. **25**, 120 (2000)
69. R.C. Woodruff, J.N. Thompson Jr., Genetica **116**, 371 (2002)
70. J.E. Haber, Ann. Rev. Genet. **32**, 561 (1998)
71. J. Kirchner, C.M. Connolly, S.B. Sandmeyer, Science **267**, 1488 (1995)
72. J.M. Kim, S. Vanguri, J.D. Boeke, A. Gabriel, D.F. Voytas, Genome Res. **8**, 464 (1998)
73. H. Eibel, P. Philippsen, Nature **307**, 386 (1984)
74. S. Zou, N. Ke, J.M. Kim, D.F. Voytas, Genes Dev. **10**, 634 (1996)
75. S.B. Sandmeyer, Proc. Natl. Acad. Sci. USA **100**, 5586 (2003)
76. W. Xie, X. Gai, Y. Zhu, D.C. Zappulla, R. Sternglanz, D.F. Voytas, Mol. Cell. Biol. **21**, 6606 (2001)
77. T.L. Singleton, H.L. Levin, Eukaryot. Cell **1**, 44 (2002)
78. N.J. Bowen, I.K. Jordan, J.A. Epstein, V. Wood, H.L. Levin, Genome Res. **13**, 1984 (2003)
79. F.D. Kelly, H.L. Levin, Cytogenet. Genome Res. **110**, 566 (2005)
80. X. Wu, Y. Li, B. Crise, S.M. Burgess, Science **300**, 1749 (2003)
81. R.S. Mitchell, B.F. Beitzel, A.R.W. Schroder, P. Shinn, H. Chen, Ch.C. Berry, J.R. Ecker, F.D. Bushman, PLoS Biol. **2**, e234 (2004)
82. A.C. Spradling, D. Stern, I. Kiss, J. Roote, T. Laverty, G.M. Rubin, Proc. Natl. Acad. Sci. USA **92**, 10824 (1995)
83. C. Hama, Z. Ali, T.B. Kornberg, Genes Dev. **4**, 1079 (1990)

84. J.A. Kassis, E. Noll, E.P. Vansickle, W.F. Odenwald, N. Perrimon, Proc. Nat. Acad. Sci. USA **89**, 1919 (1992)
85. M.O. Fauvarque, J.M. Dura, Genes Dev. **7**, 1508 (1993)
86. E. Taillebourg, J.M. Dura, Proc. Nat. Acad. Sci. USA **96**, 6856 (1999)
87. Y. Xiong, T.H. Eickbush, Cell **55**, 235 (1988)
88. W.D. Burke, H.S. Malik, J.P. Jones, T.H. Eickbush, Mol. Biol. Evol. **16**, 502 (1999)
89. M.L. Pardue, P.G. DeBaryshe, Ann. Rev. Genet. **37**, 485 (2003)
90. M. Belfort, P.S. Perlman, J. Biol. Chem. **270**, 30237 (1995)
91. G. Mohr, D. Smith, M. Belfort, A.M. Lambowitz, Genes Dev. **14**, 559 (2000)
92. M. Karberg, H. Guo, J. Zhong, R. Coon, J. Perutka, A.M. Lambowitz, Nat. Biotechnol. **19**, 1162 (2001)
93. B. Errede, T.S Cardillo, G. Wever, F. Sherman, Cold Spr. Harb. Symp. Quant. Biol. **45**, 593 (1981)
94. S.J. Gould, N. Eldredge, Nature **366**, 223 (1993)
95. Y. Zhu, J. Dai, P.G. Fuerst, D.F. Voytas, Proc. Natl. Acad. Sci. USA **100**, 5891 (2003)
96. N. Harden, M. Ashburner, Genetics **126**, 387 (1990)
97. N.H. Patel, V.E. Prince, Genome Biol. **1**, reviews 1027.1–1027.4 (2000) (For the electronic version see: http://genomebiology.com/2000/1/5/reviews/1027)
98. W.F. Doolittle, C. Sapienza, Nature **284**, 601 (1980)
99. L.E. Orgel, F.H. Crick, Nature **284**, 604 (1980)
100. A.E. Peaston, A.V. Evsikov, J.H. Graber, W.N. de Vries, A.E. Holbrook, D. Solter, B.B. Knowles, Dev. Cell **7**, 597 (2004)
101. J.R. Koza, D. Andre, in *Biocomputing: Proceedings of the 1996 Pacific Symposium*, ed. by H. Lawrence, T. Klein (World Scientific Publishing Co., Singapore, 1996) pp. 500-511
102. H. Lipson, Artif. Life **7**, 419 (2001)

7

Genomic Changes in Bacteria: From Free-Living to Endosymbiotic Life

F.J. Silva, A. Latorre, L. Gómez-Valero, and A. Moya

7.1 Introduction

Symbiosis is the association between two or more distinct organisms during at least one part of their lifecycle. Although this term is sometimes used in a narrower sense, it includes for most authors a set of different situations such as mutualism, parasitism and commensalism. Mutualism is defined as an association in which both partners derive benefit from living together. Parasitism is an association in which one of the partners benefits, while the other is harmed. Finally, commensalism is an association in which one of the two members benefits while the other is neither harmed nor obtains an advantage. In most cases, the association is established between a pluricellular eukaryote and a microorganism such as a bacterium or unicellular fungus. These microbial symbionts establish either facultative or obligate associations with their hosts. In the latter, the symbionts are always required to be together.

The terms endosymbiosis and endosymbionts are applied to those symbionts that live inside their hosts (the other partner in the association). This association may also be narrower when the endosymbiont lives inside the host's cells. Endocytobiosis is the term applied to intracellular symbiosis [1].

Many bacterial species have been reported to be able to infect and produce a pathogenic effect on insects. The pathogenic relation between the spore-forming bacterium *Bacillus thuringiensis* and many species of lepidoptera, coleoptera or diptera is well-known, but there are many other parasitic or pathogenic relations involving species of the genera *Clostridium, Serratia, Pseudomonas, Photorhabdus, Wolbachia*, etc. There are also many mutualistic associations between insects and microorganisms, and in most of them the endosymbiont adapted to live intracellularly in a specialized host cell called mycetocyte when it contains a fungal endosymbiont, and bacteriocyte when it contains a bacterial symbiont. These cells are derived from different parts and tissues of the insect such as fat body or Malpighian tubules and form organs called mycetome or bacteriome depending on the endosymbiont type. Bacteriocyte-associated endosymbionts have been described in many

Table 7.1. Examples of bacteriocyte-associated endosymbionts of insects

Endosymbiont species	Bacterial group	Host (Insect order)	Host (Common name)
Buchnera aphidicola	g-Pro	Hemiptera	Aphids
Carsonella ruddii[a]	g-Pro	Hemiptera	Psyllids
Baumannia cicadellinicola[a]	g-Pro	Hemiptera	Sharpshooters
Portiera aleyrodidarum[a]	g-Pro	Hemiptera	Whiteflies
Tremblaya princeps[a]	b-Pro	Hemiptera	Mealybugs
Wigglesworthia glossinidia	g-Pro	Diptera	Tsetse flies
Sodalis glossinidius	g-Pro	Diptera	Tsetse flies
Blochmannia floridanus[a]	g-Pro	Hymenoptera	Carpenter ants
Sithopilus oryzae primary endosymbiont (SOPE)	g-Pro	Coleoptera	Weevils
Nardonella[a]	g-Pro	Coleoptera	Weevils
Blattabacterium sp.	Flavobacterium	Dictyoptera	Termites
Blattabacterium sp.	Flavobacterium	Dictyoptera	Cockroaches

g-Pro, gamma-proteobacteria; b-Pro, beta-proteobacteria
[a]Taxonomic name designation is *Candidatus*.

insect orders such as hemiptera (aphids, psyllids or whiteflies), diptera (tsetse flies), hymenoptera (carpenter ants), dictyoptera (termites or cockroaches) or coleoptera (weevils) (Table 7.1).

Bacteriocytes are differentiated cells produced during insect development that harbour bacterial endosymbionts (Fig. 7.1). In some species, the presence of the bacterial symbiont is required for differentiation of the bacteriocyte. In others, however, the bacteriomes are formed before being inhabited by the endosymbionts [1]. Each bacteriocyte contains a large number of bacterial cells. The control of this number in the aphid endosymbiont, *Buchnera aphidicola*, is probably carried out by its own bacteriocyte. Not only have degenerating bacteria been observed inside bacteriocytes, but it has recently been demonstrated that the most abundant transcripts of the *B. aphidicola* bacteriocytes encode invertebrate type lysozymes [2]. These enzymes probably control bacterial number by degrading cell walls and lysing bacteria.

Bacteriocyte endosymbionts are typically maternally inherited. Several types of transmission have been described (see [1,3] for a review). In ants and parthenogenetic aphids, oocytes are infected by invasive bacteria that leave the bacteriomes and penetrate through the ovary [4], whereas in viviparous aphids the developing embryos may also be infected by bacteria released from mother bacteriocytes. In some beetle species such as those from the genus *Sitophilus*, there is no transmission from the bacteriome to the ovary, since endosymbionts are permanently in the ovaries and in the female germ cells. Finally, the means of transmission of *Wigglesworthia glossinidia* in tsetse flies is not transovarial, since transmission apparently takes place in the milk glands through which the mother feeds the developing larvae.

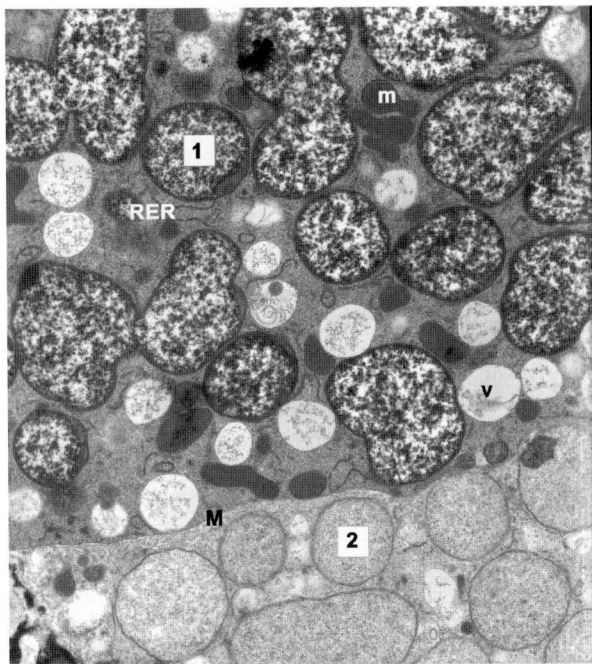

Fig. 7.1. Bacteriocytes of the aphid *Cinara cedri*. Electron microscopic image of a semithin serial section of 1.5 μm from *C. cedri*. Parts of two cells are observed separated by the cell membrane (M). Primary bacteriocyte (*up*) containing *B. aphidicola* cells (1). Secondary bacteriocyte (*down*) containing the secondary endosymbiont (R-type) (2). Other abbreviations: rough endoplasmic reticulum (RER); mitochondria (m); vacuole (v)

One of the most important characteristics affecting sequence evolution in endosymbionts is the reduction of the effective population size because of the small number of cells that are usually transmitted to the insect offspring. The number of endosymbiont cells transmitted to each sexual egg has been estimated in several *B. aphidicola* strains, ranging from 850 to 8,000 [5]. This bottleneck population structure makes endosymbiont sequences evolve completely differently than those from free-living bacteria.

Obligate maternal transmission has been reported in several bacterial endosymbionts [3]. This produces the coevolution of the host and bacterial lineages. However, there are also several examples in which endosymbiont and host phylogenies do not coincide. For example, it used to occur with the so-called secondary endosymbionts (maternally inherited but without sharing a long evolutionary history with their hosts as primary endosymbionts do), which in several insects may infect other individuals, species or even more phylogenetically distant taxa [6, 7]. The case of the genus *Wolbachia*, an intracellular gram-negative bacterium, which is found in association with a variety of invertebrate species, including insects, mites, spiders, terrestrial crustaceans and nematodes, is an example of wide taxonomic distribution.

The nature of the associations between several insect species such as aphids, carpenter ants or tsetse flies with bacterial species such as *B. aphidicola*, *Blochmannia floridanus* and *W. glossinidia*, respectively, are mainly nutritional [8–10]. These bacteria compensate for the deficient diet of the insects, synthesizing several compounds such as amino acids and vitamins (Fig. 7.2). On the other hand, these endosymbionts are unable to live outside the host cell because they have lost many capabilities for carrying out a free life and they require many different types of compounds obtained from the host cytoplasm, including several nucleotides, amino acids and vitamins.

An appealing topic in evolutionary biology is the consequences that a change in lifestyle produces in an organism. The ancestors of many bacterial obligate mutualists were probably free-living bacteria with the ability to infect insects and in some way benefit from that. At the beginning they were possibly parasites or commensalists but with time evolved to a mutualistic condition. This review will discuss several of the most important changes that the genome of these organisms experienced during the transition to the new way of life, and afterwards its adaptation to this type of life. These changes may be summarized as:

1. Changes in the pattern of sequence evolution.
2. Drastic reduction of the genome size with a strong tendency to lose genes and to lose DNA, the former being associated to the loss of many metabolic capabilities.

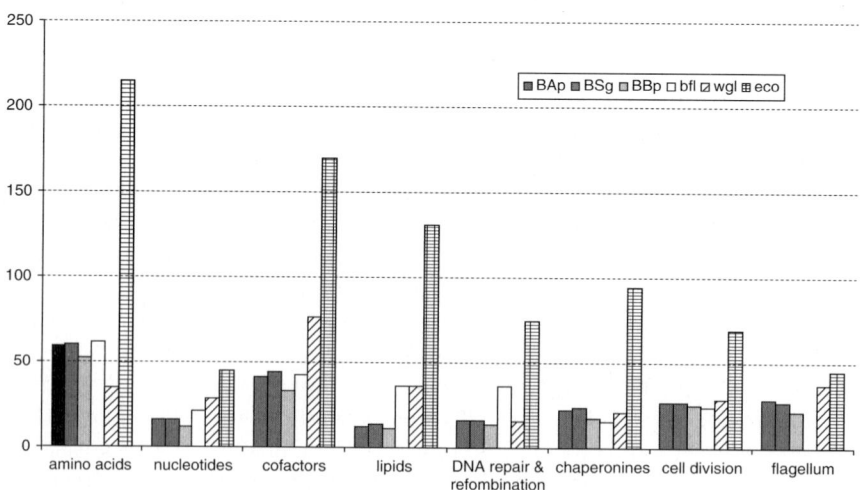

Fig. 7.2. Comparative analysis of the number of genes involved in several functional categories. Genome abbreviations: *B. aphidicola* strains (BAp, BSg and BBp); *Bl. floridanus* (bfl), *W. glossinidius* (wgl); *E. coli* K12 (eco)

3. Changes in the chromosomal rearrangement rates and the loss of the parasexual phenomenon with the loss of an efficient system for recombination, and the ability of acquired foreign DNA sequences.

7.2 Genetic and Genomic Features of Endosymbiotic Bacteria

7.2.1 Sequence Evolution in Endosymbionts

The comparative analysis of the bacterial genes and genomes has revealed several features that may probably be applied to all or most of the endosymbiont lineages. They are: (1) the increase in the AT content, (2) the acceleration of the substitution rates, (3) the loss of the synonymous codon bias selected in other organisms to increase the translation efficiency and (4) the change in the amino acid composition of the proteins.

Increase in the AT Content

The average base composition in bacterial genomes is very variable, ranging in the complete sequenced genomes from 22.5% to 72.1% GC content (see a selection of genomes in Table 7.2). In general, the GC content of intracellular bacterial symbionts is lower than that of extracellular bacteria [11]. A similar result was also reported for the GC content of obligate symbionts (including pathogen, mutualist or intracellular bacteria) when compared to those of free living bacteria [12]. These two classification systems do not overlap completely because some host-associated bacteria do not live intracellularly. To be completely sure that the increase in AT content is associated with the shift to intracellularity or obligate host-associated life, it is very important to compare each endosymbiont lineage with its closest free-living relative. These independent comparisons may permit us to know whether the decrease in endosymbiont GC content is a general feature exhibited in any bacterial group.

The first insect endosymbionts analysed and compared with phylogenetically related bacterial species belonged to the gamma-proteobacteria. The comparison of the composition of a small number of genes revealed that in most cases coding and ribosomal RNA genes presented lower GC content in the endosymbionts. The complete genome sequences of the insect endosymbionts *B. aphidicola*, *W. glossinidia* and *Bl. floridanus* [8–10, 13, 14] showed that they were among the bacterial species with the lowest genome GC contents (from 22.5% to 27.4%, see Table 7.2). Although endosymbiont phylogenies are difficult and controversial [10, 15–18], we know that species belonging to the genera *Escherichia*, *Salmonella* or *Yersinia*, which are their closest relatives, contain higher GC contents (48–52% GC). Thus, the hypothesis that

Table 7.2. GC content and genome size in a selected group of complete genomes (sorted by increase GC content)

Species	Group	Size (Mb)	GC (%)	Host	OI	Comments
Wigglesworthia glossinidia	g-Pro	0.70	22.5	I	Yes	Mutualist
Mycoplasma mobile 163 K	Firm	0.78	25.0	F	–	Pathogen
Buchnera aphidicola BBp	g-Pro	0.62	25.3	I	Yes	Mutualist
Ureaplasma parvum	Firm	0.75	25.5	H	–	Pathogen
Buchnera aphidicola BAp	g-Pro	0.66	26.2	I	Yes	Mutualist
Buchnera aphidicola BSg	g-Pro	0.64	26.3	I	Yes	Mutualist
Clostridium tetani E88	Firm	2.87	27.0	H	–	Pathogen
Blochmannia floridanus	g-Pro	0.71	27.4	I	Yes	Mutualist
Onion yellows phytoplasma OY-M	Firm	0.86	27.7	P	–	Pathogen
Rickettsia prowazekii str. Madrid E	a-Pro	1.11	29.0	H	Yes	Pathogen
Mycoplasma genitalium G-37	Firm	0.58	31.7	H	–	Pathogen
Wolbachia (end. of *Brugia malayi*)	a-Pro	1.08	34.2	N	Yes	Mutualist
Wolbachia (end. of *D. melanogaster*)	a-Pro	1.27	35.2	I	Yes	Pathogen
Bacillus thuringiensis ser. konkukian	Firm	5.31	35.4	I	–	Pathogen
Haemophilus influenzae Rd KW20	g-Pro	1.83	38.0	H	–	Pathogen
Chlamydia trachomatis D/UW-3/CX	Chla	1.04	40.0	H	Yes	Pathogen
Mycoplasma pneumoniae M129	Firm	0.82	40.0	H	–	Pathogen
Photorhabdus luminescens	g-Pro	5.69	42.8	I/N	–	Pathogen/ Mutualist
Vibrio cholerae O1 biovar eltor	g-Pro	4.03	47.0	H	–	Pathogen
Yersinia pestis CO92	g-Pro	4.83	48.0	H	–	Pathogen
Escherichia coli K12	g-Pro	4.64	50.0	H	–	Commensal
Escherichia coli O157:H7	g-Pro	5.5	50.0	H	–	Pathogen
Salmonella typhimurium LT2	g-Pro	4.95	52.0	H	–	Pathogen
Mycobacterium leprae TN	Acti	3.27	57.8	H	–	Pathogen
Xanthomonas campestris	g-Pro	5.08	64	P	–	Pathogen
Mycobacterium tuberculosis	Acti	4.4	65.6	H	–	Pathogen
Pseudomonas aeruginosa PAO1	g-Pro	6.26	67.0	H	–	Pathogen
Ralstonia solanacearum GMI1000	b-Pro	5.81	69.0	P	–	Pathogen
Thermus thermophilus HB8	Dei-Ther	2.12	69.5	–	–	–
Nocardia farcinica IFM 10152	Acti	6.29	70.7	H	–	Pathogen
Streptomyces avermitilis MA-4680	Acti	9.12	72.0	–	–	–
Streptomyces coelicolor A3(2)	Acti	9.05	72.1	–	–	–

g-Pro (gamma-proteobacteria), Firm (Firmicutes), a-Pro (alpha-proteobacteria), Chla (Chlamydiae), b-Pro (beta-proteobacteria), Acti (Actinobacteria), Dei-Ther (Deinococcus-Thermus), end (endosymbiont). I (insect), F (fish), H (human), P (plant), N (nematode), OI (Obligate intracellular).

a GC richer ancestor of these endosymbionts experienced an overall tendency to increase its AT content after the shift to endosymbiosis is possible. Natural selection is acting against this tendency by avoiding, to a variable degree, a high reduction in the gene GC content. However, when the gene is inactivated, the effect of natural selection disappears and the tendency to increase AT content becomes stronger. This is the situation observed in *B. aphidicola*, in which, when a gene is inactivated, a degenerative process produces its reduction in size and in GC content. A correlation between both parameters has been detected in *B. aphidicola*, where, on average, finally reaches 47% of the initial gene GC content [19], which is around 15% GC content for non-coding regions.

In other gamma-proteobacterial endosymbionts such as *Carsonella ruddii* (psyllid endosymbiont), the GC content may be even smaller. Thus, the GC content of a 37 kb genome segment was as low as 19.9%, with most of the coding genes with GC contents smaller than 20% and even ribosomal genes with very low values (33.1% and 35.6% for 23S and 16S, respectively) [20].

However, there are exceptions to this situation in gamma-proteobacteria, with the peculiarly high GC content of the primary endosymbiont of the weevil *Sitophilus oryzae* (SOPE), which presents an average value of 54% [11]. This exception, however, may be associated with the fact that it is in a preliminary stage of adaptation to intracellular life and, for that reason, both GC content and genome size are high and large [15]. In fact, in a comparison of the GC content of two genes in several gamma-proteobacteria, secondary endosymbiont GC contents were intermediate between those from free-living bacteria and those from primary endosymbionts [21]. Another exception to the general rule of low GC content is the beta-proteobacteria primary endosymbiont of mealybugs. It presents in 65 kb a high 57% GC content with all coding genes with values higher than 52% [22]. Because many beta-proteobacteria have even higher GC contents, a comparison with a close relative will be necessary to determine whether this bacterium is, or is not, an exception for this feature.

The analysis of the previous results leads to the general assumption that endosymbionts, especially intracellular endosymbionts, have low GC contents, and to conclude that their ancestors probably presented an equilibrium in the genome GC content, which was broken when the shift to intracellularity or host-association occurred. From that moment the tendency of the genes and genomes was to increase their AT content. Several hypotheses have been put forward to explain the low GC content of the endosymbionts. First, a mutational bias favouring higher rates of mutation from GC nucleotides to AT nucleotides would increase the AT content if mutations were fixed by genetic drift and natural selection were unable to avoid fixation, at least in non-essential genes or degenerate sites. The small effective population size of *B. aphidicola*, with strong bottlenecks in each aphid generation and the prevention of the recovery of wild type phenotypes by recombination (a phenomenon known as Muller's ratchet), would produce an increase in the rate of nucleotide substitution, especially those changes leading to the misincorporation of A and T nucleotides [3,11,23]. A neutral explanation for the AT mutational bias may be related to the loss of several DNA repair genes, for example those correcting the deamination of cytosine which, if left uncorrected, will render an uracil and finally thymine [3]. A selective explanation was also proposed recently, based on the higher energy cost and the limited availability of G and C over A and T in the obligate pathogens and symbionts [12]. The authors proposed that the higher fitness of the AT rich individuals would produce the selection of a general mutation bias, not the particular selection of each GC to AT substitution.

Acceleration of the Substitution Rates

The first wide comparative study on nucleotide substitution rates between endosymbiont and free-living bacteria showed the existence of an accelerated evolution for the former [23]. The analysis of the 16S rDNA of five independently derived endosymbiont clades (*Buchnera*, *Wigglesworthia*, *Wolbachia* and endosymbionts of mealybugs and whiteflies) showed, in every case, faster evolution in endosymbionts than in free-living relatives. The same relative rate test applied to coding genes in *Buchnera* showed an even higher acceleration in non-degenerate sites. The main difference in the rate of evolution when comparing the divergence of five coding genes between *E. coli* and *Salmonella* and two distant *B. aphidicola* strains was the extremely large increase in the non-synonymous substitution rate during the evolution of endosymbionts [23]. This accelerated evolutionary rate was also observed to be a characteristic of other very distant symbiont species such as mutualist fungi or bacterial endosymbionts of molluscan, insects or oligochaetes hosts [24–26].

Comparison of evolutionary rates between endosymbionts and free-living bacteria shows marked differences regarding the evolution of synonymous and non-synonymous sites. Thus, the number of synonymous substitutions per site (Ks) in *B. aphidicola* is homogeneous among loci, which contrasts with the heterogeneity found in free-living bacteria. In addition, the estimation of the synonymous substitution rate showed a slight increase (over twofold) in the endosymbiont over *E. coli* in genes with a low codon bias [27]. These differences are extremely high when comparing the rates of non-synonymous substitutions for the genes of *B. aphidicola* with those of free-living bacteria [13, 27, 28].

The explanation for the comparatively higher rates of substitution in endosymbionts, especially at the non-synonymous sites, was attributed to the unusual population structure of these organisms [23]. The strict vertical transmission to the host offspring, with a probable complete asexuality of the bacterial population due to the division in subpopulations and to the probable loss of efficient systems for competence and recombination [29, 30], and with continuous bottlenecks in the population size at each aphid generation [5], would favour the fixation of slightly deleterious mutations. Recently, it has been proposed that these high evolutionary rates would be mainly due to an enhanced mutation rate and to a relaxation of purifying selection [31].

Loss of the Codon Usage Bias

Many bacterial species show a selected codon usage bias throughout the genome. The strength of the selected codon usage bias varies substantially among species, and in 30% of the genomes examined, there was no significant evidence that selection had been effective [32]. This group includes obligate intracellular parasites or endosymbionts, such as species in the genera *Buchnera*, *Wigglesworthia*, *Rickettsia*, *Mycoplasma* or *Chlamydia*. Several deeper analyses of endosymbiont lineages have been carried out, finding very little variation of the synonymous codon usage across the genome in *W. glossinidia*, regardless of whether the genes have high or low expression levels [33]. On the contrary, a slight residual codon bias was detected in the three *B. aphidicola* genomes within the leading and the lagging strands [34]. Finally, it was recently reported that codon usage differs between strands of replication and between putative high and low expression genes in five insect endosymbiont genomes [35]. The strength of a general codon usage bias is correlated with several factors such as the number of rRNA operons and tRNA genes in the genome, or the level of expression of the genes. These results are consistent with the hypothesis that species exposed to selection for rapid growth have a more strongly selected codon usage bias [32].

Change in Amino Acid Composition of the Proteins

The bias to increase AT content in the endosymbiont coding genes has a strong effect on protein composition. When compared with free-living relatives, *B. aphidicola* proteins show many differences in the frequency of GC-rich encoded amino acid such as Val, Ala or Gly and AT-rich encoded amino acids such as Ile, Asn or Ser, which are decreased and increased, respectively [27, 36]. A parsimonic reconstruction of amino acid substitutions in four *B. aphidicola* strains, although putatively affected by homoplasy, suggested that in the early evolutionary stage of the lineage after the divergence from free-living enterics, a high number of amino acid replacements increasing AT took place. In its recent evolution, however, replacement of AT or GC-rich encoded amino acids is taking place at a similar rate, or there may even be those decreasing AT in a high number [27]. However, because the frequency of AT rich amino acid is very high in the current *B. aphidicola* proteins, the former observation already shows the higher tendency of GC encoded amino acids over AT encoded to mutate.

The bias to a high frequency of AT rich amino acids is also observed in other insect endosymbionts such as *W. glossinidia* [9] or *Bl. floridanus* [10]. The case of the psyllid endosymbiont *C. ruddii* is especially remarkable, because in the analysis of 32 polypeptide sequences containing an ortholog in *B. aphidicola*, *R. prowazekii* and *E. coli*, 50% of the residues consisted of the five AT-rich encoded amino acids (Phe, Ile, Lys, Asn and Tyr), which contrasted with the situation of *E. coli* (22.3%) but also with that of *R. prowazekii* (29.4%) and *B. aphidicola* (29.3%) [20].

The comparison of the amino acid changes between *E. coli* and the genomes of *W. glossinidia*, *B. aphidicola* and *Bl. floridanus* has shown that variation in amino acid usage strongly correlates with GC content and the putative expression level of genes, and that selection against costly amino acids in high expression genes is a secondary factor that explains amino acid usage [35].

The high frequency of some positively charged amino acids (especially Lys) has produced an extreme increase in the isoelectric point (pI) of the proteins (average values of 9.84, 9.6 and 8.9 for *W. glossinidia*, *B. aphidicola* and *Bl. floridanus*) over those of *E. coli* or *H. influenzae* (pI = 7.2) [8–10]. A comparative analysis of the orthologous *E. coli* and *Bl. floridanus* protein pIs has shown that most *E. coli* proteins with pIs ranging from 4 to 10 have an ortholog in *Bl. floridanus*, with a higher pI in a tendency, which reaches around 10.5 in the most extreme cases [10]. However, natural selection is acting against this change in some proteins, with 20% of the *Bl. floridanus* proteins presenting a difference of pI against *E. coli* smaller than 0.5.

Amino acid substitutions are also producing a negative effect on protein folding. The high expression levels of genes encoding chaperones, especially GroEL protein, may counteract this negative effect. In fact, the amino acid substitution rate of chaperones is very low in endosymbionts, and it has been experimentally demonstrated that the GroEL protein may act as a buffer against the effect of deleterious mutations [37]. Intracellular bacteria contain more hydrophobic proteins as a result of the increase in AT-rich encoded amino acids. It also makes them less stable against misfolding [38].

7.2.2 Reductive Evolution: DNA Loss and Genome Reduction in Obligate Bacterial Mutualists

One of the most interesting features of the obligate intracellular bacteria of insects was the extremely small size of their genomes with the five complete sequenced genomes reported to date being smaller than 800 kb (Table 7.2). In addition, other genome sizes estimated by pulsed field electrophoresis revealed a similar situation with the 680 kb of *Baumannia cicadellinicola*, the intracellular endosymbiont of sharpshooters [39], and the even more extreme situation of several *B. aphidicola* strains with sizes as small as 450 kb [40]. These small genomes are apparently derived from bacterial ancestors of larger genomes. For example, the genome of *B. aphidicola* BAp, with around 600 genes, is derived from an ancestor of at least 1,800 genes and 2 Mb [41].

Gene and DNA loss are not necessarily two simultaneous processes. For example, the genome of *Mycobacterium leprae* has experienced an important and probably recent loss of genes presenting 1,604 protein-coding genes and 1,116 pseudogenes, which contrasts with the 3,959 genes of *Mycobacterium tuberculosis* [42]. However, most of the DNA of these genes still remains in the 3,2 Mb *M. leprae* genome. Thus, we will first discuss the reasons why some genes are lost in a genome and second the reasons why the DNA of either a gene or a non-functional DNA segment (i.e. a pseudogene) is lost.

The shift of a bacterial cell to a completely host-associated life provokes natural selection to act to maintain primarily those genes with essential or very important functions for either the bacterial cell machinery or the maintenance of the symbiotic relation. Thus, on the one hand, a minimal number of around 206 genes has been proposed as necessary to maintain the main vital functions of the simplest hypothetical bacterial cell [43] and, on the other hand, insect-associated endosymbionts are required to produce several metabolites absent or deficient in insect diets. The number of genes of several functional categories in insect endosymbiont genomes is drastically reduced when compared with *E. coli* (Fig. 7.2). However, several genes involved in vitamins or amino acid synthesis have been retained because they are important to complement the insect diet. Thus, between 33 and 76 genes involved in cofactor metabolism and 35 and 62 in amino acid metabolism are present in the genomes of the three *B. aphidicola* strains, *W. glossinidia*, and *Bl. floridanus*. They act to complement the diets of the aphid, fly and ant hosts. Because of the absence of competition with other organisms and because many compounds are available in the cytoplasm of the host cell where it lives, many genes become non-essential and natural selection does not act to avoid their loss from the genome species. If the number of lost genes increases when compared to a free-living bacterium, the only way to compensate for it would be to increase gene acquisition through horizontal gene transfer. However, at a final stage the endosymbiont genomes have lost not only those genes involved in DNA uptake, but also those involved in recombination, which would favour the incorporation of foreign DNA molecules into the chromosome [8, 13, 14, 29].

Gene inactivation does not necessarily imply DNA loss. In fact, mammal pseudogenes have such a small deletion rate that the half-life of a pseudogene (the period for losing half of its nucleotides) has been estimated at 884 million years (My) [44]. The situation in other eukaryotes seems to be similar but there are important exceptions, such as *Drosophila* with a half-life of 14 My. The loss of DNA is also being studied in bacteria, and a primary observation was that the number of fixed deletions overtakes those of insertions when a pseudogene is compared with the orthologous gene of a phylogenetically related species [45]. It led to the proposal that in bacteria there is an indel bias that favours deletions over of insertions. It was proposed that this bias was associated with a mutation bias and with the random fixation of the mutations. The extreme reduction in the transition to mutualism of several bacterial endosymbionts was associated with this bias. However, it was proposed that in a final stage in *B. aphidicola*, the rate of DNA loss was as small as one nucleotide per 10,000 years [46]. One recent analysis of the 164 gene losses that took place during the recent evolution of the three *B. aphidicola* lineages showed that the half-life of a pseudogene was only 23.9 My, a non-irrelevant rate, and that the disintegration rate during the initial steps of *Buchnera* evolution would have been even higher because several mechanisms, now lost in *B. aphidicola*, would have produced faster and more drastic losses of nucleotides [19].

7.2.3 Chromosomal Rearrangements Throughout Endosymbiont Evolution

When the genomes of the *B. aphidicola* strains BAp and BSg, which diverged 50–70 My ago, were obtained [8, 13], one of the most surprising facts was that the order of the genes in both genomes was basically identical. There were no chromosomal rearrangements (inversion, translocation, duplications or horizontal gene transfer insertions) between them. The only difference was the absence of several genes in each one of the two genomes. Losses were scattered throughout the two compared chromosomes [13]. This led to the proposal that a genomic stasis takes place after the transition to intracellular life, a consequence of the loss of repeated sequences and an effective recombinational system mainly because of the absence of the recA gene. After the sequencing of the third and more divergent (86–164 My) *B. aphidicola* strain BBp, the number of rearrangements was maintained at very low levels, with only two translocations of plasmid genes to the chromosome and two small inversions affecting 1 and 6 genes, respectively [14]. The same situation was recently observed after the sequencing of a second *Blochmannia* species, *Bl. pennsilvanicus* [47], which diverged from *Bl. floridanus* (16–19 My ago) [48]. However, this chromosomal stability is not extended to plasmid genes. Thus, up to seven types of leucine-type plasmids have been observed in *B. aphidicola* strains, differing in the gene content and order [49, 50]. In addition, the transfer of the leucine cluster to the chromosome has taken place several times during the evolution of *B. aphidicola* lineages, with four different chromosomal insertion sites [51, 52]. Plasmid variability has also been observed in the tryptophan plasmid, but to a lesser extent [50].

When the genomes of either *B. aphidicola* or *Bl. floridanus* were compared with those of their close relative *E. coli*, many chromosomal rearrangements were detected [10,41], leading to the proposal that during the transition to an obligate intracellular stage many chromosomal rearrangements took place, which were associated in some cases with recombination between repeat elements such as transposable elements, phages, duplicated genes and other repeats [53]. Excluding horizontal gene transfer insertions, the most frequent chromosomal rearrangements in gamma-proteobacteria are inversions. Genome rearrangement rates have been recently estimated for a subset of genes shared by 30 gamma-proteobacterial complete genomes, observing that, in general, the progressive increase in sequence-based distances between genome pairs was associated to the increase in their rearrangement-based distances (breakpoint or inversion distances), but with several groups of distances not following this pattern. *B. aphidicola* and *W. glossinidia* lineages were evolving on average at moderately higher rates compared with free-living enteric bacteria, respectively, while *Bl. floridanus* was evolving at an almost identical rate [18]. A detailed examination of a phylogeny based on genome rearrangement distances showed that the rate of gene order evolution of this set was very heterogeneous, with periods of complete stasis such as in the case of *B. aphidicola* strains, or the *E. coli* K12, and *Salmonella typhimurium* lineages, and others of extreme acceleration such as the divergence of the two *Yersinia pestis* strains, the *Shigella flexneri* lineage, or the period of adaptation to intracellular life in *Buchnera* [18]. The comparison of two phylogenies based on genome rearrangements (inversions) and sequence (amino acid substitutions) distances (Fig. 7.3) reveals the important differences between both rates in some lineages. The extremely low sequence distances between the two *Shigella* genomes and *E. coli* (a divergence of less than 1 My) contrast with the high rearrangement distances because of the acceleration of the *Shigella*

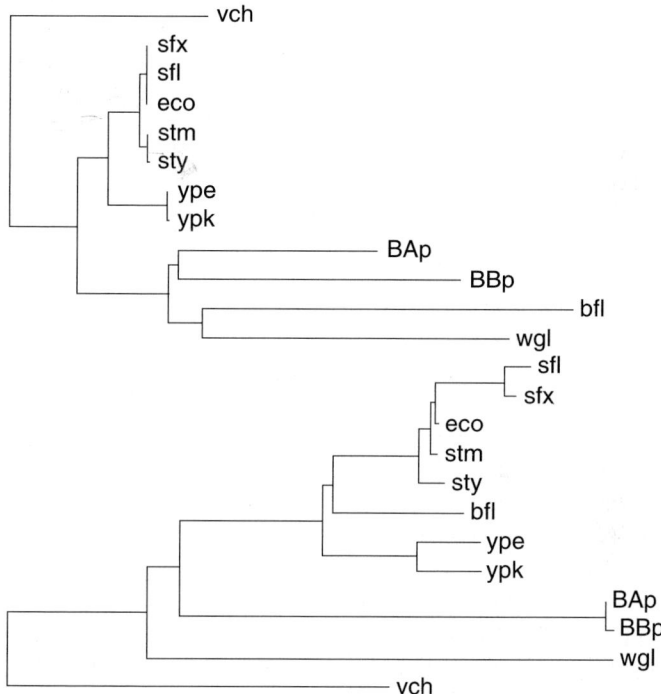

Fig. 7.3. Phylogenies of a selected group of gamma-proteobacteria based on amino acid substitution distances (*upper panel*) and genome rearrangement distances (*lower panel*). Distances were estimated in a previous paper [18]. Amino acid substitutions were obtained from a concatenated alignment of 10 well-conserved proteins involved in translation and rearrangements from the minimal number of inversions separating two genomes composed of a set of 244 genes shared by 30 completely sequenced gamma-proteobacterial genomes. Taxon names: eco (*Escherichia coli* K12); sfl (*Shigella flexneri* 2a str. 301); sfx (*Shigella flexneri* 2a str. 2457T); stm (*S. typhimurium* LT2); sty (*Salmonella enterica* subsp. enterica serovar Typhi str. CT18); vch (*Vibrio cholerae* O1 biovar eltor str. N16961); ype (*Yersinia pestis* CO92); ypk (*Yersinia pestis* KIM); BAp (*Buchnera aphidicola* BAp); BBp (*Buchnera aphidicola* BBp); BSg (*Buchnera aphidicola* BSg); bfl (*Blochmannia floridanus*); wgl (*Wigglesworthia glossinidia*)

lineage. On the contrary, the high sequence distance between *B. aphidicola* BAp and BBp (divergence 86–164 My) contrasts with the minimal rearrangement rate. The low rearrangement and sequence distances between *E. coli* K 12 and *S. typhimurium* (100 My of estimated divergence) are also remarkable. Thus, the evolution of the endosymbiont lineages passes through a period of extreme acceleration of the rearrangement rate at the initial stage and a period of complete stasis later.

Therefore, natural selection has played a very important role in the maintenance of the chromosomal architecture. Periods of relaxation of the natural selection pressure produce a strong increase in the rate of fixation of rearrangements. However, the presence of repeats and an efficient system of recombination are required to generate rearrangements and, for that reason, the endosymbiont genomes in a final stage become completely stable, even in the absence of natural selection pressure for the maintenance of the gene order.

7.3 Conclusions and Prospects

In this review, we have described the main genomic changes that have taken place during the evolution of obligate intracellular bacterial symbionts from ancestral free-living or, at least, facultative symbiont forms. Most of the changes have been observed by comparing ancient intracellular established lineages with free-living phylogenetically related ones. However, the initial steps of this transition have not received close attention. The main hypotheses consider that many changes affecting gene order, gene content, functional capabilities, etc., occurred in a very short lapse of time. To better characterize these changes it would be necessary to identify and analyse bacterial species that are in this situation at present, which is a difficult task, though several candidates are currently being analysed. The sequencing of the genomes of the secondary endosymbiont of the tsetse fly, *Sodalis glossinidius* and the primary endosymbiont of *Sitophilus oryzae* (SOPE) are in progress at the RIKEN GSC (Japan) and our group at Valencia University, respectively. The comparison of these two species possessing large genomes with those of some related free-living enterics, will show the main changes occurring at the initial stages of an obligate symbiont association.

The discovery that many insects contain, in addition to a primary bacterial endosymbiont, several other endosymbionts (secondary endosymbionts) opens the possibility of detecting a variety of situations which once analysed, may render a general image of the endosymbiont scenario during the initial stages. Especially interesting are the reasons for the coexistence of these endosymbionts in the same insect and whether they are able to form mutual stable situations among themselves and the host, or whether they are in a transitory stage that will end with the replacement of the primary endosymbiont.

One of the most interesting changes observed during the progress of the genome sequencing of SOPE was the presence of many copies of several insertion sequence elements (IS), which are responsible for several gene inactivations and are a putative source of repeat elements. Homologous recombination may use them to produce genome rearrangements including insertions, transposition, duplications and deletions.

The extent of genome reduction in these bacterial endosymbionts, and the definition of the minimal gene contents for bacterial cellular survival and the support of host life is a very interesting topic, which will be enriched with the sequencing of very small genomes. This is the case of the *B. aphidicola* strain BCc endosymbiont of the aphid *Cinara cedri*, whose genome is almost completely sequenced, and contains a chromosome of less than 420 kb and two small plasmids. Finally, the study of the insect–endosymbiont relationship requires the analysis of endosymbionts belonging to different taxonomic orders. Our group is contributing to these studies by sequencing the genome of the flavobacterium endosymbionts of some cockroaches.

Acknowledgements

This work was supported by grant BMC2003-00305 from the *Ministerio de Ciencia y Tecnología* (Spain) and grant Grupos03/204 from *Generalitat Valenciana* (Spain). L G-V was funded by a pre-doctoral fellowship from *Generalitat Valenciana* (Spain).

References

1. P. Nardon, C. Nardon, Annales de la Societé Entomologique de France **34**, 105 (1998)
2. A. Nakabachi, S. Shigenobu, N. Sakazume, T. Shiraki, Y. Hayashizaki, P. Carninci, H. Ishikawa, T. Kudo, T. Fukatsu, Proc. Natl. Acad. Sci. USA **102**, 5477 (2005)
3. J.J. Wernegreen, Nat. Rev. Genet. **3**, 850 (2002)
4. P. Buchner, in *Endosymbiosis of Animals with Plant Microorganims* (Interscience, New York, 1965)
5. A. Mira, N.A. Moran, Microb. Ecol. **44**, 137 (2002)
6. J.P. Sandstrom, J.A. Russell, J.P. White, N.A. Moran, Mol. Ecol. **10**, 217 (2001)
7. J.A. Russell, A. Latorre, B. Sabater-Muñoz, A. Moya, N.A. Moran, Mol. Ecol. **12**, 1061 (2003)
8. S. Shigenobu, H. Watanabe, M. Hattori, Y. Sakaki, H. Ishikawa, Nature **407**, 81 (2000)
9. L. Akman, A. Yamashita, H. Watanabe, K. Oshima, T. Shiba, M. Hattori, S. Aksoy, Nat. Genet. **32**, 402 (2002)
10. R. Gil, F.J. Silva, E. Zientz, F. Delmotte, F. González-Candelas, A. Latorre, C. Rausell, J. Kamerbeek, J. Gadau, B. Holldobler, R.C.H.J. van Ham, R. Gross, A. Moya, Proc. Natl. Acad. Sci. USA **100**, 9388 (2003)

11. A. Heddi, H. Charles, C. Khatchadourian, G. Bonnot, P. Nardon, J. Mol. Evol. **47**, 52 (1998)
12. E.P.C. Rocha, A. Danchin, Trends Genet. **18**, 291 (2002)
13. I. Tamas, L. Klasson, B. Canback, A.K. Naslund, A.S. Eriksson, J.J. Wernegreen, J.P. Sandstrom, N.A. Moran, S.G.E. Andersson, Science **296**, 2376 (2002)
14. R.C.H.J. van Ham, J. Kamerbeek, C. Palacios, C. Rausell, F. Abascal, U. Bastolla, J.M. Fernandez, L. Jimenez, M. Postigo, F.J. Silva, J. Tamames, E. Viguera, A. Latorre, A. Valencia, F. Moran, A. Moya, Proc. Natl. Acad. Sci. USA **100**, 581 (2003)
15. C. Lefevre, H. Charles, A. Vallier, B. Delobel, B. Farrell, A. Heddi, Mol. Biol. Evol. **21**, 965 (2004)
16. E. Lerat, V. Daubin, N.A. Moran, PLoS Biol. **1**, E19 (2003)
17. B. Canback, I. Tamas, S.G. Andersson, Mol. Biol. Evol. **21**, 1110 (2004)
18. E. Belda, A. Moya, F.J. Silva, Mol. Biol. Evol. **22**, 1456 (2005)
19. L. Gómez-Valero, A. Latorre, F.J. Silva, Mol. Biol. Evol. **21**, 2172 (2004)
20. M.A. Clark, L. Baumann, M.L.L. Thao, N.A. Moran, P. Baumann, J. Bacteriol. **183**, 1853 (2001)
21. A. Moya, A. Latorre, B. Sabater-Muñoz, F.J. Silva, J. Mol. Evol. **55**, 127 (2002)
22. L. Baumann, M.L. Thao, J.M. Hess, M.W. Johnson, P. Baumann, Appl. Environ. Microbiol. **68**, 3198 (2002)
23. N.A. Moran, Proc. Natl. Acad. Sci. USA **93**, 2873 (1996)
24. F. Lutzoni, M. Pagel, Proc. Natl. Acad. Sci. USA **94**, 11422 (1997)
25. A.S. Peek, R.C. Vrijenhoek, B.S. Gaut, Mol. Biol. Evol. **15**, 1514 (1998)
26. M. Woolfit, L. Bromham, Mol. Biol. Evol. **20**, 1545 (2003)
27. M.A. Clark, N.A. Moran, P. Baumann, Mol. Biol. Evol. **16**, 1586 (1999)
28. E.U. Brynnel, C.G. Kurland, N.A. Moran, S.G.E. Andersson, Mol. Biol. Evol. **15**, 574 (1998)
29. F.J. Silva, A. Latorre, A. Moya, Trends Genet. **19**, 176 (2003)
30. C. Dale, B. Wang, N.A. Moran, H. Ochman, Mol. Biol. Evol. **20**, 1188 (2003)
31. T. Itoh, W. Martin, M. Nei, Proc. Natl. Acad. Sci. USA **99**, 12944 (2002)
32. P.M. Sharp, E. Bailes, R. Grocock, J.F. Peden, R.E. Sockett, Nucl. Acids Res. **33**, 1141 (2005)
33. J.T. Herbeck, D.P. Wall, J.J. Wernegreen, Microbiology **149**, 2585 (2003)
34. C. Rispe, F. Delmotte, R.C.H.J. van Ham, A. Moya, Genome Res. **14**, 44 (2004)
35. J. Schäber, C. Rispe, J. Wernegreen, A. Buness, F. Delmotte, F.J. Silva, A. Moya, Gene **352**, 109 (2005)
36. S. Shigenobu, H. Watanabe, Y. Sakaki, H. Ishikawa, J. Mol. Evol. **53**, 377 (2001)
37. M.A. Fares, M.X. Ruiz-González, A. Moya, S.F. Elena, E. Barrio, Nature **417**, 398 (2002)
38. U. Bastolla, A. Moya, E. Viguera, R.C.H.J. van Ham, J. Mol. Biol. **343**, 1451 (2004)
39. N.A. Moran, C. Dale, H. Dunbar, W.A. Smith, H. Ochman, Environ. Microbiol. **5**, 116 (2003)
40. R. Gil, B. Sabater-Muñoz, A. Latorre, F.J. Silva, A. Moya, Proc. Natl. Acad. Sci. USA **99**, 4454 (2002)
41. F.J. Silva, A. Latorre, A. Moya, Trends Genet. **17**, 615 (2001)
42. S.T. Cole, K. Eiglmeier, J. Parkhill, K.D. James, N.R. Thomson, P.R. Wheeler, N. Honore, T. Garnier, C. Churcher, D. Harris, K. Mungall, D. Basham,

D. Brown, T. Chillingworth, R. Connor, R. M. Davies, K. Devlin, S. Duthoy, T. Feltwell, A. Fraser, N. Hamlin, S. Holroyd, T. Hornsby, K. Jagels, C. Lacroix, J. Maclean, S. Moule, L. Murphy, K. Oliver, M.A. Quail, M.A. Rajandream, K.M. Rutherford, S. Rutter, K. Seeger, S. Simon, M. Simmonds, J. Skelton, R. Squares, S. Squares, K. Stevens, K. Taylor, S. Whitehead, J.R. Woodward, B.G. Barrell, Nature **409**, 1007 (2001)
43. R. Gil, F.J. Silva, J. Peretó, A. Moya, Microbiol. Mol. Biol. Rev. **68**, 518 (2004)
44. D.A. Petrov, D.L. Hartl, Mol. Biol. Evol. **15**, 293 (1998)
45. A. Mira, H. Ochman, N.A. Moran, Trends Genet. **17**, 589 (2001)
46. A. Mira, L. Klasson, S.G.E. Andersson, Curr. Opin. Microbiol. **5**, 506 (2002)
47. P.H. Degnan, A.B. Lazarus, C.D. Brock, J.J. Wernegreen, Syst. Biol. **53**, 95 (2004)
48. P.H. Degnan, A.B. Lazarus, J.J. Wernegreen, Genome Res. **15**, 1023 (2005)
49. R.C.H.J. van Ham, F. González-Candelas, F.J. Silva, B. Sabater, A. Moya, A. Latorre, Proc. Natl. Acad. Sci. USA **97**, 10855 (2000)
50. A. Latorre, R. Gil, F.J. Silva, A. Moya, Heredity **95**, 339 (2005)
51. B. Sabater-Muñoz, L. Gómez-Valero, R.C.H.J. van Ham, F. J. Silva, A. Latorre, Appl. Environ. Microbiol. **68**, 2572 (2002)
52. B. Sabater-Muñoz, R.C.H.J. van Ham, A. Moya, F.J. Silva, A. Latorre, J. Bacteriol. **186**, 2646 (2004)
53. N.A. Moran, G.R. Plague, Curr. Opin. Genet. Dev. **14**, 627 (2004)

Part III

Phylogenetic Analysis

8

Molecular Phylogenetics: Mathematical Framework and Unsolved Problems

X. Xia

Phylogenetic relationship is essential in dating evolutionary events, reconstructing ancestral genes, predicting sites that are important to natural selection, and, ultimately, understanding genomic evolution. Three categories of phylogenetic methods are currently used: the distance-based, the maximum parsimony, and the maximum likelihood method. Here, I present the mathematical framework of these methods and their rationales, provide computational details for each of them, illustrate analytically and numerically the potential biases inherent in these methods, and outline computational challenges and unresolved problems. This is followed by a brief discussion of the Bayesian approach that has been recently used in molecular phylogenetics.

8.1 Introduction

Biodiversity comes in many colors and shades, and unorganized biodiversity cannot only dazzle our eyes but also confuse our minds. Phylogenetics is a special branch of science with the aim to organize biodiversity based on the ancestor–descendent relationship. Molecular phylogenetics uses molecular sequence data to achieve its three main objectives: (1) to reconstruct the branching pattern of different evolutionary lineages such as species and genes, (2) to date evolutionary events such as speciation or gene duplication and subsequent functional divergence, and (3) to understand and summarize the evolutionary processes by substitution models. With the rapid increase of DNA and protein sequence data, and with the realization that DNA is the most reliable indicator of ancestor–descendent relationships, molecular phylogenetics has become one of the most dynamic fields in biology with solid theoretical foundations [1–3] and powerful software tools [4–8]. I will not argue for the importance of molecular phylogenetics other than quoting Aristotle's statement that "He who sees things from the very beginning has the most advantageous view of them."

It is not always easy to see things from the very beginning. The evolutionary process depicted in Fig. 8.1 shows an ancestral population with a

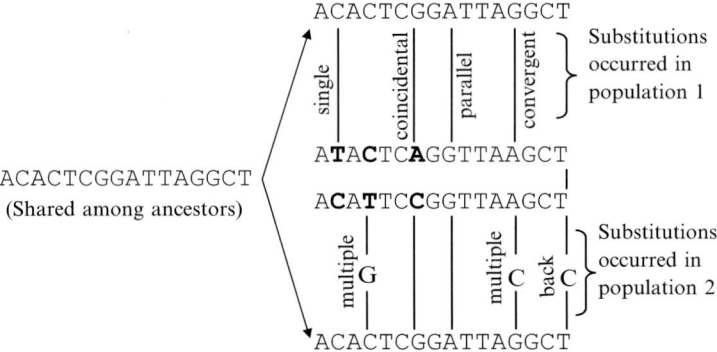

Fig. 8.1. Illustration of nucleotide substitutions and the difficulty in correcting multiple hits

single sequence shared among all individuals that have subsequently split into two populations and evolved and accumulated substitutions independently. Twelve substitutions have occurred, but only three differences can be observed between the sequences from the two extant species. The most fundamental difficulty in molecular phylogenetics is to estimate the true number of substitutions (i.e., 12) from the observed number of differences between extant sequences (i.e., 3). In short, the difficulty lies in how to correct for multiple hits.

The number of substitutions per site is known as a genetic distance. The simplest genetic distance, known as the p-distance (D_p), between two sequences is simply the number of different sites (N) divided by the sequence length (L). For the two sequences in Fig. 8.1, $D_p = 3/16$. Because D_p does not correct for multiple hits, it is typically a severe underestimation of the genetic distance and has to be corrected. In the next few sections, I will first detail commonly used substitution models, derive genetic distances based on the substitution models, and introduce the three categories of molecular phylogenetic methods: the distance-based, the maximum parsimony, and the maximum likelihood methods. This is followed by a numerical illustration of the Bayesian inference, together with a brief discussion of the Bayesian approach that has recently been used in molecular phylogenetics [9]. Potential problems with these phylogenetic methods will be highlighted.

8.2 Substitution Models

Substitution models reflect our understanding of how molecular sequences change over time. They are the theoretical foundation for computing the genetic distance in the distance-based phylogenetic method and for computing the likelihood value in the maximum likelihood method for phylogenetics.

There are three types of molecular sequences, i.e., nucleotide, amino acid, and codon sequences. Consequently, there are three types of substitution models, i.e., nucleotide-based, amino acid-based, and codon-based. We will focus on nucleotide-based substitution models, with only a brief discussion on amino acid-based and codon-based models to highlight a few potential problems.

8.2.1 Nucleotide-Based Substitution Models and Genetic Distances

Let $\boldsymbol{P}(t)$ be the vector of the four nucleotide frequencies in the order of A, G, C, T at time t. Nucleotide-based substitution models are characterized by a Markov chain of four discrete states as follows:

$$\boldsymbol{P}(t+1) = \boldsymbol{P}(t) \begin{bmatrix} M_{AA} & M_{AG} & M_{AC} & M_{AT} \\ M_{GA} & M_{GG} & M_{GC} & M_{GT} \\ M_{CA} & M_{CG} & M_{CC} & M_{CT} \\ M_{TA} & M_{TG} & M_{TC} & M_{TT} \end{bmatrix} = \boldsymbol{P}(t) \times \boldsymbol{M}, \qquad (8.1)$$

where $\boldsymbol{M}$ is the transition probability matrix and M_{ij} is the probability of changing from state i to state j in one unit of time. Three frequently used special cases of (8.1) will be detailed here: the JC69 model [10], the K80 model [11], and the TN93 model [12].

The simplest nucleotide substitution model is the JC69 one-parameter model, in which all off-diagonal elements in $\boldsymbol{M}$ are identical and designated as α. The four diagonal elements in $\boldsymbol{M}$ are $1 - 3\alpha$ constrained by the row sum equal to 1. There is a corresponding rate matrix, designated by $\tilde{\boldsymbol{Q}}$, that differs from $\boldsymbol{M}$ only in that the diagonal elements are -3α, constrained by the row sum equal to 0. It is often more convenient to derive substitution rates by using $\tilde{\boldsymbol{Q}}$ instead of $\boldsymbol{M}$, as will be clear later. Following (8.1), we have

$$\begin{aligned} P_A(t+1) &= P_A(t)(1-3\alpha) + P_G(t)\alpha + P_C(t)\alpha + P_T(t)\alpha, \\ P_G(t+1) &= P_A(t)\alpha + P_G(t)(1-3\alpha) + P_C(t)\alpha + P_T(t)\alpha, \\ P_C(t+1) &= P_A(t)\alpha + P_G(t)\alpha + P_C(t)(1-3\alpha) + P_T(t)\alpha, \\ P_T(t+1) &= P_A(t)\alpha + P_G(t)\alpha + P_C(t)\alpha + P_T(t)(1-3\alpha). \end{aligned} \qquad (8.2)$$

Arranging the left-hand side to be $\boldsymbol{P}(t+1) - \boldsymbol{P}(t)$ and then applying the continuous approximation, we have

$$\begin{aligned} \frac{\partial P_A}{\partial t} &= -P_A(t)(3\alpha) + (P_G(t) + P_C(t) + P_T(t))\alpha, \\ \frac{\partial P_G}{\partial t} &= -P_G(t)(3\alpha) + (P_A(t) + P_C(t) + P_T(t))\alpha, \\ \frac{\partial P_C}{\partial t} &= -P_C(t)(3\alpha) + (P_A(t) + P_G(t) + P_T(t))\alpha, \\ \frac{\partial P_T}{\partial t} &= -P_T(t)(3\alpha) + (P_A(t) + P_G(t) + P_C(t))\alpha. \end{aligned} \qquad (8.3)$$

Equation (8.3) is a special case of a general equation. Designate $\boldsymbol{d}$ as the vector of the four partial derivatives, the general equation is

$$\boldsymbol{d} = \boldsymbol{P}(t) \times \tilde{\boldsymbol{Q}}, \tag{8.4}$$

where $\tilde{\boldsymbol{Q}}$ is the rate matrix mentioned before.

Suppose that we start with nucleotide A, what is the probability that it will stay as A or change to one of the other three nucleotides after time t? Given the initial condition that $P_A(0) = 1$ and $P_G(0) = P_C(0) = P_T(0) = 0$ and the constraint that $P_A + P_G + P_C + P_T = 1$, (8.3) can be solved to yield

$$P_A(t) = \frac{1}{4} + \frac{3}{4}e^{-4\alpha t},$$

$$P_G(t) = P_C(t) = P_T(t) = \frac{1}{4} - \frac{1}{4}e^{-4\alpha t}. \tag{8.5}$$

The time t in (8.5) is the time from the ancestor to the present. When we compare two extant sequences, the time is $2t$, i.e., from one sequence to the ancestor and then back to the other sequence. So (8.5) has its general form as

$$M_{ii}(t) = \frac{1}{4} + \frac{3}{4}e^{-8\alpha t},$$

$$M_{ij}(t) = \frac{1}{4} - \frac{1}{4}e^{-8\alpha t}, \quad i \neq j. \tag{8.6}$$

The genetic distance (D), which is the number of substitutions per site, is defined as $D = 2t\mu$, where μ is the rate of substitution. For the JC69 mode, $\mu = 3\alpha$, so $\alpha t = D/6$. Now we can readily derive D, now designated D_{JC69}, from the probability of a site being different which is estimated by the p-distance (D_p) defined before. According to (8.6),

$$D_p = 1 - M_{ii}(t) = \frac{3}{4}\left(1 - e^{-8\alpha t}\right) = \frac{3}{4}\left(1 - e^{-4D_{JC69}/3}\right),$$

$$D_{JC69} = -\frac{3}{4}\ln\left(1 - \frac{4D_p}{3}\right). \tag{8.7}$$

For the two sequences in Fig. 8.1, $D_p = 3/16 = 0.1875$ and $D_{JC69} = 0.21576$. The equilibrium frequencies are derived by setting $(\boldsymbol{P}(t+1) - \boldsymbol{P}(t))$ in (8.3) to zero. Solving the resulting simultaneous equations with the constraint that the four frequencies sum up to 1, we have $P_A(t) = P_G(t) = P_C(t) = P_T(t) = 0.25$. In summary, the JC69 model assumes that (1) the four nucleotides can change into each other with equal probability and (2) the equilibrium frequencies are all equal to 0.25.

The variance of D_{JC69} can be obtained by using the "delta" method [13]. When a variable Y is a function of a variable X, i.e., $Y = F(X)$, the delta method allows us to obtain approximate formulation of the variance of Y if (1) Y is differentiable with respect to X and (2) the variance of X is known. The same can be extended to more variables.

The mathematical concept for the delta method is illustrated below, starting with the simplest case of $Y = F(X)$. Regardless of the functional relationship between Y and X, we always have

$$\Delta Y \approx \left(\frac{dY}{dX}\right) \Delta X, \tag{8.8}$$

$$(\Delta Y)^2 \approx \left(\frac{dY}{dX}\right)^2 (\Delta X)^2, \tag{8.9}$$

where ΔY and ΔX are small changes in Y and X, respectively. Note that the variance of Y is the expectation of the squared deviations of Y, i.e.,

$$\begin{aligned} V(Y) &= E(\Delta Y)^2, \\ V(X) &= E(\Delta X)^2. \end{aligned} \tag{8.10}$$

Replacing $(\Delta Y)^2$ and $(\Delta X)^2$ in (8.9) with $V(Y)$ and $V(X)$, we have

$$V(Y) \approx \left(\frac{dY}{dX}\right)^2 V(X). \tag{8.11}$$

This relationship allows us to obtain an approximate formulation of the variance of either Y or X if we know either $V(X)$ or $V(Y)$. For the variance of D_{JC69}, we note that D_{JC69} is a function of D_p, and the variance of D_p is known from the binomial distribution:

$$V(D_p) = \frac{D_p(1-D_p)}{L}, \tag{8.12}$$

where L is the length of the two aligned sequences. From the expression of D_{JC69} in (8.7), we have

$$\frac{\partial D_{JC69}}{\partial D_p} = \frac{1}{1 - 4D_p/3},$$

$$V(D_{JC69}) = \left(\frac{\partial D_{JC69}}{\partial D_p}\right)^2 V(D_p) = \frac{D_p(1-D_p)}{L(1-4D_p/3)^2}. \tag{8.13}$$

Kimura [11] noted that transitional substitutions typically occur much more frequently than transversions, and consequently proposed the two-parameter K80 model in which the rate of transitional substitutions ($A \leftrightarrow G$ and $T \leftrightarrow C$) is designated as α and the rate of transversion substitutions ($A \leftrightarrow T$, $A \leftrightarrow C$, $G \leftrightarrow T$ and $G \leftrightarrow C$) as β. Substituting this new $\tilde{Q}$ into (8.4) and solving the equations with the initial condition that $P_A(0) = 1$ and $P_G(0) = P_C(0) = P_T(0) = 0$ and the constraint that $P_A + P_G + P_C + P_T = 1$ as before, we have

$$P_A(t) = \frac{1}{4} + \frac{1}{4}e^{-4\beta t} + \frac{1}{2}e^{-2(\alpha+\beta)t},$$

$$P_G(t) = \frac{1}{4} + \frac{1}{4}e^{-4\beta t} - \frac{1}{2}e^{-2(\alpha+\beta)t},$$

$$P_C(t) = P_T(t) = \frac{1}{4} - \frac{1}{4}e^{-4\beta t}. \tag{8.14}$$

Note again that time t in (8.14) should be $2t$ when used between two extant sequences. So (8.14) has its general form as

$$P_s(t) = \frac{1}{4} + \frac{1}{4}e^{-8\beta t} - \frac{1}{2}e^{-4(\alpha+\beta)t},$$

$$P_v(t) = \frac{1}{2} - \frac{1}{2}e^{-8\beta t}, \tag{8.15}$$

where $P_s(t)$ and $P_v(t)$ are the probabilities that a site differs by a transition and a transversion, respectively, between two sequences that have diverged for time t, and can be estimated by the proportion of sites that differ by a transition (P) and a transversion (Q), respectively. This leads to

$$\beta t = -\frac{\ln(1-2Q)}{8},$$

$$\alpha t = -\frac{\ln(1-2P-Q)}{4} + \frac{\ln(1-2Q)}{8}. \tag{8.16}$$

Recall that the genetic distance is defined as $2t\mu$ where $\mu = \alpha + 2\beta$ for the K80 model. Therefore,

$$D_{K80} = 2\alpha t + 4\beta t = \frac{1}{2}\ln(a) + \frac{1}{4}\ln(b), \tag{8.17}$$

where $a = 1/(1-2P-Q)$ and $b = 1/(1-2Q)$.

For the two sequences in Fig. 8.1, $P = 2/16$, $Q = 1/16$, $D_{K80} = 0.22073$. The equilibrium frequencies are derived by setting $\boldsymbol{d}$ in (8.4) to the $\boldsymbol{0}$ vector. Solving the resulting simultaneous equations with the constraint that the four frequencies sum up to 1, we have $P_A(t) = P_G(t) = P_C(t) = P_T(t) = 0.25$. Thus, the K80 model shares with the JC69 model the assumption that the equilibrium frequencies are all equal to 0.25. The reader might have noticed this because nucleotide frequencies are not featured in the expression of D_{JC69} or D_{K80}.

The variance of D_{K80} can be derived by the delta method as before,

$$dD_{K80} = \left(\frac{\partial D_{K80}}{\partial P}\right)dP + \left(\frac{\partial D_{K80}}{\partial Q}\right)dQ = d_1(dP) + d_2(dQ), \tag{8.18}$$

$$\begin{aligned}(dD_{K80})^2 &= [d_1(dP) + d_2(dQ)]^2 \\ &= d_1^2(dP)^2 + 2d_1d_2(dPdQ) + d_2^2(dQ)^2 \\ &= d_1^2 V(P) + 2d_1d_2 \text{Cov}(P,Q) + d_2^2 V(Q), \\ &= \begin{bmatrix} d_1 & d_2 \end{bmatrix} \begin{bmatrix} V(P) & \text{Cov}(P,Q) \\ \text{Cov}(P,Q) & V(Q) \end{bmatrix} \begin{bmatrix} d_1 \\ d_2 \end{bmatrix}. \end{aligned} \tag{8.19}$$

Recall that P and Q stand for the proportion of sites that differ by a transitional change and by a transversional change, respectively. Designate R as the proportion of identical sites ($R = 1 - P - Q$). From the trinomial distribution of $(R + P + Q)^L$, we have

$$V(P) = \frac{P(1-P)}{L},$$
$$V(Q) = \frac{Q(1-Q)}{L},$$
$$\text{Cov}(P, Q) = -\frac{PQ}{L}. \tag{8.20}$$

Substituting these into (8.19), we have the variance of D_{K80},

$$V(D_{K80}) = (dD_{K80})^2 = \frac{a^2 P + c^2 Q - (aP + cQ)^2}{L} \tag{8.21}$$

where $c = (a + b)/2$, with a and b defined in (8.17).

Note that (8.19) is a general equation for computing the variance by the delta method. For any function $Y = F(X_1, X_2, \ldots, X_n)$, the variance of Y is obtained by the variance–covariance matrix of X_i multiplied left and right by the vector of partial derivatives of Y with respect to X_i.

Tamura and Nei [12] noticed the rate difference between $C \leftrightarrow T$ and $A \leftrightarrow G$ transitions and proposed the TN93 model with the following rate matrix:

$$\tilde{Q} = \begin{bmatrix} T & \ldots & \alpha_1 \pi_C & \beta \pi_A & \beta \pi_G \\ C & \alpha_1 \pi_T & \ldots & \beta \pi_A & \beta \pi_G \\ A & \beta \pi_T & \beta \pi_C & \ldots & \alpha_2 \pi_G \\ G & \beta \pi_T & \beta \pi_C & \alpha_2 \pi_A & \ldots \end{bmatrix}, \tag{8.22}$$

where π_i designates equilibrium nucleotide frequencies, and the diagonal (dots) is constrained by the row sum equal to 0.

Following the same protocol as before, and designating P_1, P_2, and Q as the probabilities of $C \leftrightarrow T$ transitions, $A \leftrightarrow G$ transitions and $R \leftrightarrow Y$ transversions (R means either A or G and Y means either C or T), respectively, we can obtain,

$$\begin{aligned} P_1 &= \pi_T P_{TC}(2t) + \pi_C P_{CT}(2t) \\ &= \frac{2\pi_T \pi_C}{\pi_Y} \left(\pi_Y + \pi_R e^{-2\beta t} - e^{-2(\alpha_1 \pi_Y + \beta \pi_R)t} \right), \end{aligned} \tag{8.23}$$

$$\begin{aligned} P_2 &= \pi_A P_{AG}(2t) + \pi_G P_{GA}(2t) \\ &= \frac{2\pi_A \pi_G}{\pi_R} \left(\pi_R + \pi_Y e^{-2\beta t} - e^{-2(\alpha_2 \pi_R + \beta \pi_Y)t} \right), \end{aligned} \tag{8.24}$$

$$Q = 2\pi_R \pi_y (1 - e^{-2\beta t}). \tag{8.25}$$

Solving for $\alpha_1 t$, $\alpha_2 t$, and βt from (8.23) to (8.25), we have

$$\alpha_1 t = \frac{1}{2\pi_Y}\left[-\ln\left(1 - \frac{Q}{2\pi_Y} - \frac{P_1 \pi_Y}{2\pi_T \pi_C}\right) + \pi_R \ln\left(1 - \frac{Q}{2\pi_R \pi_Y}\right)\right], \quad (8.26)$$

$$\alpha_2 t = \frac{1}{2\pi_R}\left[-\ln\left(1 - \frac{Q}{2\pi_R} - \frac{P_2 \pi_R}{2\pi_A \pi_G}\right) + \pi_Y \ln\left(1 - \frac{Q}{2\pi_R \pi_Y}\right)\right], \quad (8.27)$$

$$\beta t = -\frac{1}{2}\ln\left(1 - \frac{Q}{2\pi_R \pi_Y}\right). \quad (8.28)$$

$$\begin{aligned} D_{TN93} &= 2t\left[\pi_A(\beta\pi_T + \beta\pi_C + \alpha_2\pi_G) + \pi_C(\beta\pi_A + \beta\pi_G + \alpha_1\pi_T) + \right. \\ &\quad \left. \pi_T(\beta\pi_A + \beta\pi_G + \alpha_1\pi_C) + \pi_G(\beta\pi_T + \beta\pi_C + \alpha_2\pi_A)\right] \\ &= 4\left(\pi_R \pi_Y \beta t + \pi_A \pi_G \alpha_2 t + \pi_C \pi_T \alpha_1 t\right). \end{aligned} \quad (8.29)$$

Because we can estimate P_1, P_2, and Q by the proportion of sites with $C \leftrightarrow T$ transitions, $A \leftrightarrow G$ transitions, and $R \leftrightarrow Y$ transversions, respectively, D_{TN93} can be readily computed. For the two sequences in Fig. 8.1, $D_{TN93} = 0.2525$. The variance of D_{TN93} can be easily obtained by left- and right-multiplying the variance–covariance matrix of P_1, P_2, and Q with the vector of the three derivatives of D_{TN93} with respect to P_1, P_2, and Q in the same way as shown in the last term of (8.19). The variance and covariance of P_1, P_2 and Q can be obtained in the same way as in (8.20).

Many more substitution models and genetic distances have been proposed [14], with the number of all possible time reversible models of nucleotide substitution being 203 [15]. In addition, there are more complicated models underlying the Log-Det and the paralinear distances [16, 17] that can presumably accommodate the nonstationarity of the substitution process. Different substitution models often lead to different trees produced and constitute a major source of controversy in molecular phylogenetics [18–20].

8.2.2 Amino Acid-Based and Codon-Based Substitution Models

Amino acid-based models [21, 22] are similar in form to those nucleotide-based models in the previous section, except that the discrete states of the Markov chain will be 20 instead of only 4. Because of the large size of the transition matrix, the transition probabilities are typically derived from empirical transition matrices [23, 24].

There are three inherent difficulties with amino acid-based models. First, protein-coding genes often differ much in substitution patterns, and one can never be sure if any of the empirical transition matrices is appropriate for the protein sequences one is studying. Second, note that an amino acid replacement is effected by a nonsynonymous codon replacement. Two codons can differ by 1, 2, or 3 sites, and an amino acid replacement involving two codons differing by one site is expected to be more likely than that involving two codons differing by three sites. Only a codon-based model can incorporate this information. Third, two similar amino acids are expected to, and do, replace each other more frequently than two different amino acids [25]. However, the similarity between amino acids is difficult to define. For example, polarity may be highly conserved at some sites but not at others. Two very different amino acids rarely replace each other in functionally important domains but can replace each other frequently at unimportant segment. Moreover, the likelihood of two amino acids replacing each other also depends on neighboring amino acids [26]. For example, whether a stretch of amino acids will form an α-helix may depend on whether the stretch contains a high proportion of amino acids with high helix-forming propensity, and not necessarily on whether a particular site is occupied by a particular amino acid.

The codon-based substitution models [27, 28] were proposed to overcome some of the difficulties in amino acid-based models. These models share the third difficulty above with the amino acid-based models, and have additional problems of their own. For example, one cannot get good estimate of codon frequencies because protein-coding genes are typically very short. An alternative is to use the F3×4 codon frequency model [8, 29]. However, codon usage is affected by many factors, including differential ribonucleotide and tRNA abundance as well as biased mutation [30–32].

For example, the site-specific nucleotide frequencies are poor predictors of codon usage (Table 8.1) of protein-coding genes in *Escherichia coli* K12. The A-ending codon is used frequently for coding lysine, but the G-ending codon is used frequently for coding glutamine (Table 8.1). The reason for this is simple. Six Lys-tRNA genes in *E. coli* K12 all have anticodons being UUU which can translate the AAA lysine codon better than the AAG lysine codon. For glutamine codons, there are two copies of Glu-tRNA genes (glnX and glnV) with a CUG anticodons matching the CAG codon and another two copies (glnW and glnU) with the UUG anticodon matching the CAA codon. However, the former is more abundant than the latter in the *E. coli* cell [33], which would favor the use of CAG against the CAA codon for glutamine.

Table 8.1. Site-specific nucleotide frequencies and codon usage in two codon families

Base	CS_1	CS_2	CS_3	Codon	AA	N_{cod}
A	0.273	0.32	0.180	AAG	Lys	24
C	0.189	0.24	0.326	AAA	Lys	149
G	0.409	0.16	0.219	CAG	Gln	73
U	0.129	0.28	0.275	CAA	Gln	7

AA, amino acid; N_{cod}, number of codon; CS, codon site. Results based on eight highly expressed genes (gapC, gapA, fbaB, ompC, fbaA, tufA, groS, groL) from the *Escherichia coli* K12 genome (GenBank Accession: NC_000913).

One should expect the F3 × 4 codon frequency model to perform poorly in such a situation, which unfortunately is frequently encountered.

8.3 Tree-Building Methods

Three categories of tree-building methods are in common use: the distance-based, the maximum parsimony, and the maximum likelihood methods. These methods have their respective advantages and disadvantages, and I will provide mathematical details for the reader to understand their problems.

8.3.1 Distance-Based Methods

The distance-based methods build trees from a distance matrix, and are represented by the neighbor-joining (NJ) method [34], the Fitch–Margoliash (FM) method [35], and the FastME method [36]. The calculation of genetic distances has already been covered in previous sections. Other than the simplest UPGMA method, each tree-building method consists of two steps: (1) the evaluation of branch lengths for a given topology by the least-squares (LS) method, the NJ method or the FM method, and (2) the selection of the best tree based on either the minimum evolution (ME) criterion or the least-squares or the weighed least-squares criterion referred to hereafter as the FM criterion. One should not confuse, e.g., the FM way of evaluating branch lengths with the FM criterion for choosing the best tree.

There are many ways of evaluating branch lengths for a given tree, and I will only present the LS method here. For the three-OTU (operational taxonomic unit) tree in Fig. 8.2a, the branch lengths (x_i) can be solved uniquely by the following equations:

$$d_{12} = x_1 + x_2,$$
$$d_{13} = x_1 + x_3,$$
$$d_{23} = x_2 + x_3. \tag{8.30}$$

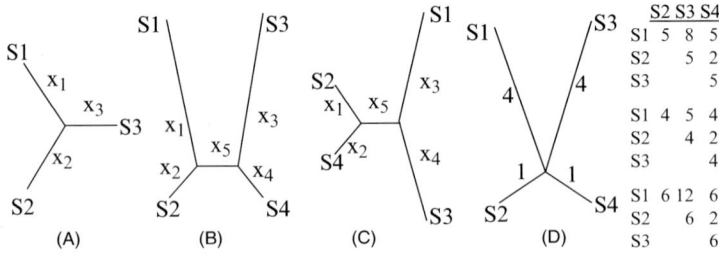

Fig. 8.2. Topologies for illustrating the distance-based methods in phylogenetic reconstruction

For the four-OTU tree in Fig. 8.2b, we can write down the equations in the same way as in (8.30), but there will be six equations for five unknowns. The LS method finds the x_i values that minimize the sum of squared deviations (SS),

$$\text{SS} = \sum (d_{ij} - \bar{d}_{ij})^2$$
$$= [d_{12} - (x_1 + x_2)]^2 + \cdots + [d_{34} - (x_3 + x_4)]^2. \tag{8.31}$$

By taking the partial derivatives with respect to x_i, setting the derivatives to zero and solving the resulting simultaneous equations, we get

$$\begin{aligned}
x_1 &= d_{13}/4 + d_{12}/2 - d_{23}/4 + d_{14}/4 - d_{24}/4, \\
x_2 &= d_{12}/2 - d_{13}/4 + d_{23}/4 - d_{14}/4 + d_{24}/4, \\
x_3 &= d_{13}/4 + d_{23}/4 + d_{34}/2 - d_{14}/4 - d_{24}/4, \\
x_4 &= d_{14}/4 - d_{13}/4 - d_{23}/4 + d_{34}/2 + d_{24}/4, \\
x_5 &= -d_{12}/2 + d_{23}/4 - d_{34}/2 + d_{14}/4 + d_{24}/4 + d_{13}/4. \tag{8.32}
\end{aligned}$$

With four OTUs, there are three unrooted trees. There are two commonly used criteria for choosing the best tree. The first is the ME criterion based on the tree length (TL), which is the summation of all x_i values. The tree with the smallest TL is chosen as the best tree. In contrast, the FM criterion chooses the tree with the smallest SS

$$\text{SS} = \sum_{i=1}^{n-1} \sum_{j=i+1}^{n} \frac{(d_{ij} - \bar{d}_{ij})^2}{d_{ij}^P}, \tag{8.33}$$

where n is the number of OTUs and P often takes the value of 0 or 2.

Whether a distance-based method will recover the true tree depends critically on the accuracy of the distance estimates. We will briefly examine this problem with both the ME criterion and the FM criterion. Let TL_B and TL_C be the tree length for Trees B and C in Fig. 8.2. Suppose that OTUs 1 and 3 have diverged from each other so much as to have experienced substitution saturation [37] to cause difficulty in estimating the true D_{13}. Let pD_{13} be the estimated D_{13}, where p measures the degree of underestimation ($p < 1$) or over-estimation ($p > 1$). Designate D_{TL} as the difference in TL between the two trees,

$$D_{\text{TL}} = \text{TL}_B - \text{TL}_C = \frac{d_{12} + d_{34} - (pd_{13} + d_{24})}{4}. \tag{8.34}$$

According to the LS method of branch evaluation, Tree B is better than Tree C if $D_{\text{TL}} < 0$, and worse than Tree C if $D_{\text{TL}} > 0$. Simple distances such as the p-distance or JC69 distance tend to have $p < 1$ and consequently increase the chance of having $D_{\text{TL}} > 0$, i.e., favoring the incorrect Tree C. This is the long-branch attraction problem, first recognized in the maximum parsimony method. Genetic distances corrected with the gamma-distributed rates over sites [12, 38–40] tend to have $p > 1$ when there is in fact no rate heterogeneity over sites, and consequently would favor Tree B over Tree C, leading to long-branch repulsion [41].

The long-branch attraction and repulsion problem is also present with the FM criterion. Let SS_B and SS_C be SS in (8.33) for Trees B and C, respectively. With $P = 0$ in (8.33) and letting $D_{SS} = SS_B - SS_C$, we have

$$4D_{SS} = (d_{13} + d_{24})^2 - (d_{12} + d_{34})^2 + 2(d_{14} + d_{23})[(d_{12} + d_{34}) - (d_{13} + d_{24})]$$
$$= x^2 - y^2 + 2z(y - x), \tag{8.35}$$

where $x = d_{13} + d_{24}$, $y = d_{12} + d_{34}$, and $z = d_{14} + d_{23}$. We now focus on Tree D, for which y is expected to equal z. Now (8.35) is reduced to

$$4D_{SS} = (x - y)^2. \tag{8.36}$$

If branch lengths are accurately estimated, then $x = y = 10$, and $D_{SS} = 0$, i.e., neither Tree B nor Tree C is favored. However, if d_{13} (i.e., the summation of the two long branches) is under- or over-estimated, then $D_{SS} > 0$ favoring Tree C. This means that both under- and over-estimation of the distance between divergence taxa will lead to long-branch attraction. This can be better illustrated with a numerical example with Tree D in Fig. 8.2, which also displays three distance matrices. The first one is accurate, the second one has genetic distances more underestimated for more divergent taxa, and the third has genetic distances more over-estimated for more divergent taxa (e.g., when gamma-distributed rates are assumed when the rate is in fact constant over sites). Note that Tree B and Tree C converge to Tree D when $x_5 = 0$. Table 8.2 shows the results by applying the ME and LS criterion in analyzing the three distance matrices.

When the distances are accurate, both the ME criterion and the FM criterion recover Tree D (the true tree) with $x_5 = 0$, $TL = 10$, and $SS = 0$. However, ME criterion favors Tree C when long branches are underestimated, and Tree B when long branches are over-estimated. In contrast, the FM criterion would favor Tree C with both under- and over-estimated distances (Table 8.2) when negative branches are allowed.

Table 8.2. Distance matrix: effect of under- and over-estimation of genetic distances

	TreeB	TreeC	TreeB	TreeC	TreeB	TreeC
TL	10	10	7.75	7.5	12.5	13
SS	0	0	0.25	0	1	0
x_5	0	0	−0.25	0.5	0.5	−1

Columns 2–3 are results from accurate genetic distances, columns 4–5 for genetic distances more underestimated for more divergent taxa, and columns 6–7 for genetic distances more over-estimated for more divergent taxa.

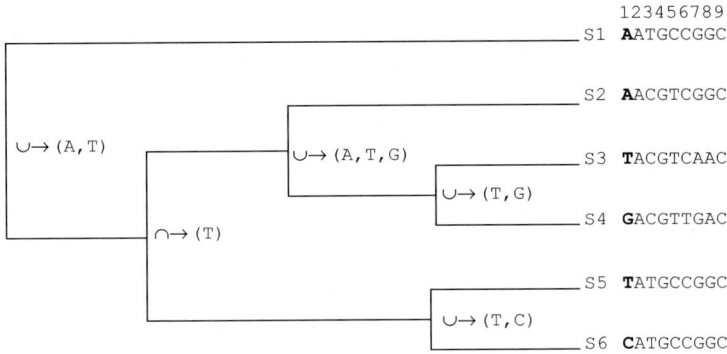

Fig. 8.3. Computing the minimum number of changes for the first site of the six alignment sequences in phylogenetic reconstruction using the maximum parsimony method

8.3.2 Maximum Parsimony Methods

In contrast to the distance-based methods, maximum parsimony (MP) and maximum likelihood methods are character-based methods. The six aligned sequences in Fig. 8.3 have nine sites, with sites 2, 4, 9 being monomorphic, and the rest of sites being polymorphic. A polymorphic site with at least two different states each represented by at least two OTUs is defined as an informative site. The MP method operates on informative sites only.

Given a topology, the minimum number of changes for each sequence site is computed, with the computation of the first site illustrated in Fig. 8.3. Each node is represented by a set of characters, with the terminal nodes (leaves) each represented by a set containing a single character. The method traverses through each internal node, starting from the node closest to the leaves. If two sets of the two daughter nodes have an empty intersection, then the node will be represented by the union of the two daughter sets, otherwise the node will be represented by the intersection. Once the operation reaches the root, then the number of union operations is the minimum number of changes needed to map the site to the tree. Site 1 in Fig. 8.3 requires four union operations, whereas sites 3, 5, and 8 each require only one union operation. Sites 6 and 7, which are polymorphic with two nucleotide states but not informative, will require one change for any topology. So the minimum number of changes, also referred to as the tree length, given the topology and the sequences in Fig. 8.3, is nine. The same computation is done for other possible topologies and the tree with the smallest tree length is taken as the MP tree.

The MP method is known to be inconsistent [42, 43], and I will provide a simple demonstration here by using trees in Fig. 8.4. With four species, we have three possible unrooted topologies, designated T_i ($i = 1, 2, 3$), with T_1 being the correct topology.

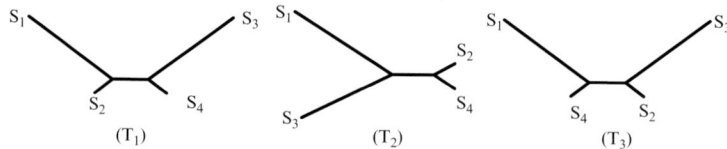

Fig. 8.4. The long-branch attraction problem in the maximum parsimony methods

Let X_{ij} be nucleotide at site j for species X_i, and L be the sequence length. For simplicity, assume that nucleotide frequencies are all equal to 0.25. Suppose that the lineages leading to X_1 and X_3 have experienced full substitution saturation, so that

$$\Pr(X_{1j} = X_{ij, i \neq 1}) = \Pr(X_{3j} = X_{ij, i \neq 3}) = 0.25 \qquad (8.37)$$

where Pr stands for probability. The lineages leading to X_2 and X_4 have not experienced substitution saturation and have

$$\Pr(X_{3j} = X_{4j}) = P, \qquad (8.38)$$

where $P > 0.25$. For simplicity, let us set $P = 0.8$, and $L = 1{,}000$.

We now consider the expected number of informative sites, designated by n_i ($i = 1, 2, 3$), favoring T_i. By definition, site j is informative and favoring T_1 if it meets the following three conditions: $X_{1j} = X_{2j}$, $X_{3j} = X_{4j}$, $X_{1j} \neq X_{3j}$. Similarly, site j favors T_2 if $X_{1j} = X_{3j}$, $X_{2j} = X_{4j}$, $X_{1j} \neq X_{2j}$. Thus, the expected numbers of informative sites favoring T_1, T_2, and T_3, respectively, are

$$\begin{aligned}
E(n_1) &= \Pr(X_{1j} = X_{2j}, X_{3j} = X_{4j}, X_{1j} \neq X_{3j})L \\
&= 0.25 \times 0.25 \times 0.75 \times 1{,}000 \approx 47, \\
E(n_2) &= \Pr(X_{1j} = X_{3j}, X_{2j} = X_{4j}, X_{1j} \neq X_{2j})L \\
&= 0.25 \times 0.80 \times 0.75 \times 1{,}000 \approx 150, \\
E(n_3) &= E(n_1) \approx 47.
\end{aligned} \qquad (8.39)$$

The equations mean that, in spite of T_1 being the true topology, we should have, on an average, only about 47 informative sites favoring T_1 and T_3, but 150 sites supporting the wrong tree T_2. This is one of the several causes for the familiar problem of long-branch attraction [44] or short-branch attraction [45]. Because it is the two short branches that contribute a large number of informative sites supporting the wrong tree, "short-branch attraction" seems a more appropriate term for the problem than "long-branch attraction."

8.3.3 Maximum Likelihood Methods

The maximum likelihood (ML) method is based on explicit substitution models. Many different types of computer simulation have demonstrated the

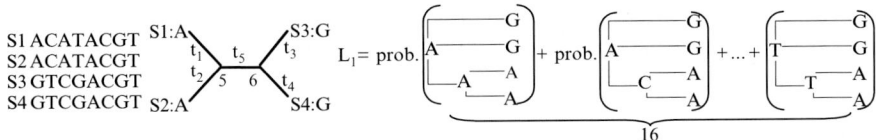

Fig. 8.5. Likelihood calculation for the first site of the four aligned sequences

superiority of the ML method in recovering the true tree. I now use the four aligned sequences in Fig. 8.5 to illustrate numerically the computation involved in the ML method based on the JC69 model. With four sequences, we have three possible unrooted topologies of which one is shown in Fig. 8.5.

The sequences have eight sites, with the first four sites sharing one site pattern and the last four sites sharing another site pattern. So we need only two site-specific likelihood functions. The likelihood function of the first site, given the topology in Fig. 8.5, is the summation of the 16 probabilities corresponding to the 16 nucleotide combinations of the two internal nodes with unknown nucleotides (Fig. 8.5). Thus, the likelihood of the first site is

$$L_1 = \pi_A M_{AA}(t_1) M_{AA}(t_2) M_{AA}(t_5) M_{AG}(t_3) M_{AG}(t_4)$$
$$+ \pi_C M_{CA}(t_1) M_{CA}(t_2) M_{CA}(t_5) M_{AG}(t_3) M_{AG}(t_4) + \cdots$$
$$+ \pi_T M_{TA}(t_1) M_{TA}(t_2) M_{TT}(t_5) M_{TG}(t_3) M_{TG}(t_4), \quad (8.40)$$

where $M_{ij}(t)$ for the JC69 model has already been given in (8.6) except that "$8\alpha t$" should be replaced by "$4\alpha t$." Note that $L_2 = L_3 = L_4 = L_1$. We can write $L_5 (= L_6 = L_7 = L_8)$ in a similar fashion.

The sequences in Fig. 8.5 allow us to simplify (8.40) greatly. Note that $S_1 = S_2$ and $S_3 = S_4$ (Fig. 8.5) so that αt_1, αt_2, αt_3, and αt_4 are all zero. Now we have

$$L_1 = 0.0625 - 0.0625 e^{-4\alpha t_5},$$
$$L_5 = 0.0625 + 0.1875 e^{-4\alpha t_5}. \quad (8.41)$$

With the assumption that all sites evolve independently, the likelihood function for all eight sites is simply

$$L = L_1^4 L_5^4,$$
$$\ln L = 4 \ln(L_1) + 4 \ln(L_5)$$
$$= 4 \ln(0.0625 - 0.0625 e^{-4\alpha t_5}) + 4 \ln(0.0625 + 0.1875 e^{-4\alpha t_5}). \quad (8.42)$$

The αt_5 value that maximizes $\ln L$ in (8.42) is $\alpha t_5 = 0.27465$, which leads to $\ln L = -21.02998$. The branch length between nodes 5 and 6 is $3\alpha t_5 = 0.82396$. We can do the same calculation for the other two possible topologies, and then choose the tree with the largest $\ln L$ value as the ML tree. In this particular example, the tree in Fig. 8.5 is the ML tree because it has the ln L value greater than that of the other two trees. One may also find that the ML tree, including its estimated branch lengths, is identical to the tree from a distance-based method such as the neighbor-joining [34], the FastME [36] or the FM method [35] as long as the JC69 distance is used.

There are two major criticisms on the ML method in phylogenetics. The first is that the application of the likelihood in phylogenetics is not really an ML method in its conventional sense because the topology is not in the likelihood function [3,46]. To see this point, we can illustrate the conventional ML method with a simple example.

Suppose we wish to estimate the proportion of males (p) of a fish population in a large lake, a random sample of N fish contains M males. With the binomial distribution, the likelihood function is

$$L = \frac{N!}{M!(N-M)!} p^M (1-p)^{N-M}. \tag{8.43}$$

The maximum likelihood method finds the value of p that maximizes the likelihood value. This maximization process is simplified by maximizing the natural logarithm of L instead:

$$\ln L = A + M \ln(p) + (N-M) \ln(1-p)$$
$$\frac{\partial \ln L}{\partial p} = \frac{M}{p} - \frac{N-M}{1-p} = 0$$
$$p = \frac{M}{N}. \tag{8.44}$$

The likelihood estimate of the variance of p is the negative reciprocal of the second derivative,

$$\text{Var}(p) = -\frac{1}{\partial^2 \ln L / \partial p^2} = -\frac{1}{-M/p^2 - (N-M)/(1-p)^2} = \frac{p(1-p)}{N}. \tag{8.45}$$

Note that, in contrast to the likelihood in (8.44), which is a function of p (the parameter to be estimated), the likelihood in (8.42) does not have the topology as a parameter. Without the convenient "$\partial \ln L / \partial \theta = 0$" formulation, we have to do either exhaustive or branch-and-bound search in order to find the topology that maximizes that likelihood. In practice, exhaustive or branch-and-bound search is rarely done, which implies that few of the published ML trees are authentic ML trees. Thus, Nei's criticism highlights more of a practical difficulty than a theoretical one because the likelihood principle does not require the parameter to be continuous and differentiable [47]. The criticism can also be applied to other phylogenetic methods. However, other methods are generally faster and can search the tree space more thoroughly than the ML method. Therefore, while it is not particularly controversial to claim that an authentic ML tree is generally better than a tree satisfying the

MP, ME, or FM criterion, it is not unreasonable for one to expect the latter to be as good as or better than an "ML" tree that is derived from searching a small subset of all possible topologies. This is particularly pertinent with reconstructing very large phylogenies [48].

The second criticism is on the assumptions shared by nearly all the substitution models currently implemented in the likelihood framework: (1) the substitutions occur independently in different lineages, (2) substitutions occur independently among sites, and (3) the process of substitution is described by a time-homogeneous (stationary) Markov process. The first assumption is false in taxa with a history of horizontal gene transfer [49–54]. The problem of the second assumption can be illustrated with the following example involving the GAT and GGT codons. Both codons end with a T. Whether a $T \to A$ substitution would occur depends much on whether the second position is an A or a G. The $T \to A$ substitution is rare when the second codon position is A because a $T \to A$ mutation in the GAT codon is nonsynonymous, but relatively frequent when the second codon position is G because such a $T \to A$ mutation in a GGT codon is synonymous. So nucleotide substitutions do not occur independently among sites. This is one of the reasons for using codon-based models but these models have their own problems as mentioned before. The third assumption is also problematic. Suppose, we wish to reconstruct a tree from a group of orthologous sequences from both invertebrate and vertebrate species, there is little DNA methylation in invertebrate genomes, but heavy DNA methylation in some vertebrate genomes. DNA methylation greatly enhanced the $C \to T$ transition (and consequently the $G \to A$ transition on the opposite strand [55]). The net result is a much elevated transition/transversion bias and increased AT% in the lineages with DNA methylation, violating the third assumption.

More complicated models have been proposed in response to our increased knowledge of the substitution process. However, such parameter-rich models require more data for reliable parameter estimation. The dilemma is that increasing the sequence length also increases the heterogeneity of substitution processes [56] including heterotachy [57] operating on different sequence segments and consequently increase the number of parameters to be estimated. Such heterogeneity over sites implies that the consistency of the ML method [47, 58] is not of much value because we cannot get long sequences for a fixed and small number of parameters. Take for example the estimation of the proportion of male fish in the lake. If we get only six male fish in a sample with no female, then the likelihood estimation of p is 1 which is worse than our wildest guess without any data.

8.3.4 Bayesian Inference

The Bayesian approach has only recently been used in phylogenetic inference [9]. Here, I illustrate the basic principle of the Bayesian approach by using the problem of estimating the proportion (p) of males when the sample of six fish being all males. For a continuous variable such as p, the Bayes' theorem is

$$f(\theta|y) = \frac{f(y|\theta)f(\theta)}{\int f(y|\theta)f(\theta)d\theta}, \qquad (8.46)$$

where θ is the parameter of interest, y is the observed sample data, $f(\theta)$ is the prior probability for incorporating our prior knowledge on θ, $f(y|\theta)$ is the likelihood, and $f(\theta|y)$ is the posterior probability. In practice, (8.46) is rarely used because the integration in the denominator is difficult unless $f(\theta)$ and $f(y|\theta)$ are very simple, although the MCMC (Markov chain Monte Carlo) approach [59,60] can alleviate the problem. Two alternative approaches have been devised to ease the computation burden, one being to use discrete approximations to continuous probability models, and the other being to use the conjugate prior distributions. For our example involving a stationary and independent Bernoulli process in estimating p, the conjugate prior distribution is the beta distribution with the following $f(p)$:

$$f(p) = \frac{(n-1)!}{(r-1)!(n-r-1)!} p^{r-1}(1-p)^{n-r-1}. \qquad (8.47)$$

Let us designate n and r as n' and r' in the prior distribution, n'' and r'' in the posterior distribution, and just as n and r in the sample. If we expect the fish species to have equal number of males and females, then for a sample of six fish ($n' = 6$), we expect r' to be 3. The prior probability can be calculated from (8.47) and shown in Fig. 8.6.

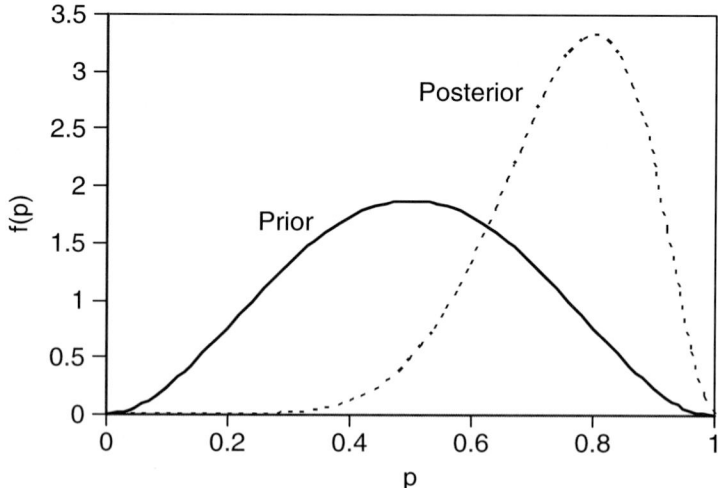

Fig. 8.6. Comparison between prior and posterior probabilities

It can be proven that, if the prior distribution of p is a beta distribution, then the posterior distribution will also be a beta distribution with two parameters computed according to (8.48). In our actual sample with six males and 0 female ($n = 6$ and $r = 6$),

$$r'' = r' + r = 3 + 6 = 9,$$
$$n'' = n' + n = 6 + 6 = 12. \quad (8.48)$$

Now the posterior probability can be calculated by using (8.47) and shown in Fig. 8.6 in comparison with the prior probability. Thus, our prior expectation of $p = 0.5$ has been revised by the actual sample to $p = 0.8$. One may note that if the population of fish is indeed made of all males, e.g., when there is a high concentration of androgen masculinizing all individuals to males [61], then the likelihood estimate of $p = 1$ is correct and the Bayesian estimate of $p = 0.8$ is wrong, and the wrong estimate may lead to our failure to identify an environmental crisis.

In Bayesian phylogenetics, θ is a collection of the tree topology, the rate matrix, and the branch lengths, and the likelihood function is formulated as in the maximum likelihood method. The main difficulty is the justification of the prior probability [1, 62, 63] which is problematic even for the simple example of estimating p.

8.4 Final Words

What I have presented is only the tip of the iceberg. One need to go down and get wet to see what is truly big (and truly messy). The models and equations are presented more for convenience than for mathematical rigor, but should work well for pedagogical purposes. I used to tell my son that his toys were alive with their own minds so he should be nice to them otherwise they would be upset and refuse to play with him. Such a lousy worldview nevertheless seemed to work perfectly well for him. I am sure that my son will grow out of this worldview, just as I am sure that the reader will grow out of the conceptual framework of molecular phylogenetics that is presented in this chapter.

Acknowledgments

I thank S. Aris-Brosou, B. Golding, D. Hickey, A. Meyer, S. Kumar, J.R. Lobry, D. Sankoff, and H. Xiong for comments, suggestions and corrections, and NSERC Discovery, RTI and Strategic grants for funding.

References

1. J. Felsenstein, *Inferring Phylogenies* (Sinauer, Sunderland, MA, 2004)
2. W.-H. Li, *Molecular Evolution* (Sinauer, Sunderland, MA, 1997)

3. M. Nei, S. Kumar, *Molecular Evolution and Phylogenetics* (Oxford University Press, New York, 2000)
4. S. Kumar, K. Tamura, I.B. Jakobsen, M. Nei, Bioinformatics **17**, 1244 (2001)
5. D.L. Swofford, *Phylogenetic Analysis Using Parsimony (* and Other Methods)*, 4th edn. (Sinauer, Sunderland, MA, 2000)
6. X. Xia, Z. Xie, J. Hered. **92**, 371 (2001)
7. X. Xia, *Data Analysis in Molecular Biology and Evolution* (Kluwer, Boston, 2001)
8. Z. Yang, *Phylogenetic Analysis by Maximum Likelihood* (PAML) (University College, London, 2002) version 3.12
9. J.P. Huelsenbeck, F. Ronquist, R. Nielsen, J.P. Bollback, Science **294**, 2310 (2001)
10. T.H. Jukes, C.R. Cantor, in *Evolution of Protein Molecules*. ed. by H.N. Munro. Mammalian Protein Metabolism (Academic, New York, 1969) pp. 21–123
11. M. Kimura, J. Mol. Evol. **16**, 111 (1980)
12. K. Tamura, M. Nei, Mol. Biol. Evol. **10**, 512 (1993)
13. M. Kimura, T. Ohta, J. Mol. Evol. **2**, 87 (1972)
14. K. Tamura, S. Kumar, Mol. Biol. Evol. **19**, 1727 (2002)
15. J.P. Huelsenbeck, B. Larget, M.E. Alfaro, Mol. Biol. Evol. **21**, 1123 (2004)
16. J.A. Lake, Proc. Natl. Acad. Sci. USA **91**, 1455 (1994)
17. P.J. Lockhart, M.A. Steel, M.D. Hendy, D. Penny, Mol. Biol. Evol. **11**, 605 (1994)
18. M.S. Rosenberg, S. Kumar, Mol. Biol. Evol. **20**, 610 (2003)
19. X. Xia, Syst. Biol. **49**, 87 (2000)
20. X. Xia, Z. Xie, K.M. Kjer, Syst. Biol. **52**, 283 (2003)
21. J. Adachi, M. Hasegawa, J. Mol. Evol. **42**, 459 (1996)
22. H. Kishino, T. Miyata, M. Hasegawa, J. Mol. Evol. **31**, 151 (1990)
23. M.O. Dayhoff, R.M. Schwartz, B.C. Orcutt, in *A Model of Evolutionary Change in Proteins*. ed. by M.O. Dayhoff. Atlas of Protein Sequence and Structure, vol. 5, suppl. 3, (National Biomedical Research Foundation, Washington DC, 1978) pp. 345–352
24. D.T. Jones, W.R. Taylor, J.M. Thornton, Comput. Appl. Biosci. **8**, 275 (1992)
25. X. Xia, W.H. Li, J. Mol. Evol. **47**, 557 (1998)
26. X. Xia, Z. Xie, Mol. Biol. Evol. **19**, 58 (2002)
27. N. Goldman, Z. Yang, Mol. Biol. Evol. **11**, 725 (1994)
28. S.V. Muse, B.S. Gaut, Mol. Biol. Evol. **11**, 715 (1994)
29. Z. Yang, R. Nielsen, Mol. Biol. Evol. **17**, 32 (2000)
30. X. Xia, Genetics **44**, 1309 (1996)
31. X. Xia, Genetics **149**, 37 (1998)
32. X. Xia, Gene **345**, 13 (2005)
33. T. Ikemura, in *Correlation Between Codon Usage and tRNA Content in Microorganisms*. ed. by D.L. Hatfield, B. Lee, J. Pirtle. Transfer RNA in Protein Synthesis (CRC, Boca Raton, FL, 1992) pp. 87–111
34. N. Saitou, M. Nei, Mol. Biol. Evol. **4**, 406 (1987)
35. W.M. Fitch, E. Margoliash, Science **155**, 279 (1967)
36. R. Desper, O. Gascuel, J. Comput. Biol. **9**, 687 (2002)
37. X. Xia, Z. Xie, M. Salemi, L. Chen, Y. Wang, Mol. Phylogenet. Evol. **26**, 1 (2003)
38. G.B. Golding, Mol. Biol. Evol. **1**, 125 (1983)

39. M. Nei, T. Gojobori, Mol. Biol. Evol. **3**, 418 (1986)
40. L. Jin, M. Nei, Mol. Biol. Evol. **7**, 82 (1990)
41. P.J. Waddell, Statistical Methods of Phylogenetic Analysis: Including Hadamard Conjugations, LogDet Transforms, and Maximum Likelihood. PhD Thesis, Massey University, New Zealand (1995)
42. J. Felsenstein, Syst. Zool. **27**, 401 (1978)
43. N. Takezaki, M. Nei, J. Mol. Evol. **39**, 210 (1994)
44. M.D. Hendy, D. Penny, Syst. Zool. **38**, 297 (1989)
45. M. Nei, Annu. Rev. Genet. **30**, 371 (1996)
46. M. Nei, *Molecular Evolutionary Genetics* (Columbia University Press, New York, 1987)
47. J.T. Chang, Math. Biosci. **137**, 51 (1996)
48. K. Tamura, M. Nei, S. Kumar, Proc. Natl. Acad. Sci. USA **101**, 11030 (2004)
49. C. Medigue, T. Rouxel, P. Vigier, A. Henaut, A. Danchin, J. Mol. Biol. **222**, 851 (1991)
50. E.V. Koonin, Mol. Microbiol. **50**, 725 (2003)
51. H. Philippe, C.J. Douady, Curr. Opin. Microbiol. **6**, 498 (2003)
52. C.G. Kurland, B. Canback, O.G. Berg, Proc. Natl. Acad. Sci. USA **100**, 9658 (2003)
53. J.R. Brown, Nat. Rev. Genet. **4**, 121 (2003)
54. J.A. Eisen, Curr. Opin. Genet. Dev. **10**, 606 (2000)
55. X. Xia, J. Mol. Evol. **57**, S21 (2003)
56. X. Xia, Mol. Biol. Evol. **15**, 336 (1998)
57. B. Kolaczkowski, J.W. Thornton, Nature **431**, 980 (2004)
58. J. Felsenstein, Annu. Rev. Genet. **22**, 521 (1988)
59. W.K. Hastings, Biometrika **57**, 97 (1970)
60. N. Metropolis, A.W. Rosenbluth, M.N. Rosenbluth, A.H. Teller, E. Teller, J. Chem. Phys. **21**, 1087 (1953)
61. D. Baron, J. Cocquet, X. Xia, M. Fellous, Y. Guiguen, R.A. Veitia, J. Mol. Endocrinol. **33**, 705 (2004)
62. D. Zwickl, M. Holder, Syst. Biol. **53**, 877 (2004)
63. K.M. Pickett, C.P. Randle, Mol. Phylogenet. Evol. **34**, 203 (2005)

9

Phylogenetics and Computational Biology of Multigene Families

P. Liò, M. Brilli, and R. Fani

This chapter introduces the study of the major evolutionary forces operating in large gene families. The reconstruction of duplication history and phylogenetic analysis provide an interpretative framework of the evolution of multigene families. We present here two case studies, the first coming from Eukaryotes (chemokine receptors) and the second from Prokaryotes (TIM barrel proteins), showing how functional and structural constraints have shaped gene duplication events.

9.1 Introduction

The genes and genomes sequencing projects have revealed that much of the increase in gene number from bacteria to humans has been mainly promoted by DNA duplication, which can lead to redundant gene copies.

In some special cases, an adaptive advantage for redundancy can be found, e.g., an increased dosage of a protein, as is the case for the developing by many insects of resistances to pesticides (see e.g., [1]). Such short-term selective advantages should not be underestimated, but in the vast majority of gene duplication events, a similar dosage effect cannot be invoked; thus, the preservation by selection of both duplicates seems to be possible only if they diverge in some way [2].

In eukaryotes, gene duplications seem to have occurred often, e.g., the olfactory [3], HOX [4], or globin genes in animals [5,6].

Moreover, the remnants of whole genome duplications have been identified, e.g., for frogs [7,8], fishes [9–11], different yeast strains [12–15], and Arabidopsis [16]; these, along with events involving DNA stretches of reduced size (from less than a gene to a few genes), lead to suppose that gene duplication is one of the most important mechanisms lying at the basis of eukaryotes evolution; the whole process can be summarized as gene duplication event and evolutionary divergence of the two paralogs; this in the long run can increase the complexity of metabolism and regulatory patterns, which have been linked to an increased morphological and/or phenotypical complexity [17].

The modalities by which gene duplication promotes complexity increases are nevertheless still an open question; in fact Dehal and Boore [18], studying the two hypothesized rounds of whole genome duplication that happened in the vertebrate lineage, noticed that the vast majority of duplicated genes that originated after the proposed whole genome duplications were subsequently lost, indicating that relatively few genes may have been responsible for the increased complexity disclosed in vertebrates [18].

The three main distinct mechanisms that generate tandem duplication of DNA stretches in eukaryotes are slipped-strand mispairing, gene conversion, and unequal recombination. Various examples have been described (see e.g., [19, 20]) and they will not be discussed here.

Different extents of duplication are possible: a small part of a gene, a module, an exon, an intron, a full coding region, a full gene, a chromosome segment containing a cluster of genes, a full chromosome, or the entire genome.

It must be stressed that not every gene duplication results in acquiring a new function by one of the two duplicates: most families, for example the globin and olfactory gene clusters, also contain many duplicates that have lost function (pseudogenes). Other duplicates can retain the original function and be maintained in a given genome for indefinite time. This could happen for the yet cited gene dosage effect, or because a certain degree of genetic redundancy increases the robustness of genetic networks, as emerged by the work of Gu et al. [21]: they performed a genome-wide survey on the phenotypic effects of *Saccharomyces cerevisiae* single-gene deletion mutants and showed that 25% of the mutants with no discernible phenotype could be explained by the presence of duplicated loci [22].

Studies concerning the rate of gene duplication confirmed that the process fulfills all the requisites for being considered an important evolutionary tool: the duplication of a single gene is an event whose average rate is on the order of 0.01 per gene per million years, ranging between 0.02 and 0.002 in different species; moreover, duplicates exhibit an average half-life of about four million years [2], a time that appear to be congruent with the evolution of novel functions or the specialization of ancestral, inefficient ones. More quantitatively, it has been estimated that half of all the genes in a genome are expected to duplicate and increase to high frequency at least once on timescales of 35–350 million years [2]. It could be interesting to remind that if these estimations are correct, then the rate of duplication of a gene is of the same order of magnitude as the rate of mutation per nucleotide site. Thus, even in the absence of direct amplification of entire genomes (polyploidization), gene duplication has the potential to generate substantial molecular substrate for the origin of evolutionary novelties [2].

As anticipated, whole genome duplications are believed to have played an important role in the evolution of yeast, plant, and vertebrate species (see e.g., [5, 23]). Consequently, the present debate on vertebrate evolution concentrates on the relative contribution of the large-scale genome duplication (the so-called big-bang model), after the echinoderms/chordates split and before the vertebrate radiation, and the continuous origin by small-scale duplications.

After duplication a gene starts diverging from the ancestral sequence, referring not only to coding sequences, but also to regulatory sites. Recently, as microarray expression data accumulate in public databases, great attention has been devoted to understand the role of expression divergence in the preservation of duplicates and the appearance of ever more complex expression patterns and genetic networks.

As a result of the different extents of DNA duplication and the possible types of evolutionary divergence of the duplicates (functional and/or regulatory), eukaryotic genomes in general and the vertebrate ones in particular, contain several large gene families, whose gene products are expressed in different tissues in different amounts, permitting the high degree of specialization of the different tissues.

The analysis of these large gene families is central to studies of mechanisms of gene duplication and duplicates preservation, but are also related to researches in genome evolutionary dynamics. Similar analyses make use of statistics and bioinformatics and while being the present chapter not intended to serve as a complete literature survey on these tools, it represents a good introduction to recent aspects of this field, including description of some statistical and bioinformatics techniques.

9.2 How Do Large Gene Families Arise?

As gene duplications take place within individuals constituting a population, it is generally assumed that the duplicated gene is neutral or immediately advantageous; otherwise it will be purged from the population; similarly, for the duplicate to be preserved, it must be fixed in the population before it becomes a functionless gene. Selection cannot identify a gene with future potential, hence the duplication of a gene has been generally considered to be followed by a race between the pseudogenization of one duplicate and its fixation in the population. This depends on the need of the cell for that gene product in that condition, i.e., it depends on the gene regulation and/or product function.

Early studies on gene duplication and duplicates preservation lead to the hypothesis that a balance between pseudogenization and fixation of the duplicate, the last happening in the case the duplicate have acquired a new and advantageous function.

9.3 The Classical Model of Gene Duplication

The classical model of gene duplication states that after the duplication event, one duplicate may become functionless, while the other copy will retain the original function [22, 24]. Complete duplicates (when the duplication involves both the coding and all the regulatory sequences of the original gene) are expected to be redundant in function (at least in the immediate beginning); in this case, one duplicate may represent a backup copy shielding the other from natural selection; this, being likely that a mutation have a negative effect on function, implies that one duplicate will probably lose its function while the other will retain it.

Very rarely, an advantageous mutation may change the function of one duplicate and both duplicates may be retained. In conclusion, the most plausible fate in the light of the classical model is the pseudogenization of one of the two duplicates, but it suffers of a number of problems when explaining evidences emerged by recent genome analyses. First of all, it fails to explain the amount of functional divergence and the long-term preservation of the large numbers of paralogous genes, which constitute most of eukaryotic multigene families, because they often retain the original function for long time. Accordingly, Walsh [25] used a mathematical model of the evolutionary fate of duplicates starting immediately after the duplication event (when the duplicates are perfectly redundant); this Markov model has two absorbing states, *fixation* and *pseudogenization*. The main result is that if the population size is large enough, the fate of most duplicated genes is to gain a new function rather than become pseudogenes.

Nadeau and Sankoff [26], studying human and mice genes, estimated that about 50% of gene duplications undergo functional divergence; other researches showed that the frequency of paralogous genes preservation following ancient polyploidization events are in the neighborhood of 30–50% over periods of tens to hundreds of millions of years [2]. To overcome these limitations, new models have been proposed which better fit empirical data.

9.4 Subfunctionalization Model

In eukaryotes, gene expression patterns are typically controlled by complex regulatory regions, which finely tune the expression of a gene in a specific tissue, developmental stage, or cell lineage. Particularly interesting is the combinatorial nature of most eukaryotic promoters, composed by different and partially autonomous regions with a positive or negative effect on downstream gene transcription with the overall expression pattern being determined by their concerted (synergistic) action.

Similarly, proteins can contain different functional and/or structural domains, which may interact with different substrates and regulatory ligands, or other proteins. Every transcriptionally important site or protein domains can be considered as a subfunctional module for a gene or protein, each one contributing to the global function of that gene or protein. Starting from this idea, Lynch and Force [27] first proposed that multiple subfunctions of the original gene may play an important role in the preservation of gene duplicates. They focused on the role of degenerative mutations in different regulatory elements of an ancestral gene expressed at rates, which depend on a certain number of different transcriptional modules (subfunctions) located in its promoter region. After the duplication event, deleterious mutations can reduce the number of active subfunctions of one or both the duplicates, but the sum of the subfunctions of the duplicates will be equal to the number of original functions before duplication (i.e., the original functions have been partitioned among the two duplicates). Similarly, considering both duplicates, they are together able to complement all the original subfunctions; moreover, they can have partially redundant functions too.

This example of subfunctionalization considers both functions affecting the expression patterns dependent on promoter sequences recognized by different transcription factors, and also "hubs" proteins with different and partially independent domains. The subfunctionalization, or duplication- degeneration-complementation model (DDC) of Lynch and Force [25, 27], differs from the classical model because the preservation of both gene copies mainly depends on the partitioning of subfunctions between duplicates, rather than the occurrence of advantageous mutations.

A limitation of the subfunctionalization model is the requirement for multiple independent regulatory elements and/or functional domains; the classical model is still valid if gene functions cannot be partitioned: for example when selection pressure acts to conserve all the subfunctions together. This is often the case when multiple subfunctions are dependent on each other.

9.5 Subneofunctionalization

A recent improvement in gene duplication models has been proposed by He and Zhang [28], starting from the results of a work concerning both yeast protein interaction data (from MIPS and [29]) and human expression data (from [30]), which have been tested both under the neofunctionalization and the subfunctionalization models. Neither models alone satisfied experimental results for duplicates. Further progress are formalized as subneofunctionalization model, which is a mix of previous models; the subfunctionalization appear to be a rapid process, while the neofunctionalization requires more time and continues even long after duplication [28].

The rapid subfunctionalization observed by He and Zhang [28] can be viewed as the acquisition of expression divergence between duplicates; after this subfunctionalization has occurred, both duplicates are essential (because only together they can maintain the original expression patterns), and hence they are maintained. Once a gene is established in a genome, it can retain its function or evolve or specialize a new one (i.e., it undergoes neofunctionalization).

9.6 Tests for Subfunctionalization

Dermitzakis and Clark [31] developed a subfunctionalization test, and identified several paralogs common to both humans and mice in which it has occurred. The basic idea is that the substitution rate in a given functional region will be low if a gene possesses a function which relies on that residues, otherwise the substitution rate in the same region will be higher. The statistical method used by the authors permits the identification of regions with significantly different substitution rates in two paralogous genes. Following Tang and Lewontin [32], Dermitzakis and Clark represented the pattern of change across a gene by the cumulative proportion of differences between a pair of orthologs (termed "pattern graph"). The pattern graph shows sharp increases when the substitution rate is high and almost no change in regions of sparse changes. Dermitzakis and Clark proposed the "paralog heterogeneity test," comparing the pattern graphs for human and mouse paralogs of several interesting genes. If one paralog has had a higher rate of substitution than the other in a given region, the difference between the pattern graphs shows a sharp rise or fall. If the paralogs have evolved at the same rate within a region, the difference between the pattern graphs will change slowly across the region.

The "paralog heterogeneity test" compares the longest stretch of increasing or decreasing difference between the two pattern graphs to what would be expected if they had evolved similarly. The null distribution is simulated by the repeated random permutation of the two genes, so neither gene has distinct regions. If a region contains significant differences when compared with the null distribution, then the paralogs have evolved at different rates in that region. If so, the subfunctionalization model predicts that it is important to the function of the paralog in which it is conserved. Subfunctionalization assumes that there is more than one region of functional importance and it is possible for more than one region to be involved in a given function. By testing the sum of the two or more largest stretches, the significance of multiple regions can be determined; this can lead to predictions of functional regions for both paralogs.

9.7 Tests for Functional Divergence After Duplication

One of the possible fates of a duplicated gene is to acquire a new function, or to specialize a pre-existing but inefficient function; again statistics permit the study of functional divergence in a rigorous manner. Changes in protein function generally lead to variations in the selective forces acting on specific residues. An enzyme can evolve arriving to recognize a different substrate, which plausibly interact with residues not involved in docking the original ligand, and leaving some of them free. Similar changes can often be detected as evolutionary rate changes at the sites in question. It is known that a good estimator of the sign and strength of the selective force acting on a coding sequence is the ratio between nonsynonymous and synonymous rates: if the value exceed 1 the gene is said to be under positive selection, on the contrary under negative selection [33].

This divergence of protein functions often is revealed by a rate change in those amino acid residues of the protein that are most directly responsible for its new function. To investigate this change in evolution, a likelihood ratio test (LRT) is developed for detecting significant rate shifts at specific sites in proteins. A slow evolutionary rate at a given site would indicate that this position is functionally important for the protein. Conversely, a high evolutionary rate would indicate that the position is not involved in an important protein function. A significant rate difference between two subfamilies at a given site would thereby mean that the function of this position is probably different in the two groups.

Recent works take into account the phylogeny and the differing substitution rates between amino acids to detect functional divergence between paralogous genes. Gu [34, 35] has developed a quantitative measure for testing the function divergence within a family of duplicated genes. The method is based on measuring the decrease in mutation rate correlation between gene clusters of a gene family; the hidden Markov model (HMM) procedures allow the amino acid residues responsible for the functional divergence to be identified [34, 35].

9.7.1 Case Study 1: Chemokine Receptors Expansion in Vertebrates

Chemokine receptors (CR) are G-coupled protein receptors that bind small peptides called chemokines (chemotactic cytochines). Chemokines and their receptors are essential components of hematopoiesis, leukocyte trafficking, organogenesis, and immuno-modulation in mammalian species [36]. They are also involved in a variety of disease processes including inflammation, allergy and neoplasia. All chemokines have four conserved cysteines linked by disulfide bonds and two of them, located in the N-terminal region, defines four chemokine subfamilies: CXC, CC, C, and CX3C [37].

To date, a number of human receptors, specific for these chemokine subfamilies, have been described, though many receptors are still unassigned. Several viruses, for example Epstein-Barr, Cytomegalovirus, and Herpes Samiri, contain functional homologs to human CRs, an indication that such viruses may use these receptors to subvert the effects of host chemokines [23]. Moreover, primate immunodeficiency retroviruses have adopted CRs as essential gateways for entry into their target cells [38].

Understanding CRs evolutionary history would help the comprehension of the evolution of immune system. The phylogenetic relationships of chemokine

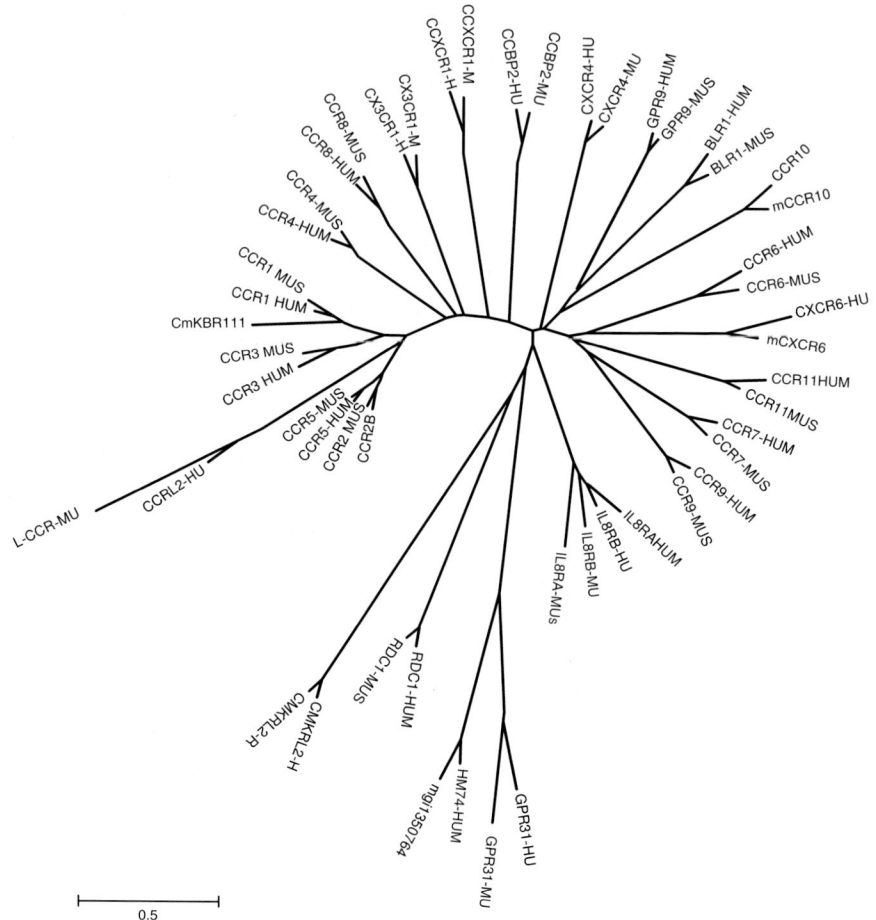

Fig. 9.1. The maximum likelihood phylogeny under the JTT+F+ Gamma model of evolution for the set of chemokine receptors amino acid sequences from human and mouse. Here, the entire sequence is considered. The scale bar refers to the branch lengths, measured in expected numbers of amino acid replacements per site

receptors using maximum likelihood [39] showed that different evolutionary models confirmed the same tree topology, with all CRs divided in four major branches. The ML tree, obtained using the JTT+F+G model of evolution, is shown in Fig. 9.1. The +F option implies that equilibrium frequencies of each state, nucleotide or amino acid, is empirically calculated from the datasets under analysis and not derived from the model.

Figure 9.2 shows the topology of chemokine receptors computed for the concatenated sequences of the external loops, which are primarily involved in the interactions with the ligands. The two topologies differ in several branches. This may imply that receptors with similar specificity do not always form

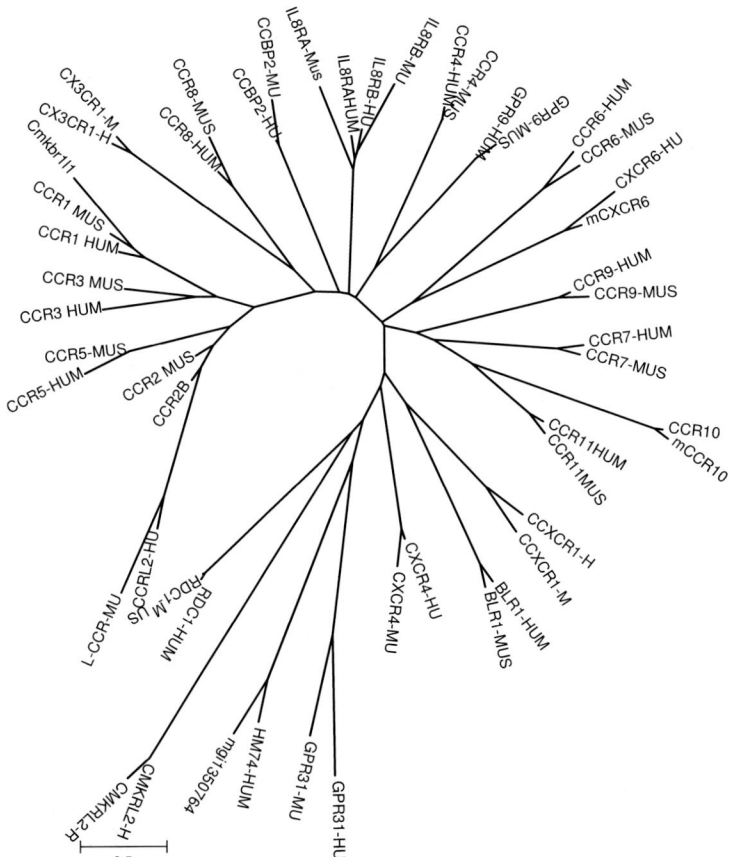

Fig. 9.2. Same as in Fig. 9.1. Here, the external loop regions are considered

monophyletic clusters (see e.g., CCR6, 1, 7, 9, and 10 with respect to the other CCR, or the CXCRs) and that the ability of a receptor to interact with a given chemokines' family could have evolved and changed different times, depending on events of gene conversion, sub and/or neofunctionalization.

9.7.2 Case Study 2: The Evolution of TIM Barrel Coding Genes

It is generally accepted that ancestral protein-encoding genes should have been relatively short sequences encoding simple polypeptides likely corresponding to functional and/or structural domains. The size and complexity of extant genes are the result of different evolutionary processes, including gene fusion, accretion of functional domains, and duplication of internal motifs (Fig. 9.3). The last mechanisms is often referred to as *gene elongation*, i.e., the increase in gene size and represents one of the most important steps in the evolution of complex genes from simple ones.

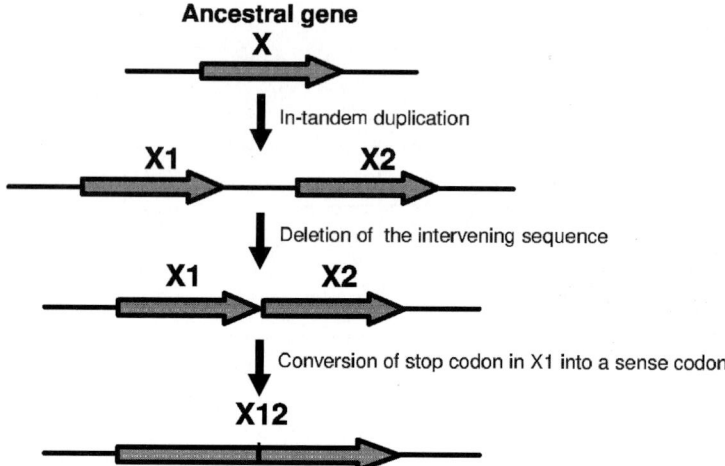

Fig. 9.3. Gene elongation: the duplication of an ancestral gene and the subsequent fusion of the two homologs to produce a longer protein

A gene elongation event might be the consequence of an in-tandem duplication of a gene; if the deletion of the intervening sequence between the two copies and a mutational change converting the stop codon of the first copy into a sense codon occurs, this might result in the elongation of the ancestral gene. Therefore, the new gene is constituted by two paralogous moieties.

The traces of an internal repetition might gradually disappear from a gene mainly because of mutations. The long-lasting evidence of an internal repetition is the presence in the protein structure of three dimensionally similar domains; Barbosa et al. [40] found an evidence for an internal gene duplication of the hisD gene which cannot be identified by analyzing the amino acid sequence. However, in other cases, the traces of the common origin of two (or more) portion within a gene (as well as of two or more genes) can be disclosed by comparing the amino acid sequence of the protein it codes for.

Particularly interesting from this point of view is the following gene family. At present, about 10% of all enzymes with known tertiary structures contains at least one domain that has the β/a-barrel fold. In spite of their structural similarity, the β/a-barrels are apparently able to support diverse functional activities. The barrel structure is composed of eight catenated strand-loop-helix-turn units. The β-strands are located in the interior of the protein, forming the staves of a barrel, whereas the a-helices pack around the exterior. The functional diversity and the poor sequence relationships among the different barrels motivated an intensive study of the evolution of their topology. Over short spans of evolution, the relationships existing between different β/a-barrel proteins can be reconstructed, but over longer periods of evolution, these relationships often become obscured. Therefore, the descent of the barrel is still under debate and arguments for both convergent and divergent evolution of the β/a-barrels have been presented. But in the last years, a large body of sequence, structural and experimental data clearly speaking toward a divergent mode of evolution of these proteins has been reported.

This accumulation of data mainly concerns the structure and the analysis of the products of hisA and hisF, which are particularly interesting from an evolutionary point of view. These two genes, whose products [N-(5′-phosphoribosyl) formimino]-5-aminoimidazole-4-carboxamide ribonucleotide isomerase and imidazole-glycerol-phosphate synthase, respectively, catalyze two central and sequential reaction in histidine biosynthesis. The comparative analysis of amino acid sequence of the proteins they code for from different archaeal, bacterial, and eukaryal organisms revealed that they are paralogous and share a similar internal organization into two paralogous modules half the size of the entire sequence.

Comparison of these modules led to the suggestion that hisA and hisF are the result of two ancient successive duplications, the first one involving an ancestral module half the size of the present-day hisA gene and leading (by a gene elongation event) to the ancestral hisA gene, which in turn underwent a duplication that gave rise to the hisF gene. Since the overall structure of the hisA and hisF genes is the same in all the (micro)organisms where they has been identified, it is quite likely that they were part of the genome of the last common ancestor and that the two successive duplication events leading to the extant hisA and hisF took place in the early stages of molecular evolution [41].

The biological significance of the subdivision of hisA and hisF into two paralogous modules half the size of the entire genes is shown in Fig. 9.4.

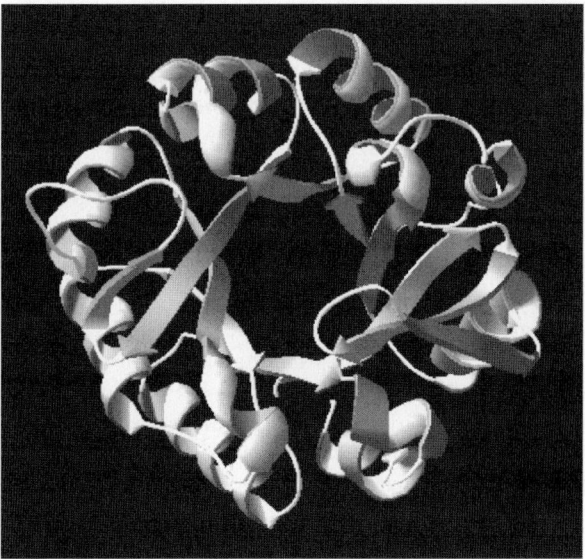

Fig. 9.4. The structure of the HisA TIM-barrel from *Thermotoga maritima* (1QO2) showing the eight internal strands and the wrapped-around helices

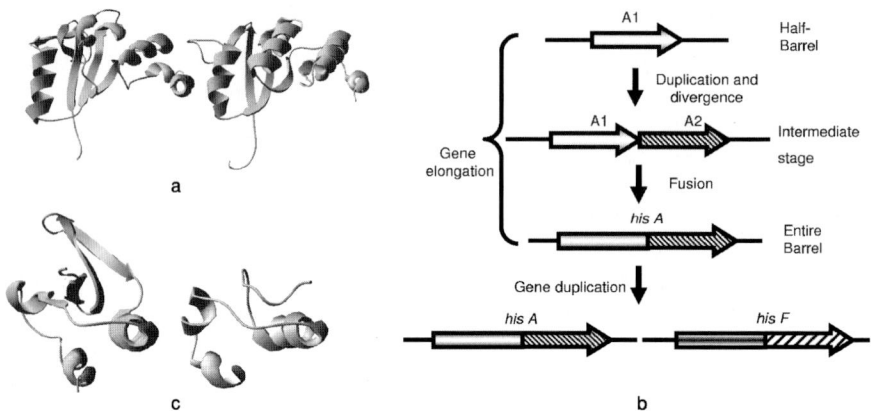

Fig. 9.5. (a) Structures of the two half-barrels corresponding to the internal repetitions of the HisA protein from *Thermotoga maritima*; (b) evolutionary model on the origin of hisA/hisF; (c) structures of the first and second quarters of the HisA protein from *Thermotoga maritima*

According to the model proposed [41,42], the ancestral gene encoded for a half-barrel protein which might undergo a homodimerization giving a functional protein. The elongation event leading to the ancestor of hisA/hisF genes, enabled the covalently fusion of two half-barrels producing a protein capable of broad enzymatic activity and whose function was refined and optimized by mutational changes occurring in time.

Over time, the whole-barrel gene might undergo a paralogous duplication event, leading to the ancestor of hisA and hisF. The model is supported not only by sequence comparative analysis [41], but also by the analysis carried out by Lang et al. [42], who studied the structure of the HisA protein by verifying that the complete barrel originated by means of a gene elongation event starting from a gene coding for a half-barrel (Fig. 9.5b). The two halves of the HisA protein were structurally superimposed (Fig. 9.5a), and they showed a maximum rms of 2.1 rÅ, while the amino acid primary sequence analysis revealed an identity of 29% for the *Methanococcus jannaschii* HisA/F proteins. In Fig. 9.5b, we describe the model introduced for the evolution of this pair of genes.

It starts with an ancestral gene coding for a half-barrel; after a gene elongation event, the gene gains four additional (β/a)-modules arriving to code for a complete barrel; this, in turn undergoes a gene duplication event leading to the present-day hisA/hisF gene pair. The analysis of the sequence and structure of HisA and HisF depicts a likely scenario for divergent evolution of (at least) some of the proteins belonging to this family.

The possibility that different β/a-barrel encoding genes might be the results of divergent evolution by one (or more) duplication of an ancestral gene received recently an elegant experimental confirmation [43] who worked with HisA and TrpF proteins. HisA and TrpF catalyze two similar reactions, an Amadori rearrangement (which is the irreversible isomerization of an aminoaldose to an aminoketose), in histidine and tryptophan biosynthesis, respectively.

Despite the lack of detectable amino acid sequence similarity, HisA and TrpF belong to the β/a-barrel family. The striking functional and structural similarities suggest that HisA and TrpF may have evolved from one common ancestral enzyme with low substrate specificity. By using random mutagenesis and selection, the authors generated several HisA variants that catalyze the TrpF reaction both in vitro and in vivo, and one of these variants retained significant HisA activity. These variants were due to single amino acid exchanges; it is possible that this "artificial" molecule probably may have the qualities of the common ancestor of HisA and TrpF.

It is also worth noting that among the TIM barrel proteins, HisA is the only one that have maintained an almost perfect subdivision in two modules half the size of the entire gene and sharing a high degree of sequence similarity. In other TIM barrels, such as HisF and TrpF, the common origin of the two halves has been obscured by point mutations and/or larger rearrangements due to functional and/or structural constraints. Therefore, it is possible that HisA might resembles the ancestral TIM barrel enzyme.

We searched for additional local internal regions of similarities and we found that they could correspond to two (β/a)-modules each (Table 9.1); being the corresponding of a quarter of barrel, we will refer to these four regions as the quarters. A structure for two of the quarters (the first and the second) is shown in Fig. 9.5c [44]. The paralogy between these two quarters is not expected by the half-barrel repetition yet known to be present in this protein (which predicts a low rms in the first vs. third quarter comparison), and it could represent the fossil of a very ancient evolutionary time, when a TIM-barrel structure was formed by a homotetramer of quarters of barrel.

Given the very ancient dating of these events, it would be not surprising if other quarters had lost their homology at the primary sequence level. Hence, the present day situation could have been reached after two gene elongation events, each one having the effect of doubling the length of the ancestral gene

Table 9.1. Percent identity (upper triangle) and similarity (lower triangle) obtained by comparing the quarters of the HisA protein from *Thermotoga maritima*

	1Q (%)	2Q (%)	3Q (%)	4Q (%)
1Q		21 (18)	26 (20)	12 (11)
2Q	36 (30)		6 (11)	9 (22)
3Q	55 (36)	29 (27)		3 (13)
4Q	44 (32)	39 (36)	25 (30)	

Given percentages are for the comparison between residues belonging to secondary structures only and considering all residues, in parentheses and the number of (β/a)-modules in the product. Then, the HisA TIM barrel would be the results of a cascade of two consecutive gene elongations.

The symmetry of the TIM-barrel structure suggests to test a further ancestral duplication step in which the minute original gene coded for a single (β/a)-module, capable of forming a homo-octamer to give the complete barrel. Although the alignment constructed from the eight single (β/a)-modules is very short, nevertheless, it still contains a certain amount of sequence and secondary structure homologies not expected from random choice. If this is so, the ancestral forms of life might have expanded their coding abilities and their genomes "simply" by duplicating a small number of mini-genes, i.e., the "starter types."

Additional studies will help in better understanding the evolutionary history of the TIM-barrel family of proteins, which are successful structures, most represented in metabolism.

References

1. T. Guillemaud, M. Raymond, A. Tsagkarakou, C. Bernard, P. Rochard, N. Pasteur, Heredity **83**, 87 (1999)
2. M. Lynch, J.S. Conery, Science **290**, 1151 (2000)
3. B.J. Trask, H. Massa, V. Brand-Arpon, K. Chan, C. Friedman, O.T. Nguyen, E. Eichler, G. van den Engh, S. Rouquier, H. Shizuya, D. Giorgi, Hum. Mol. Genet. **7**, 2007 (1998)
4. A.L. Hughes, J. da Silva, R. Friedman, Genome Res. **11**, 771 (2001)
5. T. Enver, D.R. Greaves, Curr. Opin. Biotechnol. **2**, 787 (1991)
6. D.R. Higgs, M.A. Vickers, A.O. Wilkie, I.M. Pretorius, A.P. Jarman, D.J. Weatherall, Blood **73**, 1081 (1989)
7. J. Jeffreys, V. Wilson, D. Wood, J.P. Simons, R.M. Kay, J.G. Williams, Cell **21**, 555 (1980)
8. J. Tymowska, M. Fischberg, R.C. Tinsley, Cytogenet. Cell. Genet. **19**, 344 (1977)
9. O. Jaillon, J.M. Aury, F. Brunet, J.L. Petit, N. Stange-Thomann, E. Mauceli, L. Bouneau, C. Fischer, C. Ozouf-Costaz, A. Bernot, S. Nicaud, D. Jaffe, S. Fisher, G. Lutfalla, C. Dossat, B. Segurens, C. Dasilva, M. Salanoubat, M. Levy, N. Boudet, S. Castellano, V. Anthouard, C. Jubin, V. Castelli, M. Katinka, B. Vacherie, C. Biemont, Z. Skalli, L. Cattolico, J. Poulain, V. De Berardinis, C. Cruaud, S. Duprat, P. Brottier, J.P. Coutanceau, J. Gouzy, G. Parra, G. Lardier, C. Chapple, K.J. McKernan, P. McEwan, S. Bosak, M. Kellis, J.N. Volff, R. Guigo, M.C. Zody, J. Mesirov, K. Lindblad-Toh, B. Birren, C. Nusbaum, D. Kahn, M. Robinson-Rechavi, V. Laudet, V. Schachter, F. Quetier, W. Saurin, C. Scarpelli, P. Wincker, E.S. Lander, J. Weissenbach, H. Roest Crollius, Nature **431**, 946 (2004)
10. J.S. Taylor, Y. Van de Peer, A. Meyer, Curr Biol. **11**, R1005 (2001)
11. Y. Van de Peer, J.S. Taylor, A. Meyer, J. Struct. Funct. Genomics **3**, 65 (2003)
12. F.S. Dietrich, S. Voegeli, S. Brachat, A. Lerch, K. Gates, S. Steiner, C. Mohr, R. Pohlmann, P. Luedi, S. Choi, R.A. Wing, A. Flavier, T.D. Gaffney, P. Philippsen, Science **304**, 304 (2004)

13. M. Kellis, B.W. Birren, E.S. Lander, Nature **428**, 617 (2004)
14. K.H. Wolfe, D.C. Shields, Nature **387**, 708 (1997)
15. S. Wong, G. Butler, K.H. Wolfe, Proc. Natl. Acad. Sci. USA **99**, 9272 (2002)
16. T.J. Vision, D.G. Brown, S.D. Tanksley, Science **290**, 2114 (2000)
17. F. Mazet, S.M. Shimeld, Curr. Opin. Genet. Dev. **12**, 393 (2002)
18. P. Dehal, J.L. Boore, PLoS Biol. **3**, e314 (2005)
19. S. Ohno, *Evolution by Gene Duplication* (Springer, Berlin Heidelberg New York, 1970)
20. G.P. Smith, Science **191**, 528 (1976)
21. Z. Gu, L.M. Steinmetz, X. Gu, C. Scharfe, R.W. Davis, W.H. Li, Nature **421**, 63 (2003)
22. W.H. Li, Genetics **95**, 237 (1980)
23. P.M. Murphy, Nature Immunol. **2**, 116 (2001)
24. S. Ohno, Dev. Biol. **27**, 131 (1972)
25. J.B. Walsh, Genetics **139**, 421 (1995)
26. J.H. Nadeau, D. Sankoff, Genetics **147**, 1259 (1997)
27. M. Lynch, A. Force, Genetics **154**, 459 (2000)
28. X. He, J. Zhang, Genetics **169**, 1157 (2005)
29. C. von Mering, R. Krause, B. Snel, M. Cornell, S.G. Oliver, S. Fields, P. Bork, Nature **417**, 399 (2002)
30. A.I. Su, M.P. Cooke, K.A. Ching, Y. Hakak, J.R. Walker, T. Wiltshire, A.P. Orth, R.G. Vega, L.M. Sapinoso, A. Moqrich, A. Patapoutian, G.M. Hampton, P.G. Schultz, J.B. Hogenesch, Proc. Natl. Acad. Sci. USA **99**, 4465 (2002)
31. E.T. Dermitzakis, A.G. Clark, Mol. Biol. Evol. **18**, 557 (2001)
32. H. Tang, R.C. Lewontin, Genetics **153**, 485 (1999)
33. M. Anisimova, J.P. Bielawski, Z. Yang, Mol. Biol. Evol. **18**, 1585 (2001)
34. X. Gu, Mol. Biol. Evol. **16**, 1664 (1999)
35. X. Gu, Mol. Biol. Evol. **18**, 453 (2001)
36. M. Baggiolini, Nature **392**, 565 (1998)
37. B.J. Rollins, Blood **90**, 909 (1997)
38. S.M. Owen, S. Masciotra, F. Novembre, J. Yee, W.M. Switzer, M. Ostyula, R.B. Lal, J. Virol. **74**, 5702 (2000)
39. P. Lio, M. Vannucci, Gene **317**, 29 (2003)
40. J.A. Barbosa, J. Sivaraman, Y. Li, R. Larocque, A. Matte, J.D. Schrag, M. Cygler, Proc. Natl. Acad. Sci. USA **99**, 1859 (2002)
41. R. Fani, P. Lió, I. Chiarelli, M. Bazzicalupo, J. Mol. Evol. **38**, 489 (1994)
42. D. Lang, R. Thoma, M. Henn-Sax, R. Sterner, M. Wilmanns, Science **289**, 1546 (2000)
43. C. Jurgens, A. Strom, D. Wegener, S. Hettwer, M. Wilmanns, R. Sterner, Proc. Natl. Acad. Sci. USA **99**, 9925 (2000)
44. N. Guex, M.C. Peitsch, Electrophoresis **18**, 2714 (1997)

SeqinR 1.0-2: A Contributed Package to the R Project for Statistical Computing Devoted to Biological Sequences Retrieval and Analysis

D. Charif and J.R. Lobry

The seqinR package for the R environment is a library of utilities to retrieve and analyze biological sequences. It provides an interface between: (i) the R language and environment for statistical computing and graphics, and (ii) the ACNUC sequence retrieval system for nucleotide and protein sequence databases such as GenBank, EMBL, SWISS-PROT. ACNUC is very efficient in providing direct access to subsequences of biological interest (e.g., protein coding regions, tRNA, or rRNA coding regions) present in GenBank and in EMBL. Thanks to a simple query language, it is then easy under R to select sequences of interest and then use all the power of the R environment to analyze them. The ACNUC databases can be locally installed but they are more conveniently accessed through a web server to take advantage of centralized daily updates. The aim of this chapter is to provide a handout on basic sequence analyses under seqinR with a special focus on multivariate methods.

10.1 Introduction

10.1.1 About R and CRAN

R [1, 2] is a free language and environment for statistical computing and graphics, which provides a wide variety of statistical and graphical techniques: linear and nonlinear modelling, statistical tests, time series analysis, classification, clustering, etc. Please consult the R project homepage at http://www.R-project.org/ for further information.

The Comprehensive R Archive Network, CRAN, is a network of servers around the world that store identical, up-to-date, versions of code and documentation for R. At compilation time of this document, there were 43 mirrors available from 21 countries. Please use the CRAN mirror nearest to you to minimize network load, they are listed at http://cran.r-project.org/mirrors.html.

10.1.2 About this Document

In the terminology of the R project [1,2], this document is a package *vignette*. The examples given thereafter were run under R version 2.1.0, 2005-04-18 on Wed May 4 20:29:12 2005 with Sweave [3]. The last compiled version of this document is distributed along with the seqinR package in the /doc folder. Once seqinR has been installed, the full path to the package is given by the following R code:

```
.find.package("seqinr")
```

```
[1] "/Users/lobry/Library/R/library/seqinr"
```

10.1.3 About Sequin and seqinR

Sequin is the well-known software used to submit sequences to GenBank, seqinR has definitively no connection with sequin. seqinR is just a shortcut, with no google hit, for "Sequences in R."

However, as a mnemotechnic tip, you may think about the seqinR package as the **R**eciprocal function of sequin: with sequin you can submit sequences to Genbank, with seqinR you can **R**etrieve sequences from Genbank. This is a very good summary of a major functionality of the seqinR package, to provide an efficient access to sequence databases under R.

10.1.4 About Getting Started

You need a computer connected to the Internet. First, install R on your computer. There are distributions for Linux, Mac, and Windows users on the CRAN (http://cran.r-project.org). Then, install the ape, ade4, and seqinr packages. This can be done directly in an R console with for instance the command install.packages("seqinr"). Last, load the seqinR package with:

```
library(seqinr)
```

The command lseqinr() lists all what is defined in the package seqinR:

```
lseqinr()[1:9]
```

```
[1] "AAstat"        "EXP"        "GC"
[4] "GC2"           "GC3"        "SEQINR.UTIL"
[7] "a"             "aaa"        "as.SeqAcnucWeb"
```

We have printed here only the first nine entries because they are too numerous. To get help on a specific function, say aaa(), just prefix its name with a question mark, as in ?aaa and press enter.

10.1.5 About Running R in Batch Mode

Although R is usually run in an interactive mode, some data preprocessing and analyses could be too long. You can run your R code in batch mode in a shell with a command that typically looks like:

```
unix$ R CMD BATCH input.R results.out &
```

where `input.R` is a text file with the R code you want to run and `results.out` a text file to store the outputs. Note that in batch mode, the graphical user interface is not active so that some graphical devices (e.g., `x11`, `jpeg`, `png`) are not available (see the R FAQ [4] for further details).

It is worth noting that R uses the XDR representation of binary objects in binary saved files, and these are portable across all R platforms. The `save()` and `load()` functions are very efficient (because of their binary nature) for saving and restoring any kind of R objects, in a platform independent way. To give a striking real example, at a given time on a given platform, it was about 4 minutes long to import a numeric table with 70,000 lines and 64 columns with the defaults settings of the `read.table()` function. Turning it into binary format, it was then about 8 *seconds* to restore it with the `load()` function. It is, therefore, advisable in the `input.R` batch file to save important data or results (with something like `save(mybigdata, file = "mybigdata.RData")`) so as to be able to restore them later efficiently in the interactive mode (with something like `load("mybigdata.RData")`).

10.1.6 About the Learning Curve

If you are used to work with a purely graphical user interface, you may feel frustrated in the beginning of the learning process because apparently simple things are not so easily obtained. In the long term, however, you are a winner for the following reasons.

Wheel (the)

Do not re-invent (there is a patent [5] on it anyway). At the compilation time of this document, there were 508 contributed packages available. Even if you do not want to be spoon-feed *à bouche ouverte*, it is not a bad idea to look around there just to check what is going on in your own application field. Specialists all around the world are there.

Hotline

There is a very reactive discussion list to help you, just make sure to read the posting guide at: http://www.R-project.org/posting-guide.html, before posting. Because of the high traffic on this list, we strongly suggest to answer *yes* at the question *Would you like to receive list mail batched in a daily digest?* when subscribing at https://stat.ethz.ch/mailman/listinfo/r-help. Some *bons mots* from the list are archived in the R `fortunes` package.

Automation

Consider the 178 pages of figures in the additional data file 1 http://genomebiology.com/2002/3/10/research/0058/suppl/S1) from [6]. They were produced in part automatically (with a proprietary software that is no more maintained) and manually, involving a lot of tedious and repetitive manipulations (such as italicizing species names by hand in subtitles). In few words, a waste of time. The advantage of the R environment is that once you are happy with the outputs (including graphical outputs) of an analysis for species x, it is very easy to run the same analysis on n species.

Reproducibility

If you do not consider the reproducibility of scientific results to be a serious problem in practice, then the paper by Jonathan Buckheit and David Donoho [7] is a must read. Molecular data are available in public databases, this is a necessary but not sufficient condition to allow for the reproducibility of results. Publishing the R source code that was used in your analyses is a simple way to greatly facilitate the reproduction of your results at the expense of no extra cost. At the expense of a little extra cost, you may consider to set up a RWeb server so that even the laziest reviewer may reproduce your results just by clicking on the "do it again" button in his web browser (i.e., without installing any software on his computer). For an example, involving the `seqinR` package, follow this link `http://pbil.univ-lyon1.fr/members/lobry/repro/bioinfo04/` to reproduce on-line the results from [8].

Fine Tuning

You have full control on everything, even the source code for all functions is available. The following graph was specifically designed to illustrate the first experimental evidence [9] that, on average, we have also [A]=[T] and [C]=[G] in single-stranded DNA. These data from Chargaff's laboratory (see Fig. 10.1) give the base composition of the L (Ligth) strand for seven bacterial chromosomes.

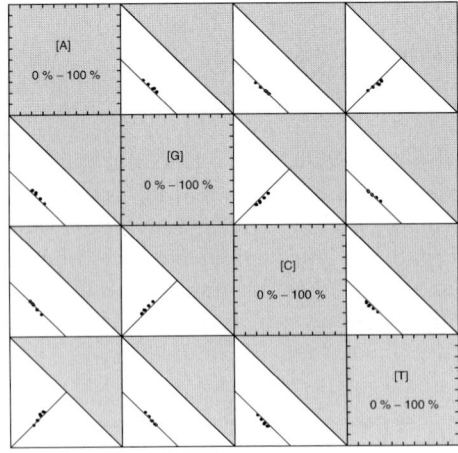

Fig. 10.1. Chargaff's data

Fig. 10.2. Genetic code number one

The graph in Fig. 10.1 is a very specialized one. The filled areas correspond to nonallowed values because the sum of the four bases frequencies cannot exceed 100%. The white areas correspond to possible values (more exactly to the projection from $\mathbb{R}^4$ to the corresponding $\mathbb{R}^2$ planes of the region of allowed values). The lines correspond to the very small subset of allowed values for which we have in addition [A]=[T] and [C]=[G]. Points represent observed values in the seven bacterial chromosomes. The whole graph is entirely defined by the code given in the example of the `chargaff` dataset (`?chargaff` to see it).

Another example of highly specialized graph is given by the function `tablecode()` to display a genetic code (Fig. 10.2) as in textbooks:

```
tablecode(dia = F)
```

It is very convenient in practice to have a genetic code at hand, and moreover here, all genetic code variants are available (Fig. 10.3):

```
tablecode(numcode = 2, dia = F)
```

Data as Fast Moving Targets

In research area, data are not always stable. Compare the following graph (Fig. 10.4):

```
dbg <- get.db.growth()
plot(x = dbg$date, y = log10(dbg$Nucl), las = 1,
     main = "The growth of DNA databases", xlab = "Year",
     ylab = "Log10 number of nucleotides")
```

with Fig. 1 in [10].

Data have been updated since then but the same R code was used to produce the figure, ensuring an automatic update. For LaTeX users, it is

Genetic code 2 : vertebrate.mitochondrial							
u u u	Phe	u c u	Ser	u a u	Tyr	u g u	Cys
u u c	Phe	u c c	Ser	u a c	Tyr	u g c	Cys
u u a	Leu	u c a	Ser	u a a	Stp	u g a	Trp
u u g	Leu	u c g	Ser	u a g	Stp	u g g	Trp
c u u	Leu	c c u	Pro	c a u	His	c g u	Arg
c u c	Leu	c c c	Pro	c a c	His	c g c	Arg
c u a	Leu	c c a	Pro	c a a	Gln	c g a	Arg
c u g	Leu	c c g	Pro	c a g	Gln	c g g	Arg
a u u	Ile	a c u	Thr	a a u	Asn	a g u	Ser
a u c	Ile	a c c	Thr	a a c	Asn	a g c	Ser
a u a	Met	a c a	Thr	a a a	Lys	a g a	Stp
a u g	Met	a c g	Thr	a a g	Lys	a g g	Stp
g u u	Val	g c u	Ala	g a u	Asp	g g u	Gly
g u c	Val	g c c	Ala	g a c	Asp	g g c	Gly
g u a	Val	g c a	Ala	g a a	Glu	g g a	Gly
g u g	Val	g c g	Ala	g a g	Glu	g g g	Gly

Fig. 10.3. Genetic code number two

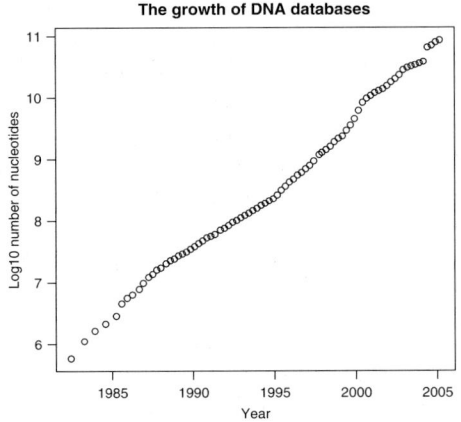

Fig. 10.4. The exponential growth of DNA databases

Table 10.1. A very simple example of amino acid counts in three proteins

	Ala	Val	Cys
1	130	70	0
2	60	40	0
3	60	35	5

worth mentioning the fantastic tool contributed by Friedrich Leish [3] called `Sweave()` that allows for the automatic insertion of R outputs (including graphics) in a LaTeX document. In the same spirit, there is a package called `xtable` to coerce R data into LaTeX tables, for instance Table 10.1 below was produced this way, enforcing a complete coherence between the R code example and the table.

10.2 How to Get Sequence Data

10.2.1 Importing Raw Sequence Data from Fasta Files

The fasta format is very simple and widely used for simple import of biological sequences. It begins with a single-line description starting with a character >, followed by lines of sequence data of maximum 80 character each. Examples of files in fasta format are distributed with the **seqinR** package in the sequences directory:

```
list.files(path = system.file("sequences", package = "seqinr"),
    pattern = ".fasta")
```
```
[1] "bb.fasta"    "ct.fasta"    "malM.fasta"    "seqAA.fasta"
```

The function `read.fasta()` imports sequences from fasta files into your workspace, for example:

```
seqaa <- read.fasta(File = system.file("sequences/seqAA.fasta",
    package = "seqinr"), seqtype = "AA")
seqaa
```
```
A06852
  [1]  "M" "P" "R" "L" "F" "S" "Y" "L" "L" "G" "V" "W" "L"
 [14]  "L" "L" "S" "Q" "L" "P" "R" "E" "I" "P" "G" "Q" "S"
 [27]  "T" "N" "D" "F" "I" "K" "A" "C" "G" "R" "E" "L" "V"
 [40]  "R" "L" "W" "V" "E" "I" "C" "G" "S" "V" "S" "W" "G"
 [53]  "R" "T" "A" "L" "S" "L" "E" "E" "P" "Q" "L" "E" "T"
 [66]  "G" "P" "P" "A" "E" "T" "M" "P" "S" "S" "I" "T" "K"
 [79]  "D" "A" "E" "I" "L" "K" "M" "M" "L" "E" "F" "V" "P"
 [92]  "N" "L" "P" "Q" "E" "L" "K" "A" "T" "L" "S" "E" "R"
[105]  "Q" "P" "S" "L" "R" "E" "L" "Q" "Q" "S" "A" "S" "K"
[118]  "D" "S" "N" "L" "N" "F" "E" "E" "F" "K" "K" "I" "I"
[131]  "L" "N" "R" "Q" "N" "E" "A" "E" "D" "K" "S" "L" "L"
[144]  "E" "L" "K" "N" "L" "G" "L" "D" "K" "H" "S" "R" "K"
[157]  "K" "R" "L" "F" "R" "M" "T" "L" "S" "E" "K" "C" "C"
[170]  "Q" "V" "G" "C" "I" "R" "K" "D" "I" "A" "R" "L" "C"
[183]  "*"
attr(,"name")
[1] "A06852"
attr(,"Annot")
[1] ">A06852                    183 residues"
attr(,"class")
[1] "SeqFastaAA"
```

A more consequent example is given in the fasta file `ct.fasta`, which contains the complete genome of *Chlamydia trachomatis* that was used in [11].

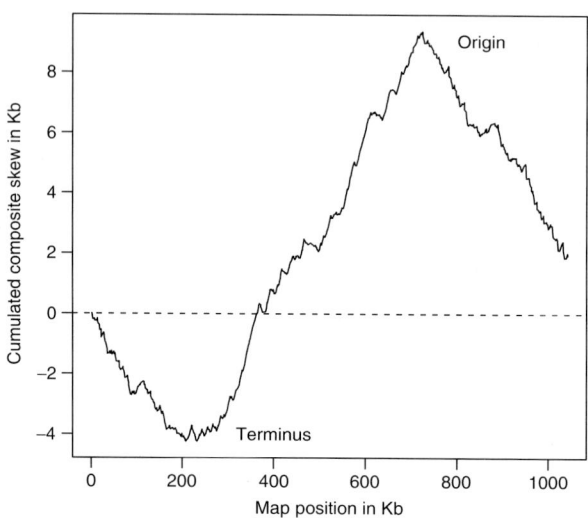

Fig. 10.5. Prediction of origin and terminus of replication with oriloc

You should be able to reproduce Fig. 1b from [11] with the following code (see Fig. 10.5):

```
out <- oriloc(seq.fasta = system.file("sequences/ct.fasta",
  package = "seqinr"), g2.coord = system.file("sequences/ct.coord",
  package = "seqinr"), oldoriloc = TRUE)
plot(out$st, out$sk/1000, type = "l", xlab = "Map position in Kb",
  ylab = "Cumulated composite skew in Kb", main = "Chlamydia trachomatis complete genome",
  las = 1)
abline(h = 0, lty = 2)
text(400, -4, "Terminus")
text(850, 9, "Origin")
```

Note that the algorithm has been improved since then and that it is more advisable to use the default option `oldoriloc = FALSE` if you are interested in the prediction of origins and terminus of replication from base composition biases (more on this at http://pbil.univ-lyon1.fr/software/oriloc.html). See also [12] for a recent review on this topic.

10.2.2 Importing Aligned Sequence Data

Aligned sequence data are very important in evolutionary studies, in this representation all vertically aligned positions are supposed to be homologous that is sharing a common ancestor. This is a mandatory starting point for comparative studies. There is a function in seqinR called `read.alignment()` to read aligned sequences data from various formats (mase, clustal, phylip, fasta, or msf) produced by common external programs for multiple sequence alignment.

Let us give an example. The gene coding for the mitochondrial cytochrome oxidase I is essential and therefore often used in phylogenetic studies because of its ubiquitous nature. Download on your local computer the following two sample tests of aligned sequences of this gene (extracted from ParaFit [13]), this can be done directly at R prompt in the R console with:

```
download.file(url = "http://pbil.univ-lyon1.fr/software/SeqinR/Datasets/louse.fasta",
    destfile = "louse.fasta")
download.file(url = "http://pbil.univ-lyon1.fr/software/SeqinR/Datasets/gopher.fasta",
    destfile = "gopher.fasta")
```

The eight genes of the first sample are from various species of louse (insects parasitics on warm-blooded animals) and the eight genes of the second sample are from their corresponding gopher hosts (a subset of rodents):

```
l.names <- readLines("http://pbil.univ-lyon1.fr/software/SeqinR/Datasets/louse.names")
l.names
```
```
[1] "G.chapini "  "G.cherriei"  "G.costaric"  "G.ewingi"
[5] "G.geomydis"  "G.oklahome"  "G.panamens"  "G.setzeri"
```

```
g.names <- readLines("http://pbil.univ-lyon1.fr/software/SeqinR/Datasets/gopher.names")
g.names
```
```
[1] "G.brevicep"  "O.cavator"   "O.cherriei"  "O.underwoo"
[5] "O.hispidus"  "G.burs1"     "G.burs2"     "O.heterodu"
```

```
louse <- read.alignment("louse.fasta", format = "fasta")
gopher <- read.alignment("gopher.fasta", format = "fasta")
```

The aligned sequences are now imported in your R environment. **SeqinR** has very few methods devoted to phylogenetic analyses but many are available in the **ape** package. This allows for a very fine tuning of the graphical outputs of the analyses thanks to the power of the R facilities. For instance, a natural question here would be to compare the topology of the tree of the hosts and their parasites to see if we have congruence between host and parasite evolution. In other words, we want to display two phylogenetic trees face to face. This would be tedious with a program devoted to the display of a single phylogenetic tree at time, involving a lot of manual copy/paste operations, hard to reproduce, and then boring to maintain with data updates.

How does it looks under R? First, we need to *infer* the tree topologies from data. Let us try as an *illustration* the famous neighbor-joining tree estimation of Saitou and Nei [14] with Jukes and Cantor's correction [15] for multiple substitutions.

```
library(ape)
louse.JC <- dist.dna(x = lapply(louse$seq, s2c),
    model = "JC69")
gopher.JC <- dist.dna(x = lapply(gopher$seq, s2c),
    model = "JC69")
l <- nj(louse.JC)
g <- nj(gopher.JC)
```

Now, we have an estimation for *illustrative* purposes of the tree topology for the parasite and their hosts. We want to plot the two trees face to face, and for this we must change R graphical parameters. The first thing to do is to save the current graphical parameter settings so as to be able to restore them later:

```
op <- par(no.readonly = TRUE)
```

The meaning of the `no.readonly = TRUE` option here is that graphical parameters are not all settable, we just want to save those we can change at will. Now, we can play with graphics (see Fig. 10.6):

```
g$tip.label <- paste(1:8, g.names)
l$tip.label <- paste(1:8, l.names)
layout(matrix(data = 1:2, nrow = 1, ncol = 2),
    width = c(1.4, 1))
par(mar = c(2, 1, 2, 1))
plot(g, adj = 0.8, cex = 1.4, use.edge.length = FALSE,
    main = "gopher (host)", cex.main = 2)
plot(l, direction = "l", use.edge.length = FALSE,
    cex = 1.4, main = "louse (parasite)", cex.main = 2)
```

We now restore the old graphical settings that were previously saved:

```
par(op)
```

OK, this may look a little bit obscure if you are not fluent in programming, but please try the following experiment. In your current working directory, that is in the directory given by the `getwd()` command, create a text file called `essai.r` with your favorite text editor, and copy/paste the previous R commands, that is

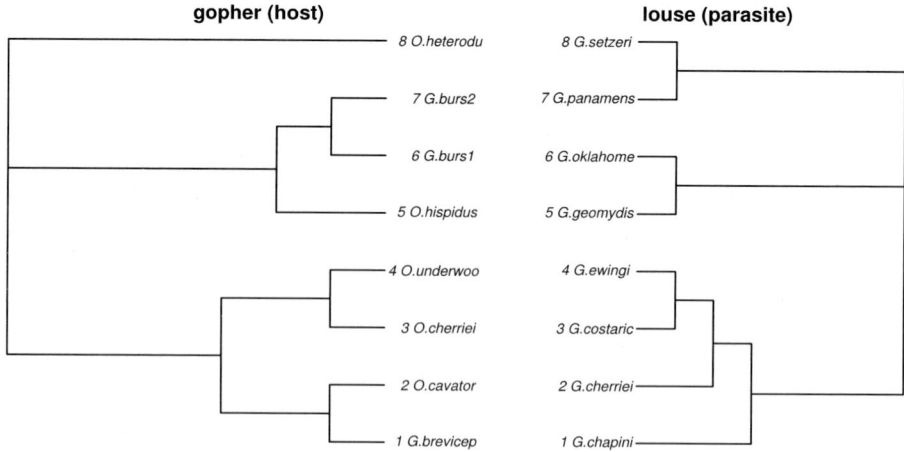

Fig. 10.6. Comparison of the phylogenies of hosts and parasites with the Jukes and Cantor model of DNA evolution

```
l.names<-readLines("http://pbil.univ-lyon1.fr/software/SeqinR/Datasets/louse.names")
g.names<-readLines("http://pbil.univ-lyon1.fr/software/SeqinR/Datasets/gopher.names")
louse <- read.alignment("louse.fasta",format="fasta")
gopher <- read.alignment("gopher.fasta",format="fasta")
louse.JC <- dist.dna(x = lapply(louse$seq, s2c), model = "JC69" )
gopher.JC <- dist.dna(x = lapply(gopher$seq, s2c), model = "JC69" )
l <- nj(louse.JC)
g <- nj(gopher.JC)
g$tip.label <- paste(1:8, g.names)
l$tip.label <- paste(1:8, l.names)
layout(matrix(data = 1:2, nrow = 1, ncol = 2), width=c(1.4, 1))
par(mar=c(2,1,2,1))
plot(g, adj = 0.8, cex = 1.4, use.edge.length=FALSE,
  main = "gopher (host)", cex.main = 2)
plot(l,direction="l", use.edge.length=FALSE, cex = 1.4,
  main = "louse (parasite)", cex.main = 2)
```

Make sure that your text has been saved and then go back to R console to enter the command:

```
source("essai.r")
```

This should reproduce the previous face-to-face phylogenetic trees in your R graphical device (we have assumed here that the files louse.fasta and gopher.fasta are present in your local working directory). Now, your boss is unhappy with working with the Jukes and Cantor's model [15] and wants you to use the Kimura's 2-parameters distance [16] instead. Go back to the text editor to change model = "JC69" by model = "K80", save the file, and in the R console source("essai.r") again, you should obtain the following graph (Fig. 10.7):

Nice congruence, isn't it? Now, something even worst, there was an error in the aligned sequence set: the first base in the first sequence in the file

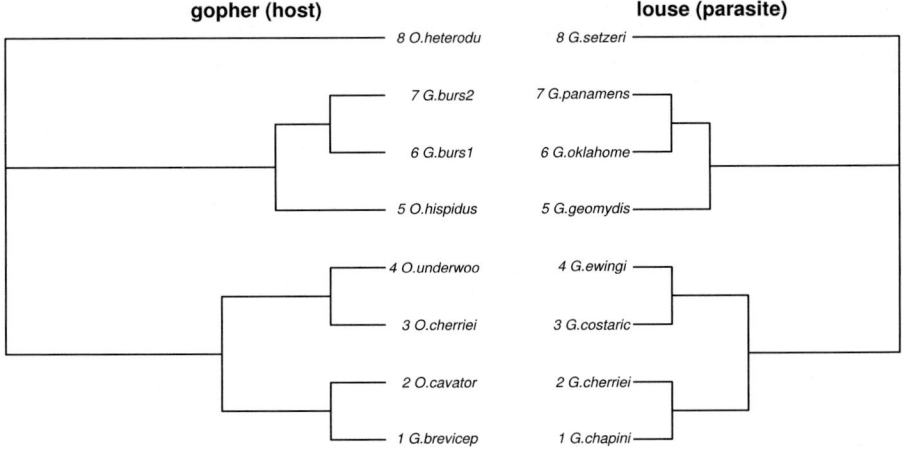

Fig. 10.7. Comparison of the phylogenies of hosts and parasites with Kimura's 2-parameters model of DNA evolution

louse.fasta is not a C but a T. Open the louse.fasta in your text editor, fix the error, go back to the R console to source("essai.r") again. That is all, your graph is now consistent with the updated dataset.

10.2.3 Complex Queries in ACNUC Databases

As a rule of thumb, after compression one nucleotide needs one octet of disk space storage (because you need also the annotations corresponding to the sequences), so that most likely you will not have enough space on your computer to work with a local copy of a complete DNA database. The idea is to import under R only the subset of sequences you are interested in. This is done in three steps:

1. Select the database you want to work with the choosebank() function. This function initiates remote access to an acnuc database. Called without arguments, choosebank() gives the list of available databases:

   ```
   choosebank()
    [1] "genbank"    "embl"        "emblwgs"    "swissprot"
    [5] "ensembl"    "emglib"      "nrsub"      "nbrf"
    [9] "hobacnucl"  "hobacprot"   "hovernucl"  "hoverprot"
   [13] "hogennucl"  "hogenprot"   "hoverclnu"  "hoverclpr"
   [17] "HAMAPnucl"  "HAMAPprot"   "hoppsigen"  "nurebnucl"
   [21] "nurebprot"  "taxobacgen"  "greview"
   ```

   If you want to work with GenBank, for instance, you call choosebank() with "genbank" as an argument and store the result in a variable in the workspace:

   ```
   mybank <- choosebank("genbank")
   str(mybank)
   List of 5
    \$socket  :Classes 'sockconn', 'connection'  int 6
    \$bankname: chr "genbank"
    \$totseqs : chr "46590656"
    \$totspecs: chr "293711"
    \$totkeys : chr "1307328"
   ```

   The list returned by choosebank() here means that in the database called genbank at the compilation time of this document there were 46,590,656 sequences from 293,711 species and a total of 1,307,328 keywords. For the following, the most important item is the first one of the list, mybank$socket, which contains all the required details of the socket connection.

2. Then, you have to say what you want, that is to compose a query to select the subset of sequences you are interested in. The way to do this is documented under ?query, we just give here a simple example. In the query below, we want to select all the coding sequences (t=cds) from cat (sp=felis catus) that are not (et no) partial sequences (k=partial). We want the result to be stored in an object called list1.

```
query(socket = mybank\$socket, listname = "list1",
      query = "sp=felis catus et t=cds et no k=partial",
      invisible = TRUE)
```

Now, there is in the workspace an object called `list1`, which does not contain the sequences themselves but the *sequence names* that fit the query. They are stored in the `req` component of the object, let us see the first ten of them:

```
sapply(list1\$req[1:10], getName)
 [1] "AB000483.PE1"    "AB000484.PE1"    "AB000485.PE1"
 [4] "AB004237"        "AB004238"        "AB009279.PE1"
 [7] "AB009280.PE1"    "AB010872.UGT1A1" "AB011965.SDF-1A"
[10] "AB011966.SDF-1B"
```

The first sequence that fit our request is `AB000483.PE1`, the second one is `AB000484.PE1`, and so on. Note that the sequence name may have an extension, this corresponds to *subsequences*, a specificity of the ACNUC system that allows to handle easily a subsequence with a biological meaning, typically a gene.

Note that the component `call` of `list1` keeps a trace of the way we have selected the sequences. At this stage, you can quit your R session saving the workspace image. The next time an R session is opened with the workspace image restored, there will be an object called `list1`, and looking into its `call` component will tell you that it contains the names of complete coding sequences from *Felis catus*.

In practice, queries for sequences are rarely done in one step and are more likely to be the result of an iterative, progressively refining, process. An important point is that a list of sequences can be reused. For instance, we can reuse `list1` get only the list of sequences that were published in 2004:

```
query(socket = mybank\$socket, listname = "list2",
      query = "list1 et y=2004", invisible = TRUE)
length(list2\$req)
[1] 41
```

Hence, there were 41 complete coding sequences in 2004 for *Felis catus* in GenBank.

3. The sequence itself is obtained with the function `getSequence()`. For example, the first 50 nucleotides of the first sequence of our request are

```
myseq <- getSequence(list1\$req[[1]])
myseq[1:50]
 [1] "a" "t" "g" "a" "a" "t" "c" "a" "a" "g" "g" "a" "g" "c"
[15] "c" "g" "t" "t" "t" "t" "t" "a" "g" "g" "c" "a" "c" "c"
[29] "t" "g" "c" "t" "c" "c" "t" "g" "g" "t" "g" "c" "t" "g"
[43] "c" "a" "g" "c" "t" "g" "g" "t"
```

They can also be coerced as string of character with the function `c2s()`:

```
c2s(myseq[1:50])
[1] "atgaatcaaggagccgttttaggcacctgctcctggtgctgcagctggt"
```

Note that what is done by getSequence() is much more complex than a substring extraction because subsequences of biological interest are not necessarily contiguous or even on the same DNA strand. Consider for instance the following coding sequence from sequence AE003734:

```
AE003734.PE35      Location/Qualifiers      (length=1833 bp)
       CDS         join(complement(162997..163210),
                   complement(162780..162919),complement(161238..162090),
                   146568..146732,146806..147266)
                   /gene="mod(mdg4)"
                   /locus_tag="CG32491"
                   /note="CG32491 gene product from transcript CG32491-RT;
                   trans-splicing"
```

To get the coding sequence manually you would have join five different pieces from AE003734 and some of them are in the complementary strand. With getSequence() you do not have to think about this:

```
query(socket = mybank\$socket, listname = "list3",
    query = "N=AE003734.PE35", invisible = TRUE)
transspliced <- getSequence(list3\$req[[1]])
tsaa <- getTrans(transspliced)
tsaa[1:50]
 [1] "M" "A" "D" "D" "E" "Q" "F" "S" "L" "C" "W" "N" "N" "F"
[15] "N" "T" "N" "L" "S" "A" "G" "F" "H" "E" "S" "L" "C" "R"
[29] "G" "D" "L" "V" "D" "V" "S" "L" "A" "A" "E" "G" "Q" "I"
[43] "V" "K" "A" "H" "R" "L" "V" "L"
```

10.3 How to Deal with Sequence

10.3.1 Sequence Classes

There are at present three classes of sequences, depending on the way they were obtained:

- **seqFasta** is the class for the sequences that were imported from a fasta file
- **seqAcnucWeb** is the class for the sequences coming from an ACNUC database server
- **seqFrag** is the class for the sequences that are fragments of other sequences

10.3.2 Generic Methods for Sequences

All sequence classes are sharing a common interface, so that there are very few method names we have to remember. In addition, all classes have their specific as.ClassName method that return an instance of the class, and is.ClassName method to check whether an object belongs or not to the class. Available methods are listed in Table 10.2.

Table 10.2. Available methods for sequences

Methods	Result	Type of result
getFrag	A sequence fragment	A sequence fragment
getSequence	The sequence	Vector of characters
getName	The name of a sequence	String
getLength	The length of a sequence	Numeric vector
getTrans	Translation into amino acids	Vector of characters
getAnnot	Sequence annotations	Vector of string
getLocation	Position of a sequence on its parent sequence	List of numeric vector

10.3.3 Internal Representation of Sequences

The current mode of sequence storage is done with vectors of characters instead of strings. This is very convenient for the user because all R tools to manipulate vectors are immediately available. The price to pay is that this storage mode is extremely expensive in terms of memory. They are two utilities called s2c() and c2s() that allows to convert strings into vector of characters, and vice versa, respectively.

Sequences as Vectors of Characters

In the vectorial representation mode, all the very convenient R tools for indexing vectors are at hand.

1. Vectors can be indexed by a vector of *positive* integers, saying which elements are to be selected. As we have already seen, the first 50 elements of a sequence are easily extracted thanks to the binary operator from:to, as in

    1:50
    ```
    [1]   1  2  3  4  5  6  7  8  9 10 11 12 13 14 15 16 17 18
    [19] 19 20 21 22 23 24 25 26 27 28 29 30 31 32 33 34 35 36
    [37] 37 38 39 40 41 42 43 44 45 46 47 48 49 50
    ```
    myseq[1:50]
    ```
    [1]  "a" "t" "g" "a" "a" "t" "c" "a" "a" "g" "g" "a" "g" "c"
    [15] "c" "g" "t" "t" "t" "t" "t" "a" "g" "g" "c" "a" "c" "c"
    [29] "t" "g" "c" "t" "c" "c" "t" "g" "g" "t" "g" "c" "t" "g"
    [43] "c" "a" "g" "c" "t" "g" "g" "t"
    ```
    The seq() function allows to build more complex integer vectors. For instance in coding sequences, it is very common to focus on third codon positions where selection is weak. Let us extract bases from third codon positions:
    ```
    tcp <- seq(from = 3, to = length(myseq), by = 3)
    tcp[1:10]
    [1]  3  6  9 12 15 18 21 24 27 30
    ```

```
myseqtcp <- myseq[tcp]
myseqtcp[1:10]
[1] "g" "t" "a" "a" "c" "t" "t" "g" "c" "g"
```

2. Vectors can also be indexed by a vector of *negative* integers saying which elements have to be removed. For instance, if we want to keep first and second codon positions, the easiest way is to remove third codon positions:

```
-tcp[1:10]
[1]  -3  -6  -9 -12 -15 -18 -21 -24 -27 -30
myseqfscp <- myseq[-tcp]
myseqfscp[1:10]
[1] "a" "t" "a" "a" "c" "a" "g" "g" "g" "c"
```

3. Vectors are also indexable by a vector of *logicals* whose TRUE values say which elements to keep. Here is a different way to extract all third coding positions from our sequence. First, we define a vector of three logicals with only the last one true:

```
ind <- c(F, F, T)
ind
[1] FALSE FALSE  TRUE
```

This vector seems too short for our purpose because our sequence is much more longer with its 1,425 bases. But under R vectors are automatically *recycled* when they are not long enough:

```
(1:30)[ind]
[1]  3  6  9 12 15 18 21 24 27 30
myseqtcp2 <- myseq[ind]
```

The result should be the same as previously:

```
identical(myseqtcp, myseqtcp2)
[1] TRUE
```

This recycling rule is extremely convenient in practice but may have surprising effects if you assume (incorrectly) that there is a stringent dimension control for R vectors as in linear algebra.

Another advantage of working with vector of characters is that most R functions are vectorized so that many things can be done without explicit looping. Let us give some very simple examples:

```
tota <- sum(myseq == "a")
```

The total number of a in our sequence is 350. Let us compare graphically the different base counts (Figs. 10.8–10.10) in our sequence:

```
basecount <- table(myseq)
myseqname <- getName(list1\$req[[1]])
dotchart(basecount, xlim = c(0, max(basecount)),
    pch = 19, main = paste("Base count in", myseqname))
```

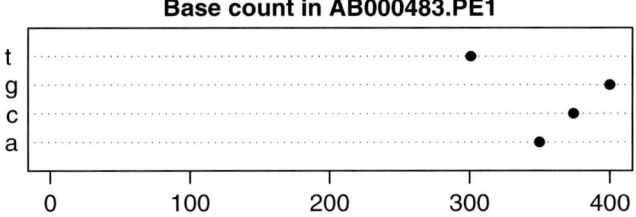

Fig. 10.8. Example of a dotchart representation of base counts in a DNA sequence

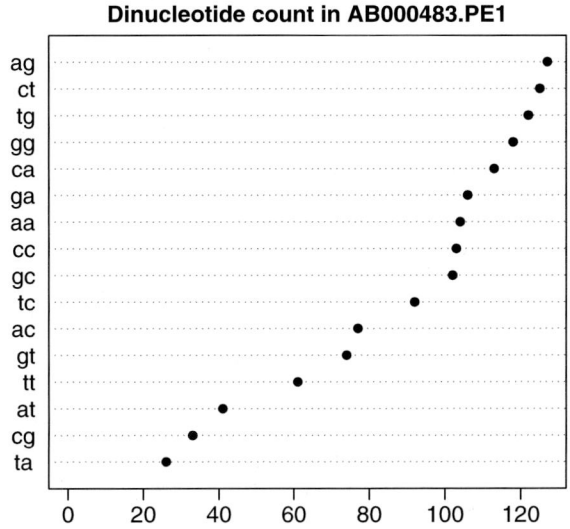

Fig. 10.9. Example of a dotchart representation of dinucleotide counts in a DNA sequence

```
dinuclcount <- count(myseq, 2)
dotchart(dinuclcount[order(dinuclcount)], xlim = c(0,
    max(dinuclcount)), pch = 19, main = paste("Dinucleotide
    count in", myseqname))

codonusage <- uco(myseq)
dotchart.uco(codonusage, main = paste("Codon usage in",
    myseqname))
```

Sequences as Strings

If you are interested in (fuzzy) pattern matching, then it is advisable to work with sequence as strings to take advantage of *regular expression* implemented

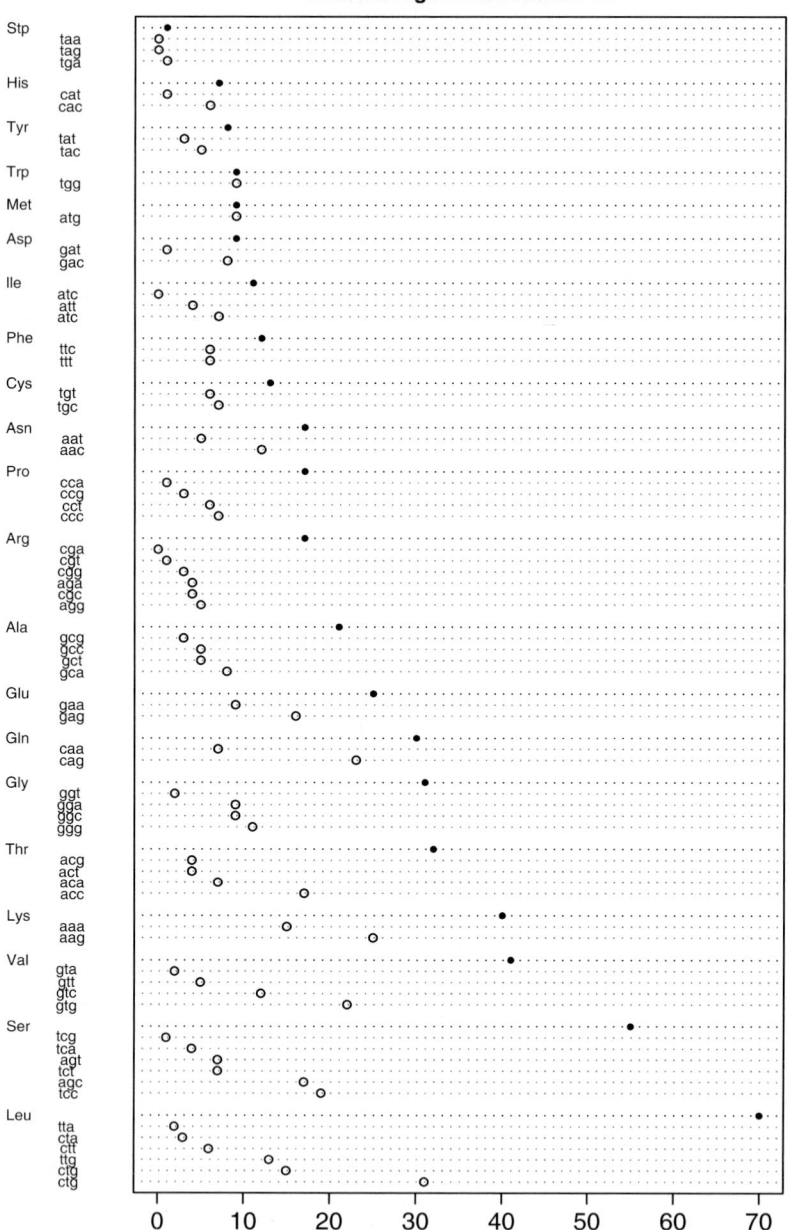

Fig. 10.10. Example of a dotchart representation of codon and amino acid counts in a coding sequence

in R. The function `words.pos()` returns the positions of all occurrences of a given regular expression. Let us suppose we want to know where are the trinucleotides "cgt" in a sequence, that is, the fragment CpGpT in the direct strand:

```
mystring <- c2s(myseq)
words.pos("cgt", mystring)

[1]   15  854  909  919  987 1248
```

We can also look for the fragment CpGpTpY to illustrate fuzzy matching because Y (IUPAC code for pyrimidine) stands C or T:

```
words.pos("cgt[ct]", mystring)

[1]  15 909 919
```

To look for all CpC dinucleotides separated by three or four bases:

```
words.pos("cc.{3,4}cc", mystring)

 [1]   27  121  152  278  431  437  471  476  477  492  555
[12]  618  722  788  809  885  886  939 1043 1046 1190 1220
[23] 1263
```

Virtually, any pattern is easily encoded with a regular expression. This is especially useful at the protein level because many functions can be attributed to short linear motifs.

10.4 Multivariate Analyses

10.4.1 Correspondence Analysis

This is the most popular multivariate data analysis technique for amino acid and codon count tables, its application, however, is not without pitfalls [17]. Its primary goal is to transform a table of counts into a graphical display, in which each gene (or protein) and each codon (or amino acid) is depicted as a point. Correspondence analysis (CA) may be defined as a special case of principal components analysis (PCA) with a different underlying metrics. The interest of the metrics in CA, that is the way we measure the distance between two individuals, is illustrated below with a very simple example (Table 10.1 inspired from [18]) with only three proteins having only three amino acids, so that we can represent exactly on a map the consequences of the metric choice.

```
df <- data.frame(matrix(c(130, 60, 60, 70, 40,
    35, 0, 0, 5), nrow = 3))
names(df) <- c("Ala", "Val", "Cys")
df
```

```
   Ala Val Cys
1  130  70   0
2   60  40   0
3   60  35   5
```

Let us first use the regular Euclidean metrics between two proteins i and i',

$$d^2(i, i') = \sum_{j=1}^{J} (n_{ij} - n_{i'j})^2 \qquad (10.1)$$

to visualize this small data set (Fig. 10.11):

```
library(ade4)
pco <- dudi.pco(dist(df), scann = F, nf = 2)
myplot <- function(res, ...) {
    plot(res\$li[, 1], res\$li[, 2], ...)
    text(x = res\$li[, 1], y = res\$li[, 2], labels = 1:3,
        pos = ifelse(res\$li[, 2] < 0, 1, 3))
    perm <- c(3, 1, 2)
    lines(c(res\$li[, 1], res\$li[perm, 1]), c(res\$li[,
        2], res\$li[perm, 2]))
}
myplot(pco, main = "Euclidean distance", asp = 1,
    pch = 19, xlab = "", ylab = "", las = 1)
```

From this point of view, the first individual is far away from the two others. But thinking about it, this is a rather trivial effect of protein size:

```
rowSums(df)

  1   2   3
200 100 100
```

With 200 amino acids, the first protein is two times bigger than the others so that when computing the Euclidean distance (10.1) its n_{ij} entries are on

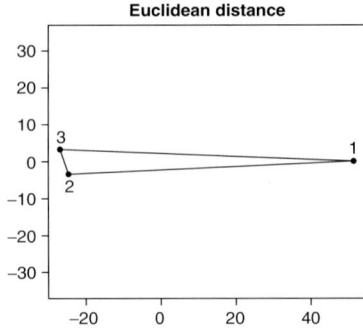

Fig. 10.11. Relationships between the three proteins with the Euclidean distance

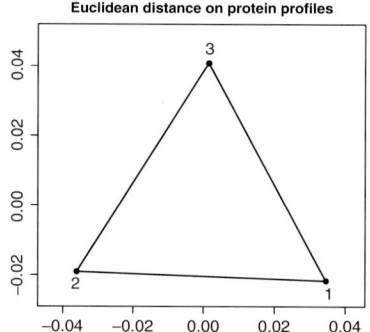

Fig. 10.12. Relationships between the three proteins with the Euclidean distance on protein profiles

average bigger, sending it away from the others. To get rid of this trivial effect, the first obvious idea is to divide counts by protein lengths so as to work with *protein profiles* (Fig. 10.12). The corresponding distance is

$$d^2(i, i') = \sum_{j=1}^{J} \left(\frac{n_{ij}}{n_{i\bullet}} - \frac{n_{i'j}}{n_{i'\bullet}} \right)^2 \qquad (10.2)$$

where $n_{i\bullet}$ and $n_{i'\bullet}$ are the total number of amino acids in protein i and i', respectively.

```
df1 <- df/rowSums(df)
df1
```

```
    Ala  Val  Cys
1  0.65 0.35 0.00
2  0.60 0.40 0.00
3  0.60 0.35 0.05
```

```
pco1 <- dudi.pco(dist(df1), scann = F, nf = 2)
myplot(pco1, main = "Euclidean distance on protein profiles",
    asp = 1, pch = 19, xlab = "", ylab = "", ylim = range(pco1\$li[,
    2]) * 1.2)
```

The pattern is now completely different with the three proteins being equally spaced. This is normal because in terms of relative amino acid composition they are all differing two-by-two by 5% at the level of two amino acids only. We have clearly removed the trivial protein size effect, but this is still not completely satisfactory. The proteins are differing by 5% for all amino acids but the situation is somewhat different for Cys because this amino acid is very rare. A difference of 5% for a rare amino acid has not the same significance than a difference of 5% for a common amino acid such as Ala in our example. To cope with this, CA make use of a variance-standardizing technique to compensate for the larger variance in high frequencies and the smaller variance in low frequencies. This is achieved with the use of the *chi-square distance*

(χ^2), which differs from the previous Euclidean distance on profiles (10.2) in that each square is weighted by the inverse of the frequency corresponding to each term

$$d^2(i, i') = \sum_{j=1}^{J} \frac{1}{n_{\bullet j}} \left(\frac{n_{ij}}{n_{i\bullet}} - \frac{n_{i'j}}{n_{i'\bullet}} \right)^2 \quad (10.3)$$

where $n_{\bullet j}$ is the total number of amino acid of kind j. With this point of view, the map is now like this (Fig. 10.13):

```
$coa <- dudi.coa(df, scann = FALSE, nf = 2)
myplot(coa, main = expression(paste(chi^2, " distance")),
    asp = 1, pch = 19, xlab = "", ylab = "")$
```

The pattern is completely different with now protein number 3, which is far away from the others because it is enriched in the rare amino acid Cys when compared with others.

The purpose of this small example was to demonstrates that the metric choice is not without significant effects on the visualization of data. Depending on your objectives, you may agree or disagree with the χ^2 metric choice, that is not a problem, the important point is that you should be aware that there is an underlying model there, *chacun a son goût* ou *chacun à son goût*, it is up to you.

Now, if you agree with the χ^2 metric choice, there is a nice representation that may help you for the interpretation of results. This is a kind of "biplot" representation in which the lines and columns of the dataset are simultaneously

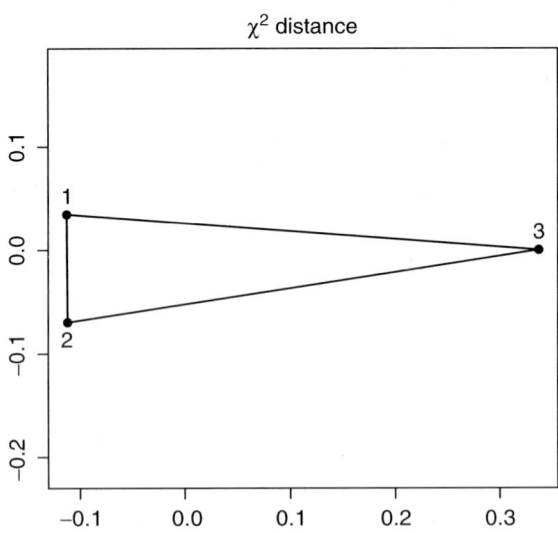

Fig. 10.13. Relationships between the three proteins with the χ^2 distance

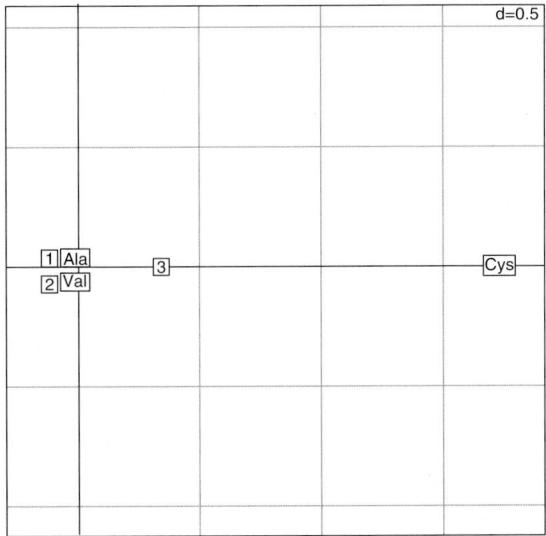

Fig. 10.14. First factorial map of correspondence analysis of the three proteins

represented, in the right way, that is as a graphical *translation* of a mathematical theorem, but let us see how does it look like in practice (Fig. 10.14):

```
scatter(coa, clab.col = 0.8, clab.row = 0.8, posi = "none")
```

What is obvious is that the Cys content has a major effect on protein variability here, no scoop. Please note how the information is well summarized here: protein number 3 differs because it is enriched in in Cys; protein number 1 and 2 are almost the same but there is a small trend that protein number 1 to be enriched in Ala. When compared with Table 10.1, this graph is of poor information here, so let us try a more big-room-sized example (with 20 columns so as to illustrate the dimension reduction technique).

Data are from [19], a sample of the proteome of *Escherichia coli*. According to the title of this chapter, the most important factor for the between-protein variability is hydrophilic–hydrophobic gradient. Let us try to reproduce this assertion (Fig. 10.15):

```
download.file(url = "ftp://pbil.univ-lyon1.fr/pub/datasets/NAR94/data.txt",
    destfile = "data.txt")
ec <- read.table(file = "data.txt", header = TRUE,
    row.names = 1)
ec.coa <- dudi.coa(ec, scann = FALSE, nf = 1)
F1 <- ec.coa$li[, 1]
hist(F1, proba = TRUE, xlab = "First factor for amino acid variability",
    col = grey(0.8), border = grey(0.5), las = 1,
    ylim = c(0, 6), main = "Protein distribution on first factor")
lines(density(F1, adjust = 0.5), lwd = 2)
```

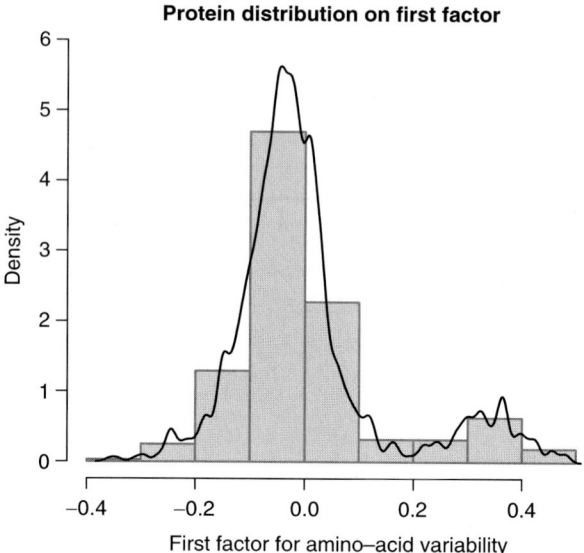

Fig. 10.15. The bimodal distribution of proteins on the first factor of variability for amino acid composition

10.4.2 Synonymous and Nonsynonymous Analyses

Genetic codes are surjective applications from the set codons ($n = 64$) into the set of amino acids ($n = 20$) (Fig. 10.16).

Two codons encoding the same amino acid are said synonymous, while two codons encoding a different amino acid are said nonsynonymous. The distinction between the synonymous and nonsynonymous level are very important in evolutionary studies because most of the selective pressure is expected to work at the nonsynonymous level, because the amino acids are the components of the proteins, and therefore more likely to be subject to selection.

K_s and K_a are an estimation of the number of substitutions per synonymous site and per nonsynonymous site, respectively, between two protein-coding genes [21]. The $\frac{K_a}{K_s}$ ratio is used as tool to evaluate selective pressure (see [22] for a nice back to basics). Let us give a simple illustration with three orthologous genes of the thioredoxin familiy from *Homo sapiens*, *Mus musculus*, and *Rattus norvegicus* species:

```
download.file(url = "http://pbil.univ-lyon1.fr/software/SeqinR/Datasets/ortho.fasta",
    destfile = "ortho.fasta")
ortho <- read.alignment("ortho.fasta", format = "fasta")
kaks.ortho <- kaks(ortho)
kaks.ortho$ka/kaks.ortho$ks
```

```
            AK002358.PE1 HSU78678.PE1
HSU78678.PE1    0.1249730
RNU73525.PE1    0.1428173    0.1363095
```

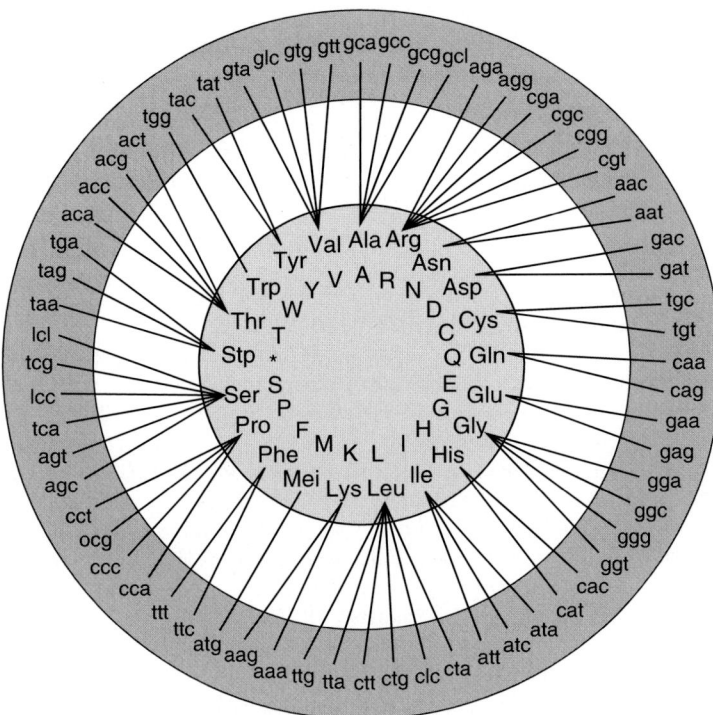

Fig. 10.16. A visual representation of the surjective nature of genetic codes, genetic code number 1. Adapted from [20]

The $\frac{K_a}{K_s}$ ratios are less than 1, suggesting a selective pressure on those proteins during evolution.

For transversal studies (i.e., codon usage studies in a genome at the time it was sequenced), there is little doubt that the strong requirement to distinguish between synonymous and an nonsynonymous variability was the source of many mistakes [17]. We have just shown here with a scholarship example that the metric choice is not neutral. If you consider that the χ^2 metric is not too bad, with respect to your objectives, and that you want to quantify the synonymous and an nonsynonymous variability, please consider reading this chapter [20], and follow this link http://pbil.univ-lyon1.fr/members/lobry/repro/jag03/ for on-line reproducibility.

Acknowledgments

Please enter `contibutors()` in your R console.

References

1. R. Ihaka, R.Gentleman, J. Comp. Graph. Stat. **3**, 299 (1996)
2. R Development Core Team, *R: A language and environment for statistical computing* (ISBN 3-900051-00-3, 2004) http://www.R-project.org
3. F. Leisch, Proceedings in Computational Statistics, 575 (2002) (ISBN 3-7908-1517-9)
4. K. Hornik, *The R FAQ* (ISBN 3-900051-08-9, 2005) http://CRAN.R-project.org/doc/FAQ/
5. J. Keogh, Australian Patent Office application number *AU 2001100012 A4* (2001). www.ipmenu.com/archive/AUI_2001100012.eps
6. J.R. Lobry, N. Sueoka, Genome Biology **3**, research0058.1 (2002) http://genomebiology.com/2002/3/10/research/0058
7. J. Buckheit, D.L. Donoho, in *Wavelets and Statistics*, ed. by A. Antoniadis (Springer, Berlin, New York, 1995)
8. D. Charif, J. Thioulouse, J.R. Lobry, G. Perrière, Bioinformatics **21**, 545 (2005); http://pbil.univ-lyon1.fr/members/lobry/repro/bioinfo04/
9. R. Rudner, J.D. Karkas, E. Chargaff, Proc. Natl. Acad. Sci. USA **63**, 152 (1969)
10. J.R. Lobry, Lecture Notes Comput. Sci. **3039**, 679 (2004). http://pbil.univ-lyon1.fr/members/lobry/repro/lncs04/
11. A.C. Frank, J.R. Lobry, Bioinformatics **16**, 560 (2000)
12. P. Mackiewicz, J. Zakrzewska-Czerwińska, A. Zawilak, M.R. Dudek, S. Cebrat, Nucleic Acids Res. **32**, 3781 (2004)
13. P. Legendre, Y. Desdevises, E. Bazin, Syst. Biol. **51**, 217 (2002)
14. N. Saitou, M. Nei, Mol. Biol. Evol. **4**, 406 (1984)
15. T.H. Jukes, C.R. Cantor, in *Mammalian Protein Metabolism*, ed. by H.N. Munro (Academic, New York, 1969) pp. 21–132
16. M. Kimura, J. Mol. Evol. **16**, 111 (1980)
17. G. Perrière, J. Thioulouse, Nucleic Acids Res. **30**, 4548 (2002)
18. C. Gautier, Ph.D. Thesis (1987), Université Claude Bernard - Lyon I
19. J.R. Lobry, C. Gautier, Nucleic Acids Res. **22**, 3174 (1994). http://pbil.univ-lyon1.fr/members/lobry/repro/nar94/
20. J.R. Lobry, D. Chessel, J. Appl. Genet. **44**, 235 (2003). http://jay.au.poznan.pl/html1/JAG/pdfy/lobry.eps
21. W.-H. Li, J. Mol. Evol. **36**, 96 (1993)
22. L.D. Hurst, Trends Genet. **18**, 486 (2002)

Part IV

Networks

11

Evolutionary Genomics of Gene Expression

I.K. Jordan and L. Mariño-Ramírez

The study of evolution at the molecular level has focused primarily on changes in gene (protein) sequences overtime [1]. Of course, phenotype is influenced not only by the sequence of genes but also by their expression patterns, i.e., the amplitude, timing, and spatial distribution of transcription. Thus, changes in gene expression are quite likely to be equally as important as sequence changes in evolution; indeed, the significance of gene expression divergence to the evolutionary process has been recognized for some time [2–4]. However, gene expression data have only recently accumulated to the levels needed for systematic evolutionary studies. This has been due to the application of new high-throughput techniques that measure gene expression levels for thousands of genes simultaneously [5–7], as well as the development of database resources needed to handle such data [8–10]. The availability of these expression data, together with the long standing interest in the evolutionary significance of gene expression, has stimulated numerous recent studies on gene expression divergence.

This chapter will provide a guide for the study of gene expression divergence. The emphasis will be placed on an integrated approach to the study of evolutionary genomics that considers both gene sequence and gene expression divergence and explores the relationship between those two aspects of the evolutionary process. The body of the chapter will be broken down into three sections. The first two body sections, on sequence divergence and on gene expression divergence, will be tutorial in nature and cover specific methodological techniques involved in the study of gene sequence and gene expression divergence. In most cases, descriptions of methods will focus on the most straightforward and widely available techniques. The third body section, on integrated analysis, will be more conceptual in nature and deal with selected examples of how gene sequence and gene expression divergence analyses have been used to address fundamental evolutionary questions.

11.1 Sequence Divergence

Methods of gene (protein) sequence analysis have been covered in great detail elsewhere. Here, some of the basic techniques and issues related to sequence analysis will be covered. A key concept for sequence analysis is that of functional, and/or selective, constraint. A functionally constrained gene is one that encodes a protein that performs a critical function for the organism, and a functionally constrained residue (site) is one that plays an important role in the function of the molecule. Changes to the sequence of such genes, or sites, are likely to be deleterious, i.e., they will reduce the fitness of the organism, and therefore will be removed by natural selection. Thus, genes, or specific sites within a gene, that are under greater selective constraint evolve more slowly, while genes (sites) that are subject to relatively less constraint evolve more rapidly. As such, comparisons of gene divergence levels can be used to make inferences about the strength of functional constraint and the relative action of natural selection.

The availability of complete genome sequences provides great utility for the comparative analysis of gene divergence levels because it allows for the controlled comparison of divergence levels for thousands of genes. To compare gene divergence levels between complete genome sequences one needs to: (1) identify orthologous genes, (2) align gene (protein) sequences, and (3) calculate substitution rates. Each of these tasks will be briefly covered below.

11.1.1 Ortholog Identification

Genes that share a common ancestor are said to be homologous, and homologous genes can be defined as orthologous or paralogous [11]. Orthologs are genes that diverged due to a speciation event, while paralogs are genes that diverged due to a gene duplication event. Considering two species, such as human and mouse, orthologs can be thought of colloquially as the pair corresponding genes, i.e., those that perform the same function, in each genome. If both genomes are completely sequenced, then pairs of orthologous genes can be identified using sequence similarity. This is the so-called "reciprocal best hits" approach [12, 13]. To implement this approach, each protein sequence encoded by genome A is compared individually to the entire set of protein sequences encoded by genome B using a sequence similarity comparison tool, such as BLAST [14, 15] or FASTA [16]. Protein sequences are typically used because they are more sensitive for sequence similarity comparisons. Then, for each protein from genome A, the protein with the highest similarity from genome B is recorded as its best hit. The same process is repeated in the opposite direction, with each protein from genome B compared individually to all proteins from genome A and the best hits recorded. Orthologous pairs are then identified as those pairs of proteins that are each other's best hits in the reciprocal sequence similarity searches.

This simple approach works quite well for closely related genomes. However, there are some caveats to ortholog identification that one should be aware of. Ortholog identification between distantly related genomes is less accurate owing because of to problems with sequence similarity comparisons as well as the phenomenon of gene loss where the corresponding member of an orthologous pair is lost in one lineage. Furthermore, the reciprocal best-hits approach defined above identifies one-to-one orthologs. However, orthologous relationships may also be one-to-many or many-to-many because of lineage-specific gene duplications that occur subsequent to the speciation event under consideration. These complex relationships can confound attempts to identify orthologs using only reciprocal best hits. Ortholog databases, such as the Clusters of Orthologous Groups database [17], use complex algorithms that post process reciprocal best hits between multiple complete genomes and allow for the representation of many-to-many orthologous relationships. A recently developed method for pairwise genome comparison, the reciprocal smallest distance algorithm, has been shown to identify many orthologs missed by reciprocal best hits [18]. In some rare cases, lineage-specific gene duplication followed by differential loss of alternate paralogous copies in the different genomes can result in erroneous identification of orthologs [19]. One way to avoid this problem is to use an ad hoc approach whereby the distribution of divergence levels between orthologs is considered and the most divergent outliers are removed.

11.1.2 Sequence Alignment

Once orthologous gene (protein) pairs are identified they need to be aligned before divergence levels can be calculated. The Clustal series is the most widely used group of programs for sequence alignment [20]. Clustal uses a heuristic approach for building multiple sequence alignments, but the initial pairwise alignment step is based on dynamic programming algorithm that guarantees an optimal solution. Given its ready availability and the reliability of its pairwise alignment step, Clustal is a good choice for aligning pairs of orthologs.

11.1.3 Sequence Distance Calculation

Gene (protein) divergence levels are calculated as sequence distances, which measure the numbers of differences between sequences normalized by their lengths. The simplest sequence distance measure is the p-distance. The p-distance is simply the proportion of differences between any two gene sequences and it is defined as

$$p = \frac{n_\mathrm{d}}{n}, \qquad (11.1)$$

where n_d is the number of differences between the sequence and n is the number of sites being compared. Alignment sites that contain gaps are usually ignored when calculating p-distances. The problem with the p-distance is that it tends to undercount the number of changes that have occurred between sequences. For example, multiple changes at a single site will only be counted as one difference. Parallel substitutions that lead to the same residue will not be counted at all. A number of different sequence distance measures have been developed that attempt to account for multiple substitutions and give a more accurate measure of sequence divergence. Examples of a few of these will be covered below for nucleotide and protein sequences.

Nucleotide Sequences

Estimates of nucleotide divergence levels that account for multiple substitutions are based on mathematical models of the substitution process. The simplest such model is the Jukes–Cantor model. In this case, it is assumed that nucleotide frequencies are equal and that nucleotide substitutions are all equally probable. The Jukes–Cantor nucleotide distance (d) [21] can be calculated simply from the p-distance,

$$d = -\frac{3}{4}\ln[1 - (4/3)p]. \tag{11.2}$$

Nucleotide substitution models become progressively more complicated by separately parameterizing different aspects of the substitution process. For instance, the Kimura two-parameter (K2P) method [22] accounts for the fact that transitions, changes from purine-to-purine or from pyrimidine-to-pyrimidine, occur at different rates than transversions, changes between purines and pyrimidines. The K2P distance can be calculated as

$$d = -\frac{1}{2}\ln(1 - 2P - Q) - \frac{1}{4}\ln(1 - 2Q), \tag{11.3}$$

where P and Q are the proportional differences between the sequences due to transitions (P) and transversions (Q). The Felsenstein (F81) model extends Jukes–Cantor by allowing for unequal nucleotide frequencies [23]. Both of these methods are merged in the Hasegawa, Kishino, and Yano (HKY85) model that allows for unequal base frequencies as well as different transition and transversion rates [24]. The most nuanced nucleotide substitution model is the general reversible model where nucleotide frequencies are unequal and all six pairs of substitution rates are free to vary [25,26]. More detailed expositions of nucleotide substitutions models can be found in [27,28].

Synonymous vs. Nonsynonymous Substitutions

One of the great advantages of using nucleotide sequence distances is the information that they can provide regarding the action of natural selection. The effects of selection on nucleotide coding sequences (CDSs) can be gleaned by comparing levels of synonymous (S) vs. nonsynonymous (N) sequence divergence [29]. S changes are substitutions in the CDS that do not change the encoded amino acid sequence, while N changes are CDS substitutions that result in amino acid differences. Thus, N changes may change the structure and/or function of the encoded protein, while S changes are largely silent. Since natural selection exerts its influence based on detectable phenotypic differences, N changes are subject to the effects of selection while S differences are, for the most part, invisible to natural selection. As such, S changes tend to be freer to accumulate than N changes. The proportion of synonymous (p_S) and nonsynonymous (p_N) differences can be calculated as the number of S (S_d), or the number of N (N_d), differences normalized by the number of S, or N, sites

$$p_S = \frac{S_d}{S} \tag{11.4a}$$

$$p_N = \frac{N_d}{N}. \tag{11.4b}$$

These measures are also needed to account for multiple substitutions to be more accurate. The p_S and p_N values can be used with the Jukes–Cantor model (11.1) to calculate d_S and d_N, respectively. This approach is employed in the Nei–Gojobori method [30] for calculating d_S and d_N. Other methods take into account factors such as transition vs. transversion differences as well as the nuances of the genetic code to try and achieve the most accurate S and N distance measures possible. Several of these methods are implemented in the program MEGA [31]. The program PAML [32] also calculates d_S and d_N using a maximum likelihood-based approach. Users should note that depending on the method employed, d_S may be referred to as K_s and d_N may be referred to as K_a.

In general, when sequence pairs are compared, $d_N/d_S \ll 1$ is indicative of purifying selection, or removal of deleterious changes, $d_N/d_S \approx 1$ suggest the absence of natural selection, and $d_N/d_S \gg 1$ is indicative of adaptive or diversifying selection, based on the fixation of beneficial N sequence changes.

Protein Sequences

As with nucleotide sequences, protein sequence divergence levels can be calculated using the proportion of differences (p-distance) between sequences, but this measure will underestimate the true amount of divergence for all but the most closely related sequences. One correction for multiple amino acid substitutions is the Poisson corrected (PC) distance [27] where the probability of k amino acid substitutions at a given site is considered to follow a Poisson distribution. PC can be calculated from the p-distance as

$$d = -\ln(1-p). \tag{11.5}$$

However, most models of the amino acid substitution process are empirical rather than analytical. Empirical models take advantage of the availability of many related protein sequences that can be reliably aligned. Given a set of protein sequence alignments, the relative probabilities of exchange between any two amino acid residues can be calculated. These probabilities can be placed into a substitution matrix, which can be employed in the calculation of distances between protein sequences. Commonly employed empirical models are based on the PAM [33] and JTT [34] amino acid substitution matrices. Distances based on empirical models are more accurate than those obtained with simple analytical models like the PC distance, because they more closely reflect biological reality.

Another important consideration when calculating divergence levels between protein sequences is the fact that changes may accumulate at vastly different rates in different regions of the sequence. As described previously, sites along a sequence that are more functionally constrained will change more slowly than sites that are less constrained. The gamma distance correction is one way to account for this rate variation across sites. The gamma correction is based on the observation that the substitution rate across sites can be considered to vary according to a gamma distribution [35]. The shape of this distribution is governed by a single parameter α. The gamma distance (d_G) can be calculated from the p-distance as

$$d_G = \alpha \left[(1-p)^{-1/\alpha} - 1 \right] . \tag{11.6}$$

The lower the value of α, the more severe the correction for multiple substitutions is. A gamma distance correction can also be used together with an empirical-based model of amino acid substitution. For instance, the JTT model can be used together with a gamma correction and this is represented as JTT+Γ. Several protein divergence calculation methods are implemented in the program MEGA.

11.2 Gene Expression Divergence

High-throughput techniques for measuring gene expression levels, such as microarray and serial analysis of gene expression (SAGE) approaches, have resulted in an explosion of gene expression data. This section will focus on how microarray data collected from different species can be used to calculate gene expression divergence. In order to illustrate the methods that can be used in the evolutionary analysis of gene expression data, examples will be taken from our own work based on the analysis of the Novartis Foundations' mammalian gene expression atlas [36, 37]. The gene expression atlas reports expression levels for thousands of human and mouse genes based on the use of Affymetrix microarray experiments. There are two versions of the atlas and this chapter focuses primarily on the second and more recent version, sometimes referred to as GNF2. In GNF2, the results of replicate experiments across a wide variety of tissue and cell-line samples are reported. The human data set has expression level measurements for 44,744 probes over 79 distinct tissue samples, while the mouse data set includes expression level measurements for 36,181 probes over 61 distinct tissue samples; in both data sets, each tissue sample is represented by two replicate experiments. Given this abundance of comparative expression data, GNF2 is a phenomenal resource for the analysis of gene expression divergence. Used together with human–mouse comparative sequence analysis, investigation of the GNF2 data can reveal much about the evolution of gene expression and the relationship between gene sequence and gene expression divergence.

11.2.1 Database Sources

There are a number of database sources available for extracting gene expression data. The Novartis Foundation hosts its own database [38] that allows for targeted querying of the gene expression atlas along with downloading of entire expression datasets. The Novartis site also provides valuable information concerning the expression atlas including useful tips for the analysis and interpretation of their data as well as information on the source of tissue samples used.

The UCSC genome browser [39] has integrated the gene expression atlas data into its browser tracks for the human and mouse genomes. As with everything on their site, the primary expression data is available for download either as SQL tables or as tabulator delimited text files. In addition, one particularly nice feature provided by the UCSC genome browser is the mapping of Affymetrix probes identifiers to Genbank accessions for known human and mouse genes. The UCSC genome browser also provides a number of other canonical gene expression datasets for other model organisms such as yeast and *Drosophila*.

The Gene Expression Omnibus (GEO) database [40] at the National Center for Biotechnology Information (NCBI) also provides the GNF2 data. This data can be queried with search terms, and users can also download the primary expression data as tabulator delimited text files. GEO is a vast repository of gene expression data and one of its strengths is the many ways that the data can be queried. For instance, users can identify all datasets generated from a specific microarray platform. This can be quite useful for extracting datasets that can be readily compared. GEO also provides nice graphical views of expression data results as well as targeted comparisons between user selected samples from a given dataset.

There are a couple of other database resources of note that store and disseminate microarray expression data including the Stanford Microarray Database [41] and the Human Gene Expression Index [42].

11.2.2 Probe-to-Gene Mapping

One of the first challenges in using GNF2, or any results based on the Affymetrix platform for that matter, is probe-to-gene mapping. Affymetrix microarray technology is based on oligonucleotide probes that are designed to correspond to specific genes, and each probe is labeled with a unique identifier (id). The primary expression data are generally distributed along with these Affymetrix ids. Since comparative analysis of gene expression data, between species, involves a gene-centric approach, the user is faced with the task of mapping these probe ids to gene ids such as Genbank or Refseq accessions. Fortunately, Affymetrix probe-to-gene mapping keys can be downloaded from the Affymetrix Web site [43]. Alternatively, the UCSC Genome Browser provides probe-to-gene mapping for GNF2. Finally, since the sequences of the probes are often provided along with expression data, users could always do their own probe-to-gene mapping but this would be somewhat labor intensive.

Users should be aware that sometimes multiple probes will map to a single gene. In such cases, the user may decide to average the expression values for the multiple probes to come up with one set of gene-specific values. Another, somewhat arbitrary, approach that is sometimes used is to take the probe that yields the highest expression value as the best one for a given gene. Far more problematic is the fact that, in some cases, a single probe will map to multiple unique genes; because of their inherent ambiguity, these data should not be used for the analysis of gene expression divergence.

11.2.3 Structure of the Data

The GNF2 data, as well as many other microarray datasets, are structured as a table with probe-specific expression data in rows and experiment (sample)-specific data in the columns. Thus, for any particular probe, or gene, the expression data consist of an array of values, one for each experimental condition. These are gene expression profiles, and they can be thought of as vectors in n-dimensional space, where n is the number of distinct samples in the data set. For instance, for any gene i, with expression levels recorded across n samples, its expression profile can be represented as

$$\text{gene}_i = [X_{i1}, X_{i2}, X_{i3}, \ldots, X_{in}], \tag{11.7}$$

where X_{ij} is the expression value for gene i in the experiment j. It is these gene-specific expression profiles that can be compared within or between species to measure gene expression divergence. When comparing profiles between species, it is essential that the identity j and number n of the samples is identical in the vectors. For example, in GNF2, the human and mouse datasets share 28 common tissue samples. These 28 samples can be arranged in the same order across the gene expression profiles so that vectors can be meaningfully compared between species.

11.2.4 Transformation and Normalization

Two critical issues with respect to microarray data are transformation and normalization. The details on transformation and normalization are outside the scope of this chapter but they have been treated in depth elsewhere [44–46]; here, these matters will be covered briefly. For microarray data, transformation generally involves taking the logarithm to the base 2 ($\log_2$) of the expression value. This procedure is typically used to transform ratios as in the case where the expression data are represented as a ratio of one experimental condition over a reference experimental condition. In this case, $\log_2$ transformation treats up-regulation and down-regulation equally and represents them in a very regular and intuitive way. For example, a ratio of 1 would mean no change in expression across conditions and the $\log_2$ value would be 0. A ratio of 4 would yield a $\log_2$ value of 2, while a ratio of $1/4$ gives a $\log_2$ value of -2. Thus, using log transformation, the magnitude of the deviations in up- and down-regulation are symmetrical around 0. The GNF2 data, however, consist of absolute expression value measurements as opposed to ratios. Nevertheless, the relative levels of gene expression for each sample j across a gene i expression profile is often represented as a ratio of the absolute expression value j over the median value of all n expression values in the profile. The use of $\log_2$ transformation for these kinds of profile median ratios has the same useful effect of mapping changes in gene-specific relative expression in a symmetrical and continuous way around 0. The basic idea behind normalization is to control for systematic variation between experimental conditions that affect the recorded levels of expression. This is particularly important when comparing expression levels, or profiles, of orthologous genes between species. Since the conditions under which the experiments for different species were conducted are sure to differ, it is essential that relative, as opposed to absolute, expression values are compared between experiments. One straightforward way to do this is to mean, or median, center the results for each microarray. This consists of dividing each individual expression by the mean, or median, expression value for the entire array.

11.2.5 Measuring Divergence

Expression Level and Breadth

Gene expression divergence can be measured in several different ways. Perhaps the most intuitive of these methods is to compare differences in gene expression level or amplitude. When comparing between species, it is important to ensure that the same tissues or samples are being compared and that the data between species has been appropriately normalized. Given a gene-specific profile of expression levels across a set of different tissues or samples, the expression level can be taken as the maximum or the average of the tissue-specific expression values. While both of these measures fairly represent the amplitude of gene expression, using the average (or the sum), over all tissue-specific values has the disadvantage of conflating gene expression level with gene expression breadth. Gene expression breadth is a measure of how widely a gene is expressed and can be counted simply as the number of tissues in which a gene is expressed at or above some threshold. Obviously, if expression breadth is to be accurately compared between species, then it helps to ensure that the same set of tissues is being compared for each species. For example, the GNF2 has a total of 28 tissues that are shared between the human and mouse experiments. Expression breadth would simply be measured as the number of j tissue samples out of 28 where gene i is expressed at or above some threshold. GNF2 provides absolute expression levels in arbitrary units that are referred to as signal intensity values. For the GNF2 data, a signal intensity value ≥ 350 can be taken to (approximately) indicate that a gene i is expressed in a tissue j. In addition to signal intensity values (i.e., absolute expression levels), the Novartis site also provides presence/absence calls for each gene i and condition j (X_{ij}). These calls simply indicate whether or not a gene i can be considered to be expressed in tissue j with a certain level of statistical confidence. Thus, another approach to determine expression breadth is simply to use the presence/absence calls for all X_{ij}.

Gene Expression Profiles

As described earlier, a gene expression profile represents the levels of expression for gene i over all experiments (tissues) j. Most comparisons of gene expression patterns consist of quantitative measures of the similarity or difference between gene expression profile vectors. Two of the most commonly used metrics for comparing expression profile vectors are the Euclidean distance and the Pearson correlation coefficient. The Euclidean distance is geometric measure of the straight line distance of two points. The higher the Euclidean distance, the more different the gene expression profiles are. For instance, if comparing two genes A and B that have two-dimensional gene expression profile vectors $A = [a_1, a_2]$ and $B = [b_1, b_2]$, the Euclidean distance d_E would be calculated as

$$d_E = \sqrt{(a_1 - b_1)^2 + (a_2 - b_2)^2}. \quad (11.8)$$

For an expression profile vector of n-dimensions, the Euclidean distance d_E would be calculated as

$$d_E = \sqrt{\sum_{j=1}^{n}(a_j - b_j)^2}. \quad (11.9)$$

The Euclidean distance is particularly sensitive to changes in the magnitude of gene expression. Comparison of genes with identical relative expression levels across n tissues may actually yield quite large Euclidean distances if their absolute expression levels differ substantially. One way to get around this is to use relative expression levels by mean or median, centering the expression levels for each gene-specific vector.

The Pearson correlation coefficient is also widely used in comparing gene expression profile vectors, and it measures the strength of the linear relationship between the vectors being compared. Pearson correlation coefficient values scale from -1 to $+1$, where -1 would correspond to the exact opposite expression pattern and $+1$ would indicate an identical expression pattern. Use of the Pearson correlation coefficient assumes that the data are normally distributed so log transformation of expression data is advised when comparing profiles with this method. It is also important to note that the Pearson correlation coefficient works best when comparing genes that are differentially expressed, i.e., when there are substantial differences in expression levels across the n samples being considered. Genes that are ubiquitously expressed, such as housekeeping genes, can obviously be considered to have very similar expression patterns. However, because there may be no discernible linear relationship between up and down expression across tissues for such evenly expressed genes, comparison of these genes using the Pearson correlation coefficient will often result in values around 0 indicating no correlation. On the other hand, the Pearson correlation coefficient is very good at identifying genes with similar tissue-specific expression patterns. There are many forms for the Pearson correlation coefficient r. Given two n-dimensional gene expression profiles vectors for genes A and B, where $A = [a_1, a_2, \ldots, a_n]$ and $B = [b_1, b_2, \ldots, b_n]$, r can be calculated as

$$r = \frac{\sum_{j=1}^{n} a_j b_j - \frac{1}{n} \sum_{j=1}^{n} a_j \sum_{j=1}^{n} b_j}{\sqrt{\sum_{j=1}^{n} a_j^2 - \frac{1}{n} \left(\sum_{j=1}^{n} a_j\right)^2} \sqrt{\sum_{j=1}^{n} b_j^2 - \frac{1}{n} \left(\sum_{j=1}^{n} b_j\right)^2}}. \quad (11.10)$$

Two other useful distance measures that are often employed include the Hamming distance and mutual information both of which are useful for considering expression data that has been rendered discrete such as presence/absence calls that can be represented as binary expression profiles.

11.2.6 Clustering and Visualization

Clustering and visualization are important components of gene expression divergence analysis, which will nevertheless be treated in only the most cursory manner here. For more detailed treatment of these issues, users can consult [44, 47–49]. The idea behind clustering is simply to group genes with similar expression profiles together. Clustering approaches can be classified as hierarchical or nonhierarchical. Hierarchical methods group profiles into clusters and also specify the relationships among the profiles within clusters, while nonhierarchical methods simply define clusters of related expression profiles with no specification of the within group relationships. Hierarchical clustering methods can be agglomerative, where profiles are successively joined until they are all connected, or divisive, where the entire set of profiles is considered as a single cluster that is progressively broken down. Nonhierarchical clustering, on the other hand, starts with a predefined number of groups and then proceeds to partition the profiles into these discrete groups. Examples of nonhierarchical clustering are K-means clustering and self-organizing maps.

Visualization provides a very intuitive way for the user to identify similarly expressed genes. In visualization, each sample j in a gene profile i is assigned a color that indicates its relative level of expression. A typical color scheme that is employed in visualization is to label relatively high-expression levels (or up-regulated) as red and relatively low expression levels (or down-regulated) as green. This allows for ready identification of genes that have similar patterns of up and down expression across their respective profiles. Visualization is often combined with clustering techniques to define related sets of genes. There are many software packages that combine clustering and visualization techniques. One freely available program that we have found to be quite useful is the TIGR Multiexperiment Viewer (MEV) [50].

11.3 Integrated Analysis

This section will treat a few selected examples from the literature that illustrate how integrated gene sequence and gene expression divergence analyses can be used to address fundamental evolutionary questions. This survey highlights new findings regarding the evolution of gene expression as well as some of the open questions that have been raised in this relatively new area of inquiry.

11.3.1 Sequence vs. Expression Divergence

We have explored the intersection of gene expression and gene sequence divergence in two recent publications of our own [51, 52]. Both of these articles dealt with mammalian evolution and combined genomic sequence analysis with analysis of gene expression data from the Novartis gene expression atlas. The first of these studies took a network-based approach to the study of gene coexpression [52]. Human gene expression profiles were compared and genes that were found to be coexpressed were linked in a network. The topology of the resulting human gene coexpression network was shown to have scale-free properties that imply evolutionary self-organization via preferential node attachment. When rates of sequence evolution between human and mouse orthologs were overlayed on the coexpression network, genes with numerous coexpressed partners, so-called "hubs" of the network, were found to evolve more slowly, on average, than genes with fewer coexpressed partners. Furthermore, coexpressed genes were demonstrated to have coevolved in the sense that they have similar rates of evolution. These observations indicate that the strength of selective constraints on gene sequences is strongly influenced by the topology of the gene coexpression network. This connection is strong for the coding regions and 3′ untranslated regions (UTRs), but the

5′ UTRs appear to evolve under a different regime. An interesting exception to this trend was found for the relationship between gene sequence divergence and gene expression profile divergence. We found no correlation between the rate of gene sequence divergence and the extent of gene expression profile divergence between human and mouse. This suggests that distinct modes of natural selection might govern sequence vs. expression divergence. Our current work is focused on the possibility that the evolution of gene expression may be driven by adaptation-driven divergence characterized convergent evolution of gene expression patterns.

In a related study, two different aspects of gene expression divergence were related to gene sequence divergence [51]. Changes in the expression level, or the amplitude of expression, between human and mouse orthologs were shown to be correlated with levels of gene sequence divergence that are determined largely by purifying selection. However, consistent with the previously described work, evolutionary changes of tissue-specific gene expression profiles did not show such a correlation with sequence divergence. This is despite the fact that divergence of both gene expression levels and profiles were significantly lower for orthologous human–mouse gene pairs than for pairs of randomly chosen human and mouse genes. Together, these findings indicate that while purifying selection is acting to constrain gene expression divergence, there is also likely to be a neutral component in evolution of gene expression. This may be particularly true for tissues where the expression of a given gene is low and functionally irrelevant. Neutral evolution of gene expression is explored in more detail in Sect. 11.3.2. One prediction of the neutral model of gene expression divergence is a regular, clock-like accumulation of gene expression changes. Relative rate tests of the gene expression divergence among human–mouse–rat orthologous gene sets did reveal clock-like evolution for gene sequence divergence, and to a lesser extent for gene expression level divergence, but not for the divergence of tissue-specific gene expression profiles. These results suggest that the evolution of tissue-specific expression profiles may be influenced by adaptively driven changes that tend to accumulate at an uneven tempo overtime.

11.3.2 Neutral Changes in Gene Expression

Neutral evolutionary changes are those that do not confer any selective advantage or disadvantage. The neutral theory of molecular evolution holds that most changes between gene sequences are neutral, with respect to organismic fitness, and accumulate because of the random fixation of variants. The relative influence of adaptively driven changes vs. neutral evolution of gene sequences was a historically contentious issue that led to many fruitful areas of inquiry. It appears that a similar debate in the literature is emerging over the relative contributions of these two evolutionary modes – selection driven vs. neutral – to the evolution of gene expression.

Only very recently, due to the systematic analysis of high-throughput gene expression data sets, has the neutral frame of reference started to be applied in earnest to the evolution of gene expression patterns. In one particularly provocative study, Khaitovich et al. evaluated the divergence of gene expression patterns within and between several mammalian species and concluded that the evolution of gene expression patterns is largely neutral [53]. They based their conclusion on several observations. First of all, expression levels between species were found to accumulate approximately linearly as a function of time; this pattern held for comparisons of both primate species and of mouse species. Second, divergence of expression levels between human and chimp were found to be the same for pseudogenes, which evolve under no selective constraint, as for (intact) nonpseudogenes. Finally, expression level differences within species were shown to be strongly correlated with expression level differences between species, suggesting that the same neutral evolutionary process are involved in the evolution of gene expression both within and between species. The conclusion that gene expression divergence is primarily neutral has substantial implications for the study of biological evolution and function. For instance, the clock-like accumulation of gene expression changes may allow for detailed inferences on the evolution of different tissues and organ systems based on changes in gene expression patterns among them. On the other hand, if gene expression changes are neutral then the application of expression data to functional inferences may be limited.

Another recent study, by Yanai et al., also concluded that gene expression divergence between species may be dominated by neutral evolution [54]. This work took advantage of the first mammalian gene expression atlas provided by Novartis to assess the rate of gene expression pattern divergence between mammalian species. Orthologous pairs of human–mouse genes were identified and their expression patterns across multiple tissues were quantified. Expression patterns were represented as profiles that reflect the relative expression levels in different tissues, and both distance measures and correlations between profiles were measured. Several surprising results came out of this analysis. None of the gene expression profiles that were most similar between human and mouse corresponded to orthologous genes. In fact, expression profiles between orthologous genes were found to diverge so rapidly that their differences are comparable to those seen between duplicated genes (paralogs) and between random gene pairs. Even the corresponding tissues between the two species did not show similar patterns of gene-specific expression levels; all human tissues were more similar to one another as were all mouse tissues. Such rapid divergence in gene expression can be attributed to the effects of natural selection based on adaptively beneficial functional differences, so-called positive selection, or to random drift based on functionally indistinguishable (i.e., neutral) differences. The authors favor the neutral model for several reasons including the presence of orthologous gene pairs that are not presumed to have changed function and the even distribution of expression differences across tissues.

Contrary to the results suggesting neutral evolution of gene expression, a few other recent studies have pointed to an important role for natural selection in constraining, and perhaps driving, gene expression divergence. Fraser et al. focused on random fluctuations in gene expression that produce noise in protein levels [55]. They investigated the biological significance of this noise, specifically asking whether fluctuations in gene expression are biologically relevant and thus subject to natural selection. To investigate this issue, they tested two specific hypotheses, namely that two classes of genes, (1) essential genes and (2) genes that encode members of multisubunit protein complexes, should both be particularly sensitive to random fluctuations (noise) in gene expression levels. Combined computational analyses of yeast gene and protein expression levels, gene knock-out effects, and protein–protein interaction data were used toward this end. The rationale behind the test was that essential genes and genes that encode members of protein complexes should be particularly subject to the effects of natural selection – indeed these classes of genes tend to show reduced rates of evolution consistent with strong purifying selection – and if fluctuations in expression levels are significant, these too should be under strong selection for the gene classes in question. They found that, for both tests and over a large range of protein production levels, these two classes of genes show significantly and substantially lower levels of noise in protein expression than other genes. From this, it was concluded that noise in gene expression is biologically relevant and is subject to the effects of natural selection. This conclusion seemingly stands in stark contrast to those of the two studies summarized above, both of which conclude that gene expression levels are neutral with respect to organismic fitness. However, the results summarized here may not be inconsistent with a neutral model of gene expression. Indeed, natural selection does have an important role under the neutral model of evolution, but its effect is to reduce, rather than enhance, levels of diversity. Thus, genes (or positions in gene sequences) that are more functionally constrained are expected to evolve more slowly than those that are less functionally constrained and this prediction has been born out time and time again. The noise in protein expression, while biologically relevant in the sense that it is deleterious, conforms to the neutral pattern with genes that are presumably more functionally constrained showing less variation in protein production.

Yet another recent study examined patterns of gene expression polymorphism (within species changes) and gene expression divergence (between species changes) in a number of datasets from different eukaryotic species including *Drosophila*, mice, and primates [56]. Here again, the evolution of gene expression was considered with respect to the neutral model of change. Using a number of different measures, the authors found that gene expression levels tend to be evolutionarily stable in that they change very little overtime. This stability strongly implies that gene expression levels are subject to selective constraint. However, there are substantial differences in the rates at which gene expression changes, and different functional classes of genes were shown to have distinct characteristic levels of change. The authors also put forward a model that explains how changes in gene expression could be driven by directional selection.

11.3.3 Evolutionary Conservation of Gene Expression

The accumulation of genome scale expression data sets is beginning to provide opportunities for cross-species comparisons of gene expression patterns. One recent report provides an example of how such studies can reveal patterns of evolutionary conservation of gene expression [57]. The authors of this work compiled large-scale gene expression data sets from six diverse species: *Escherichia coli*, *Saccharomyces cerevisiae*, *Arabidposis thaliana*, *Caenorhabditis elegans*, *Drosophila melanogaster*, and *Homo sapiens*. Their analysis included expression data for more than 40,000 genes under 2,000 experimental conditions, and the study is also notable for its combination of gene expression and sequence data analysis. Correlations between condition-specific gene expression patterns were used to identify coexpressed genes within species. Coexpression networks, where genes are the nodes and they are connected if significantly coexpressed, were found to have similar connectivity across species. The distributions of network connectivity were found to follow a power-law similar to other biological networks such as protein–protein interaction and metabolic networks. The scale-free nature of these distributions, along with their conservation between species, suggests that a fundamental mechanism is involved in the evolution of gene expression in different domains of life. Pairwise correlations between genes were also compared to the correlations between homologous gene pairs in different species and a significant fraction was found to be similar. The utility of this homologous coexpression similarity with respect to functional annotation of genes was demonstrated. In addition, sets of highly connected genes were enriched for genes that are essential and posses a high number of homologous sequences in other organisms. Expression data were broken down into modules that consist of coexpressed genes and the particular expression conditions that give rise to their coregulation. While some groups of functionally related genes show up as conserved modules in multiple species, many of the expression modules vary widely across species and so probably contribute to evolutionary diversification.

Stuart et al. conducted a similar study of the cross-species conservation of gene expression and identified pairs of genes that are coexpressed across more than 3,000 microarray experiments conducted for humans, flies, worms, and yeast [58]. A total of 22,163 evolutionarily conserved coexpression relationships were identified in this way. Links between coexpressed genes were used to build a coexpression network and this approach revealed network components that were specific for different levels of diversification, such as ancient vs. more recently evolved connections. The conservation of expression patterns suggests that coexpression gene pairs are functionally related and the functional similarity provides the mechanistic basis of the selection for maintained coexpression. In light of this finding, the authors demonstrate how the coexpression relationships can be used to provide evidence for the involvement of new genes in specific cellular functions including cell cycle, secretion, and protein expression. Notably, a few specific predictions generated based on conserved coexpression were tested and confirmed.

References

1. W.H. Li, *Molecular Evolution* (Sinauer Associates, Sunderland, 1997)
2. R.J. Britten, E.H. Davidson, Science **165**, 349 (1969)
3. R.J. Britten, E.H. Davidson, Q. Rev. Biol. **46**, 111 (1971)
4. M.C. King, A.C. Wilson, Science **188**, 107 (1975)
5. V.E. Velculescu, L. Zhang, B. Vogelstein et al, Science **270**, 484 (1995)
6. M. Schena, D. Shalon, R.W. Davis et al., Science **270**, 467 (1995)
7. M.D. Adams, J.M. Kelley, J.D. Gocayne et al., Science **252**, 1651 (1991)
8. D. Karolchik, R. Baertsch, M. Diekhans et al., Nucleic Acids Res. **31**, 51 (2003)
9. R. Edgar, M. Domrachev, A.E. Lash, Nucleic Acids Res. **30**, 207 (2002)
10. J. Gollub, C.A. Ball, G. Binkley et al., Nucleic Acids Res. **31**, 94 (2003)
11. W.M. Fitch, Syst. Zool. **19**, 99 (1970)
12. M.C. Rivera, R. Jain, J.E. Moore et al., Proc. Natl. Acad. Sci. USA **95**, 6239 (1998)
13. R.L. Tatusov, E.V. Koonin, D.J. Lipman., Science **278**, 631 (1997)
14. S.F. Altschul, W. Gish, W. Miller et al., J. Mol. Biol. **215**, 403 (1990)
15. S.F. Altschul, T.L. Madden, A.A. Schaffer et al., Nucleic Acids Res. **25**, 3389 (1997)
16. W.R. Pearson, D.J. Lipman: Proc. Natl. Acad. Sci. USA **85**, 2444 (1988)
17. E.V. Koonin, N.D. Fedorova, J.D. Jackson et al., Genome Biol. **5**, R7 (2004)
18. D.P. Wall, H.B. Fraser, A.E. Hirsh: Bioinformatics **19**, 1710 (2003)
19. I.K. Jordan, Y.I. Wolf, E.V. Koonin: BMC Evol. Biol. **4**, 22 (2004)
20. R. Chenna, H. Sugawara, T. Koike et al: Nucleic Acids Res. **31**, 3497 (2003)
21. T.H. Jukes, C.R. Cantor., in *Mammalian Protein Metabolism*, ed. by H.D. Munro (Academic, New York, 1969)
22. M. Kimura, J. Mol. Evol. **16**, 111 (1980)
23. J. Felsenstein, J. Mol. Evol. **17**, 368 (1981)
24. M. Hasegawa, H. Kishino, T. Yano, J. Mol. Evol. **22**, 160 (1985)
25. F. Rodriguez, J.L. Oliver, A. Marin et al., J. Theor. Biol. **142**, 485 (1990)
26. Z. Yang, N. Goldman, A. Friday, Mol. Biol. Evol. **11**, 316 (1994)
27. M. Nei, S. Kumar, *Molecular Evolution and Phylogenetics* (Oxford University Press, New York, 2000)
28. R.D.M. Page, E.C. Holmes, *Molecular Evolution: A Phylogenetic Approach* (Blackwell Science, Malden, 1998)
29. L.D. Hurst: Trends Genet. **18**, 486 (2002)
30. M. Nei, T. Gojobori: Mol. Biol. Evol. **3**, 418 (1986)
31. S. Kumar, K. Tamura, M. Nei: Brief. Bioinform. **5**, 150 (2004)
32. Z. Yang, Comput. Appl. Biosci. **13**, 555 (1997)

33. M.O. Dayhoff, R.M. Schwartz, B.C. Orcutt, in *Atlas of Protein Sequence and Structure*, ed. by M.O. Dayhoff (National Biomedical Research Foundation Washington, DC, 1978)
34. D.T. Jones, W.R. Taylor, J.M. Thornton, Comput. Appl. Biosci. **8**, 275 (1992)
35. T. Uzzell, K.W. Corbin, Science **172**, 1089 (1971)
36. A.I. Su, T. Wiltshire, S. Batalov et al., Proc. Natl. Acad. Sci. USA **101**, 6062 (2004)
37. A.I. Su, M.P. Cooke, K.A. Ching et al., Proc. Natl. Acad. Sci. USA **99**, 4465 (2002)
38. http://symatlas.gnf.org/SymAtlas/
39. http://genome.ucsc.edu/
40. http://www.ncbi.nlm.nih.gov/geo/
41. http://genome-www5.stanford.edu/
42. http://df216w01.mgh.harvard.edu/hio/
43. http://www.affymetrix.com/support/index.affx
44. M.M. Babu, in *Computational Genomics: Theory and Application*, ed. by R.P. Grant (Horizon Bioscience, Norwich, 2004)
45. H.C. Causton, J. Quackenbush, A. Brazma, *Microarray Gene Expression Data Analysis: A Beginner's Guide* (Blackwell Science, Malden, 2003)
46. J. Quackenbush: Nat Genet. **32** Suppl, 496 (2002)
47. R. Xu, D. Wunsch, IEEE Trans. Neural Netw. **16**, 645(2005)
48. D.R. Gilbert, M. Schroeder, J. van Helden, Trends Biotechnol. **18**, 487 (2000)
49. R.B. Altman, S. Raychaudhuri: Curr. Opin. Struct. Biol. **11**, 340 (2001)
50. http://www.tm4.org/mev.html
51. I.K. Jordan, L. Marino-Ramirez, E.V. Koonin, Gene **345**, 119 (2005)
52. I.K. Jordan, L. Marino-Ramirez, Y.I. Wolf et al., Mol. Biol. Evol. **21**, 2058 (2004)
53. P. Khaitovich, G. Weiss, M. Lachmann et al., PLoS Biol. **2**, E132 (2004)
54. I. Yanai, D. Graur, R. Ophir., Omics **8**, 15 (2004)
55. H.B. Fraser, A.E. Hirsh, G. Giaever et al., PLoS Biol. **2**, e137 (2004)
56. B. Lemos, C.D. Meiklejohn, M. Caceres et al., Evol. Int. J. Org. Evol. **59**, 126 (2005)
57. S. Bergmann, J. Ihmels, N. Barkai, PLoS Biol. **2**, E9 (2004)
58. J.M. Stuart, E. Segal, D. Koller et al., Science **302**, 249 (2003)

12
From Biophysics to Evolutionary Genetics: Statistical Aspects of Gene Regulation

M. Lässig

12.1 Introduction

Genomic functions often cannot be understood at the level of single genes but require the study of gene networks. This systems biology credo is nearly commonplace by now. Evidence comes from the comparative analysis of entire genomes: current estimates put, for example, the number of human genes at around 22,000, hardly more than the 14,000 of the fruit fly, and not even an order of magnitude higher than the 6,000 of baker's yeast. The complexity and diversity of higher animals, therefore, cannot be explained in terms of their gene numbers. If, however, a biological function requires the concerted action of several genes, and conversely, a gene takes part in several functional contexts, an organism may be defined less by its individual genes but by their interactions. The emerging picture of the genome as a strongly interacting system with many degrees of freedom brings new challenges for experiment and theory, many of which are of a statistical nature. And indeed, this picture continues to make the subject attractive to a growing number of statistical physicists.

Genes encode proteins, and proteins perform functions in the cell. Hence, a gene takes part in a biological function only if it is *expressed*, i.e., if the protein produced from it is present in the cell. Genes interact by *regulation*: the protein of one gene can influence the production of protein from another gene. Gene regulation can take place during *transcription*, the process by which the cell reads the information contained in a gene and copies it to messenger RNA (which is subsequently used to make a functional protein). This is the most fundamental level of interactions between genes: the transcription of one gene may be enhanced or reduced by the expression of other genes. Transcriptional regulation is thus a good starting point for theory. We should keep in mind, however, that it is not the only mode of gene interactions. Especially in eukaryotes, additional regulation mechanisms involving histones, chromatin, micro-RNAs, etc., become relevant, which are just entering the stage of model building. An excellent introduction to the biology of regulation can be found in [1].

This article is a primer on theoretical aspects of gene interactions, and we limit ourselves to transcriptional regulation. Clearly, the subject has rather diverse aspects:

1. Transcription is a *biophysical* process, which involves the interaction of DNA and proteins. Its regulation takes place through the binding of proteins to DNA at specific loci in the vicinity of the gene to be regulated. Already at this level, this process is rather complex and not yet fully understood. What enables the protein to find one or a few specific functional sites in a genome of up to billions of base pairs, bind there with sufficient strength to influence transcription, and leave again once its task is performed?
2. Given the protein can find its functional sites, can we as well? If that is possible, we can predict the specific gene interactions building regulatory networks from sequence data. The analysis of regulatory DNA is a major topic of research in *bioinformatics*, with the aim of identifying statistical characteristics of functional loci and of building search algorithms.
3. Regulation is also becoming an important part of *evolutionary biology* [2, 3]. If regulatory networks are to explain the differentiation of higher animals, there must be efficient modes of evolution for the interactions between genes. At the level of regulatory DNA, these modes remain largely to be explored. It is clear, however, that the underlying evolutionary dynamics is the basis of a quantitative understanding of regulatory networks.

All three aspects of regulation contribute to a unified theoretical picture. Key concepts such as the biophysical binding energy, the bioinformatic scoring function, and the evolutionary fitness turn out to be rather deeply related. We will focus on these crosslinks between different fields, which are quite likely to become important for future research. A challenge for an introductory presentation is the diversity of relevant background material, only a rather eclectic account of which can be presented here. Yet, I hope it transpires even from this short introduction that present quantitative genomics is an area of science shaped by a remarkable confluence of ideas from different disciplines.

12.2 Biophysics of Transcriptional Regulation

The fundamental step in the regulatory interaction between two genes is a binding process: the protein produced by the first gene acts as a *transcription factor* for the second gene, i.e., it binds to a functional site on the DNA close to the second gene and thereby enhances or suppresses its transcription. Binding sites are short segments of typically 10–15 base pairs in prokaryotes and even shorter segments in eukaryotes. They are primarily located in

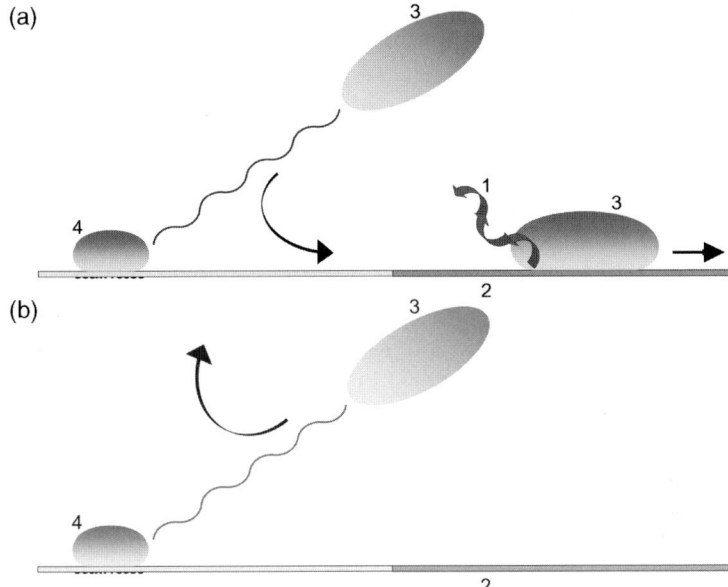

Fig. 12.1. Transcriptional regulation: Transcription is the synthesis of messenger RNA (1) whose genetic code is a copy of the coding DNA (2) of a gene, by means of RNA polymerase (3). A transcription factor (4) bound to a DNA target site interacts with RNA polymerase molecules, (**a**) enhancing or (**b**) reducing the transcription rate of a nearby gene

the *cis-regulatory region* of a gene, which lies just upstream of its protein-coding sequence and extends over hundreds of base pairs in prokaryotes and over thousands of base pairs in eukaryotes. The scenario of transcriptional regulation is sketched in Fig. 12.1. A transcription factor bound to a functional binding site regulates the downstream gene by recruiting or repelling RNA polymerase. This protein–protein interaction catalyzes or suppresses the process of transcription of the gene. All these binding processes should not be understood as on or off; they happen with certain probabilities, which are determined by the binding energies and the numbers of the molecules involved.

12.2.1 Factor-DNA Binding Energies

The interaction of a transcription factor protein with DNA is twofold: there is a position-unspecific attraction with energy E_u and a specific interaction, whose energy depends on the particular locus where the factor binds. The unspecific part is the electrostatic interaction between the positively charged protein and the negatively charged DNA backbone, while the specific part involves hydrogen bonds between the binding domain of the protein and the nucleotides of the binding locus. A locus is specified by its starting position

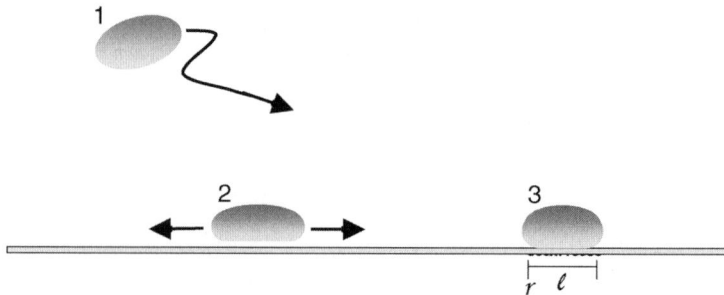

Fig. 12.2. Thermodynamic states of a transcription factor. (1) Unbound state, with three-dimensional diffusion. (2) Unspecific bound state, with one-dimensional diffusion along the DNA backbone. (3) Specific bound state. The binding energy depends on the genotype at the binding locus, which has length ℓ and whose position is specified by the coordinate r

r and its length ℓ (with relevant values ℓ of order 10). The specific binding energy $E(r)$ depends on ℓ consecutive nucleotides $\mathbf{a} = (a_1, \ldots, a_\ell)$ counted downstream from the starting position, the *sequence state* or *genotype* of that locus. Switching between unspecific and specific binding takes place via a conformation change of the factor protein. As a result of these interactions, the factor protein can be in three thermodynamic states as shown in Fig. 12.2: unbound (i.e., freely diffusing), unspecifically bound (i.e., diffusing along the DNA backbone), and specifically bound.

The biophysics of factor-DNA binding has been established in a series of seminal papers [4–7]. More recently, the characteristics of specific binding have been measured for some bacterial transcription factors [8–12]. These can be summarized as follows:

1. The single nucleotides of a binding locus $\mathbf{a} \equiv (a_1, \ldots, a_\ell)$ give approximately independent contributions to the binding energy,

$$E(\mathbf{a}) = \sum_{i=1}^{\ell} \epsilon_i(a_i). \qquad (12.1)$$

2. At each position i, there is typically one preferred nucleotide a_i^* with $\epsilon_i(a_i^*) = \min_a \epsilon_i(a)$. Hence, there is a unique "ground state" sequence $\mathbf{a}^* = (a_1^*, \ldots a_\ell^*)$ with minimal binding energy $E^* \equiv E(\mathbf{a}^*)$, i.e., with strongest binding.
3. Mismatches with respect to the minimum-energy sequence involve energy costs $\epsilon_i(a) - \epsilon_i(a_i^*) \approx 1 - 3\ k_\mathrm{B}T$ per nucleotide.
4. There is an energy difference $E_u - E^* \sim 15\ k_\mathrm{B}T$ between unspecific and strongest specific binding.

Experimental data for the binding energies $\epsilon_i(a)$ are known only for a few transcription factors. Approximate values for these energies can also be inferred from nucleotide frequencies in functional binding sites [10]. For order-of-magnitude estimates, one often uses the so-called two-state approximation [7], which is homogeneous in the nucleotide positions and distinguishes only between match and mismatch:

$$\epsilon_i(a) - \epsilon_i(a_i^*) = \begin{cases} \epsilon & \text{if } a_i \neq a_i^* \\ 0 & \text{if } a_i = a_i^* \end{cases} \quad (12.2)$$

with $\epsilon \approx 2k_B T$. In this approximation, the binding energy of a sequence **a** is simply related to the *Hamming distance* $d(\mathbf{a}, \mathbf{a}^*)$ (see also Chap. 14 by Jain and Krug), i.e., the number of nucleotide mismatches between **a** and $\mathbf{a}^*$,

$$E(\mathbf{a}) = E^* + \epsilon \cdot d(\mathbf{a}, \mathbf{a}^*). \quad (12.3)$$

12.2.2 Energy Distribution in the Genome

Figure 12.3a shows the sequence of energy values $E(r)$ found in a segment of the *E. coli* genome for a specific transcription factor, the cAMP response protein (CRP). This "energy landscape" looks quite random, i.e., consecutive energy values are approximately uncorrelated. The distribution $W_{\text{dat}}(E)$ of energies over the entire noncoding part of the *E. coli* genome is shown in Fig. 12.3b. We can compare this with the distribution $W_0(E)$ obtained from a random sequence with the same nucleotide frequencies (i.e., from a scrambled genome). The distribution $W_0(E)$ is approximately Gaussian as expected for a sum of independent random variables ϵ_i according to (12.1). The actual distribution $W_{\text{dat}}(E)$ is indeed of the same form as $W_0(E)$ for most energies. However, a closer look at the low-energy tail of the distribution shows

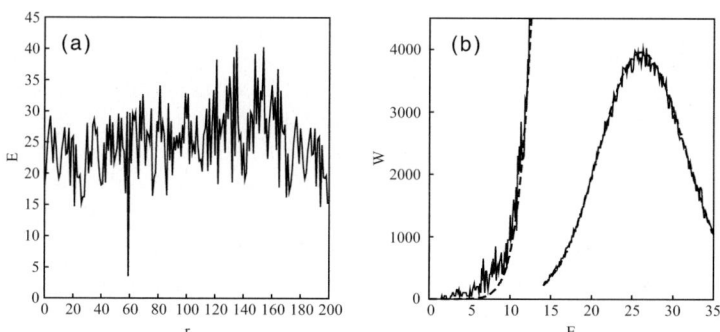

Fig. 12.3. Transcription factor binding energies of the *E. coli* genome. (a) Energy "landscape" $E(r)$ for specific binding of the CRP factor at 200 consecutive positions r in an intergenic region, with a binding site at position 59. (b) Count histogram $W_{\text{dat}}(E)$ with energy bins of width 0.1 obtained from all intergenic regions, together with the distribution $W_0(E)$ for a random sequence (*dashed line*, shown with a 30-fold zoom into the region $E < 14$; adapted from [15])

that there are significantly more strong binding sites than expected from a random sequence [13–15]. So at least some of them are there not by chance but for a reason.

12.2.3 Search Kinetics

All three thermodynamic modes of a factor molecule – free diffusion, unspecific binding, and specific binding – are important for the search kinetics toward a functional site [4–6]. The unspecific attraction causes the transcription factor to be bound to DNA with a finite probability, i.e., a given molecule spends about equal amounts of time on and off the DNA backbone. Hence, the search process is a mixture of effectively one-dimensional diffusion along the DNA backbone and three-dimensional diffusion in the surrounding medium. This proves more efficient than purely one- or three-dimensional diffusion. In the 1D mode, the factor diffuses in a flat energy landscape if it is in the conformation of unspecific binding, or in the landscape $E(r)$ if it is in the conformation of specific binding. In this way, it can sample the low-energy part of the landscape $E(r)$ while avoiding its barriers. The main obstacles on its way to a functional site are spurious binding sites, which have a low energy $E(r)$ by chance and act as traps. We lack a completely satisfactory picture of the search kinetics, which is an area of current research [13, 16]. However, this process proves to be remarkably fast. Typical search times are less than a minute, i.e., substantially shorter than typical functional intervals in a cell cycle of at least minutes. Therefore, the regulatory effect of a site is related to its probability of binding a factor molecule at equilibrium, which can be evaluated by standard thermodynamics.

12.2.4 Thermodynamics of Factor Binding

We start with the idealized but instructive problem of a single factor protein interacting with a genome of length $L \gg 1$, which contains a single functional site, while the rest of the sequence is random. Since the protein is bound to the DNA with a probability of about 1/2, we neglect the unbound state for the subsequent probability estimates and study only the bound protein, which is at equilibrium between specific and unspecific binding. At each position r, the likelihood of these two states is given by the Boltzmann factors $\exp[-E(r)/k_\mathrm{B}T]$ and $\exp[-E_u/k_\mathrm{B}T]$, respectively. Hence, the partition function for a single protein has the form

$$Z = \sum_{r=1}^{L} e^{-E(r)/k_\mathrm{B}T} + L\, e^{-E_u/k_\mathrm{B}T} . \tag{12.4}$$

The functional site, which is assumed to be positioned at $r = r_f$, must have a low specific binding energy $E \equiv E(r_f)$. We now single out this position and write

$$Z = e^{-E/k_BT} + \sum_{r \neq r_f} e^{-E(r)/k_BT} + L e^{-E_u/k_BT}$$
$$\approx e^{-E/k_BT} + Z_0, \tag{12.5}$$

where Z_0 is the partition function of a completely random sequence. The probability of the factor being bound specifically at the functional site is then

$$p(E) = \frac{e^{-E/k_BT}}{Z} = \frac{1}{1 + e^{(E-F_0)/k_BT}}, \tag{12.6}$$

where $F_0 = -k_B T \log Z_0$ is the free energy for a random genome. Thus, the binding probability depends on the binding energy in a sigmoid way, with a threshold energy $E = F_0$ between strong and weak binding. This strongly nonlinear dependence is known to physicists as a Fermi function.

It is easy to generalize the thermodynamic formalism to more than one factor molecule. Ignoring the overlap between close sites, each position r can be empty or be occupied either by an unspecifically or by a specifically bound factor. Using a chemical potential σ, the many-factor partition function can hence, be written as

$$Z(\sigma) = \prod_{r=1}^{L} Z(\sigma, r), \tag{12.7}$$

where $Z(\sigma, r)$ is a sum over the three thermodynamic states at position r,

$$Z(\sigma, r) = 1 + e^{\sigma - E(r)/k_BT} + e^{\sigma - E_u/k_BT}. \tag{12.8}$$

The chemical potential σ is determined by the number of factor molecules, n, via the relation $n = (d/d\sigma) \log Z(\sigma)$. For actual transcription factor numbers, which are of order $1 - 10^4$, this relation is well approximated by [13]

$$\sigma = \frac{F_0}{k_BT} + \log n. \tag{12.9}$$

The functional site is now occupied by a specifically bound factor with probability

$$p(E) = \frac{e^{\sigma - E/k_BT}}{Z(\sigma, r_f)} = \frac{1}{1 + e^{(E-F_0)/k_BT - \log n}}. \tag{12.10}$$

The binding probability – and hence the effects of the functional site on the regulated gene – are thus determined by the binding energy, the number of factor molecules, and on the genomic background (via the free energy F_0). The dependence $p(E)$ is a Fermi function with threshold energy $E = F_0 + k_BT \log n$, which is shifted with respect to the single-molecule case. Clearly, p is also a Fermi function of $\log n$ at fixed binding energy, with a threshold at $\log n = (E - F_0)/k_BT$.

12.2.5 Sensitivity and Genomic Design of Regulation

The regulatory machinery can be very efficient: in bacteria, it has been shown that single factor molecules can have regulatory effects. We can use (12.6) to enquire how the cell can reach this high level of sensitivity, following mostly [13]. We assume a minimal genome, which has a single functional site of maximum binding strength E^* and is otherwise random. If a single factor molecule is to affect regulation, its binding to the functional site must not be overwhelmed by the remainder of the genome. This leads to a criterion on the signal-to-noise ratio of regulatory interactions,

$$F_0 \gtrsim E^*, \tag{12.11}$$

which in turn imposes a number of constraints on the design of regulatory DNA:

1. In a random genome, there must be at most of order one minimum-energy binding sites, i.e., $L(1/4)^\ell \lesssim 1$. This gives a lower bound on the site length, $\ell \gtrsim \log L / \log 4$. For a bacterial genome ($L \sim 10^6$), we obtain $\ell \gtrsim 10$, which gives the right length of functional binding sites. However, this bound is not fulfilled in eukaryotes. Indeed, eukaryotic genomes use a different design with groups of adjacent binding sites.
2. For each minimum-energy site, there are ℓ suboptimal sites of Hamming distance 1 from the minimum-energy sequence. These must not suppress the binding to the minimum-energy site, i.e., $\exp(-E^*/k_B T) \gtrsim \ell \exp[-(E^* + \epsilon)/k_B T]$ in the two-state approximation. This gives a lower bound on the binding energy per nucleotide, $\epsilon/k_B T \gtrsim \log \ell \approx 2 - 3$.
3. Finally, the unspecific binding in the entire genome must not suppress the specific binding to a minimum-energy site, i.e., $\exp(-E^*/k_B T) \gtrsim L \exp(-E_u/k_B T)$. This produces a lower bound on the energy gap between unspecific and optimal specific binding, $(E_u - E^*)/k_B T \gtrsim \log L \approx 15$.

Quite remarkably, these bounds are fulfilled as approximate equalities in bacteria. Hence, the machinery of transcriptional regulation operates just at the threshold of single-molecule sensitivity, i.e, $F_0 \approx E^*$.

12.2.6 Programmability and Evolvability of Regulatory Networks

Of course, not every regulatory interaction is equally sensitive. To switch genes on or off, the cell uses the dependencies of the binding probability both on factor numbers and on binding energies. During the cell cycle, the level of n can vary over several orders of magnitude, say, between a few and tens of thousands of molecules. At a given value of n, the effects on the regulated genes differ since their functional sites have different values of E. The binding energies can change on evolutionary time scales by mutations of the site sequence, which leads to regulatory differences between individuals and, ultimately, between species. Both parameters are thus necessary to encode pathways in regulatory networks. This is most flexible if minimum-energy sites are indeed sensitive to a single factor molecule as discussed above. Differential *programmability* as a network design principle [13] thus favors complicated molecular structures with longer binding sites and larger binding energies. However, this competes with the *evolvability* of the system by a stochastic evolution process [17]. We have seen that the single-molecule sensitivity is just marginally reached in bacteria. This indicates that the actual machinery may result from a compromise between programmability and evolvability: binding sites are just complicated enough to work. It also indicates that genomic structures can only be understood from their evolution; this aspect will be developed further in Sect. 12.4.

12.3 Bioinformatics of Regulatory DNA

Predicting regulatory interactions between genes is clearly a key problem in bioinformatics, which is as important as the analysis of individual genes and proteins. It is not surprising that this problem is very difficult since, as we have discussed in Sect. 12.2, targeting regulatory input in a large genome is a tremendous signal-to-noise problem even for the cell itself. Its solution via the analysis of regulatory DNA requires finding statistical criteria to distinguish between functional binding sites and background sequence. A general introduction to the relevant sequence statistics can be found in [18].

12.3.1 Markov Model for Background Sequence

We begin by specifying a stochastic model for the nonfunctional segments of intergenic DNA. These are assumed to be Markov sequences with uniform single-nucleotide frequencies $p_0(a)$ ($a = A, C, G, T$). Hence, the probability of finding a given sequence has the factorized form

$$P_0(a_1, \ldots, a_k) = \prod_{i=1}^{k} p_0(a_i). \tag{12.12}$$

This assumption should not be taken too literally. The term "nonfunctional" refers to binding of a particular transcription factor. Intergenic DNA contains plenty of nonrandom elements with other functions (e.g., binding sites for other factors) or without known function (such as repeat elements). The salient point is, however, that most of intergenic DNA is well approximated by a Markov sequence with respect to binding of a given transcription factor. To make this more precise, we project the distribution $P_0(\mathbf{a})$ for segments of length ℓ onto the binding energy E as independent variable. Denoting the projected distribution for simplicity with the same letter P_0, we have

$$P_0(E) \equiv \sum_{\mathbf{a}} P_0(\mathbf{a})\, \delta(E - E(\mathbf{a})). \tag{12.13}$$

This distribution is close to the actual genomic distribution $W_{\text{dat}}(E)$ for most values of E, as we have seen in Fig. 12.3. It is possible to improve the background model by introducing small frequency couplings between neighboring letters [14, 15].

12.3.2 Probabilistic Model for Functional Sites

The sequences $\mathbf{a} = (a_1, \ldots, a_\ell)$ at functional sites of a given transcription factor are assumed to be drawn from a different distribution $Q(\mathbf{a})$. We write this distribution in the form

$$Q(\mathbf{a}) = P_0(\mathbf{a}) \exp[S(\mathbf{a})]. \qquad (12.14)$$

The quantity $S(\mathbf{a})$, which is called the *relative log likelihood score* of the distributions P_0 and Q, will turn out to have an important evolutionary meaning as well.

The single-nucleotide distribution $q_i(a)$ at a given position i within functional loci is obtained by summing the full distribution Q over all other positions

$$q_i(a) = \sum_{a_1, \ldots, a_{i-1}, a_{i+1}, \ldots, a_\ell} Q(\mathbf{a}). \qquad (12.15)$$

The set of these marginal distributions, $q_i(a)$ ($i = 1, \ldots, \ell$; $a = A, C, G, T$) is called the *position weight matrix* for binding sites of a given factor [19]. If the score function is additive in the nucleotide positions, $S(\mathbf{a}) = \sum_{i=1}^{\ell} s_i(a_i)$, the Q distribution has a factorized form, $Q(\mathbf{a}) = \prod_{i=1}^{\ell} q_i(a_i)$ with

$$q_i(a) = p_0(a) \exp[s_i(a)]. \qquad (12.16)$$

This additivity assumption is made in most of the existing literature since the position weight matrix (12.15) can be inferred from a sample of known functional site sequences, which in turn determines directly the single nucleotide scores (12.16). This scoring is the basis for a number of site prediction methods in single species and by cross-species analysis; see e.g., [19–23].

Here, we treat functional sites as coherent statistical units and do not make the assumption of additivity of the score function [15]. As will be discussed in the Sect. 12.4, functionality imposes correlations between the nucleotide frequencies within a functional site, preventing factorization of the Q distribution. Of course, it is not possible to reconstruct the full distribution $Q(\mathbf{a})$, which lives on a 4^ℓ-dimensional sequence space, from a limited sample of experimentally known functional sites. However, we can again project this distribution onto the binding energy as independent variable, $Q(E) \equiv \sum_{\mathbf{a}} Q(\mathbf{a})\delta(E - E(\mathbf{a}))$. Since all regulatory effects of a functional site depend on its sequence $\mathbf{a}$ only via the binding energy, we can also write the score as a function of the energy, $S(\mathbf{a}) = S(E(\mathbf{a}))$ (this will become obvious in

Sect. 12.4). Hence, the relationship (12.14) has the same form for the projected distributions,
$$Q(E) = P_0(E) \exp[S(E)]. \tag{12.17}$$

12.3.3 Bayesian Model for Genomic Loci

Assuming that functional loci are distributed randomly with a small probability λ, we now combine the models for background sequence and for functional sites into a model for the full distribution of sequences $\mathbf{a}$ in intergenic DNA,
$$W(\mathbf{a}) = (1 - \lambda) P_0(\mathbf{a}) + \lambda Q(\mathbf{a}). \tag{12.18}$$

(At the moment, we are ignoring the possible overlap between functional sites). In the language of statistics, this is a probabilistic model with *hidden variables*. The output of this model consists of pairs $(m, \mathbf{a})$: First, the model variable $m \in \{f, 0\}$ is drawn with probabilities λ and $1 - \lambda$ (i.e., a locus is labeled as nonfunctional or functional), then the sequence is drawn from the corresponding distribution $P_0(\mathbf{a})$ or $Q(\mathbf{a})$. However, only the sequence counts $\mathbf{a}$ are available data. The "hidden" variable m can be inferred from the data in a probabilistic way using Bayes' formula, which expresses the joint probability distribution of data and model in terms of its conditional and its marginal distributions
$$\text{prob}(\mathbf{a}, m) = \text{prob}(\mathbf{a}|m) \text{prob}(m) = \text{prob}(m|\mathbf{a}) \text{prob}(\mathbf{a}) \tag{12.19}$$
with $\text{prob}(\mathbf{a}) = \sum_m \text{prob}(\mathbf{a}|m) \text{prob}(m)$. We can solve for the conditional probability of the model for given data $\mathbf{a}$,
$$\text{prob}(m|\mathbf{a}) = \frac{\text{prob}(\mathbf{a}|m) \text{prob}(m)}{\sum_m \text{prob}(\mathbf{a}|m) \text{prob}(m)}. \tag{12.20}$$

For the probability of functionality, $\rho_f(\mathbf{a}) \equiv \text{prob}(f|\mathbf{a})$, this formula reads
$$\rho_f(\mathbf{a}) = \frac{\lambda Q(\mathbf{a})}{W(\mathbf{a})} = \frac{1}{1 + \exp[-S(\mathbf{a}) + \log \frac{1-\lambda}{\lambda}]}. \tag{12.21}$$

The dependence on S has again the form of a Fermi function. Its threshold value $S = \log[(1 - \lambda)/\lambda]$ separates sequences that are more likely to be functional or more likely to be background.

The full Bayesian model (12.18) can again be projected onto the energy variable,
$$W(E) = (1 - \lambda) P_0(E) + \lambda Q(E). \tag{12.22}$$

In this form, it can be tested against genomic data [15]. To plot the distributions P_0, Q, and W as functions of E, we use (12.1) with an energy matrix $\epsilon_i(a) = \epsilon_0 \log[q_i(a)/p_0(a)]$ estimated from the position weight matrix up to an overall constant ϵ_0 [10]. For our example of the CRP transcription factor, the distribution $Q(E)$ can be estimated from the about 50 known binding sites in

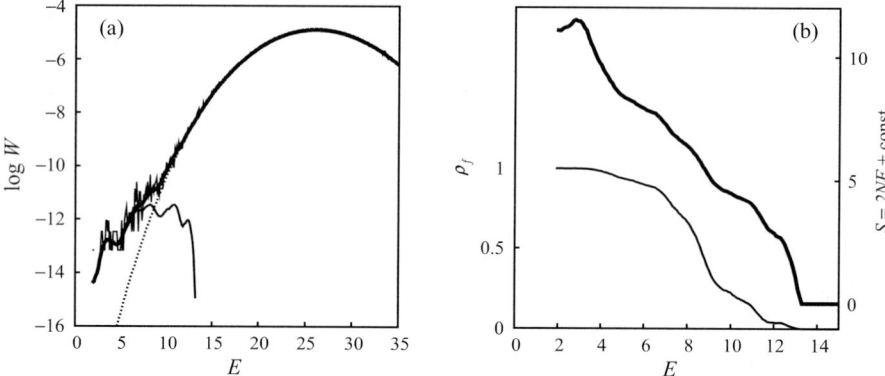

Fig. 12.4. Bayesian model for regulatory DNA and score function. (a) Energy count histogram $W_{\text{dat}}(E)$ for CRP sites in *E. coli* as in Fig. 12.3 (log scale), model distribution $W(E)$ (*thick line*), and its decomposition (12.22) into background component $(1-\lambda)P_0(E)$ (*thin dashed line*) and component $\lambda Q(E)$ ($E < E_s \approx 13$) of functional sites (*thin solid line*). (b) Log-likelihood score $S(E) = \log[Q(E)/P_0(E)]$ (shifted by a constant, *thick line*) and probability of functionality $\rho_f(E)$ (*thin line*) (adapted from [15])

the *E. coli* genome. Using this Q distribution and a probability of functionality $\lambda \approx 6 \times 10^{-4}$, the full distribution $W(E)$ produces an excellent fit of the count histogram $W_{\text{dat}}(E)$ over the entire range of energies; see Fig. 12.4a. The log likelihood score function $S(E) = \log[Q(E)/P_0(E)]$ is shown in Fig. 12.4b, shifted such that the curve has its zero at a point $E_s \approx 13$ beyond which binding becomes negligible.

The resulting probability of functionality $\rho_f(E)$ as given by (12.21) is also shown in Fig. 12.4b. This indicates the dilemma for the prediction of individual binding sites based on sequence data from a single species. Many functional sites have energies in the "twilight" region between the ensembles λQ and $(1-\lambda)P_0$, where ρ_f takes values around $1/2$. Hence, depending on the energy cutoff chosen, any prediction is torn between many false negatives or many false positives.

12.3.4 Dynamic Programming and Sequence Analysis

It is straightforward to generalize the Bayesian approach to longer segments of intergenic DNA, which are covered by an unknown number s of nonoverlapping functional sites as shown in Fig. 12.5 [21]. The hidden variables are now the sequence of left initial positions $\mathbf{r}_f \equiv (r_1, \ldots, r_s)$ of the functional sites (with the no-overlap constraint $r_{\nu+1} \geq r_\nu + \ell$ for $\nu = 1, \ldots, s-1$). The full sequence distribution in a segment of length L has the form

$$W_L(a_1, \ldots, a_L) = Z^{-1} \sum_{\mathbf{r}_f} \tilde{\lambda}^s W_L(a_1, \ldots, a_L | \mathbf{r}_f), \qquad (12.23)$$

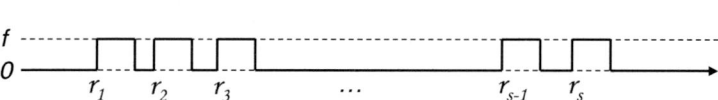

Fig. 12.5. Analysis of regulatory sequences. A configuration of s nonoverlapping binding sites is given by the sequence of left initial positions $\mathbf{r}_f = (r_1, \ldots, r_s)$ (with $r_{\nu+1} - r_\nu \geq \ell$ for $\nu = 1, 2, \ldots, s-1$). It can be associated with a path $m(r)$ which takes the values $m = f$ at the nucleotide positions of binding sites and $m = 0$ elsewhere. Dynamic programming algorithms based on a Bayesian model (12.24) of genomic sequences assign to each site configuration a probability of occurrence $\rho(\mathbf{r}|a_1, \ldots, a_L)$ for given sequence data $a_1, \ldots, a_L$; see (12.26)

where Z is a normalization factor, $\tilde{\lambda} = \lambda + O(\lambda^2)$ is a weight factor for each functional locus (the negligible correction terms originate from the no-overlap constraint), and $W_L(a_1, \ldots, a_L|\mathbf{r}_f)$ is the sequence distribution for given positions of functional loci,

$$W_L(a_1, \ldots, a_L|\mathbf{r}_f) =$$
$$p_0(a_1) \ldots p_0(a_{r_1-1}) \prod_{\nu=1}^{s} Q(a_{r_\nu}, \ldots, a_{r_\nu+\ell-1}) p_0(a_{r_\nu+\ell}) \ldots p_0(a_{r_{\nu+1}-1}) =$$
$$p_0(a_1) \ldots p_0(a_L) \exp\left[\sum_{\nu=1}^{s} S(a_{r_\nu}, \ldots, a_{r_\nu+\ell-1})\right] \quad (12.24)$$

with $r_{n+1} \equiv L + 1$. The sum over sequences $\mathbf{r}_f$ of arbitrary length s seems formidable at first, but W_L is easy to compute from the recursion

$$W_r(a_1, \ldots, a_r) = (1 - \hat{\lambda}) p_0(a_r) W_{r-1}(a_1, \ldots, a_{r-1})$$
$$+ \hat{\lambda} Q(a_{r-\ell+1}, \ldots, a_r) W_{r-\ell}(a_1, \ldots, a_{r-\ell}) \quad (12.25)$$

with the initial condition $W_0 = 1$ and $\hat{\lambda} = \tilde{\lambda} + O(\tilde{\lambda}^2)$. This type of recursion relation is usually called a *dynamic programming algorithm* in computer science. In physics, it is known as a *transfer matrix*, and the sum (12.24) is recognized as the corresponding discrete path integral in imaginary time r, if we interpret $\mathbf{r}_f$ as encoding a path $m(r)$ that takes the value $m = f$ at the nucleotide positions $r_\nu, \ldots, r_\nu + \ell - 1$ ($\nu = 1, \ldots, s$) within functional loci and $m = 0$ otherwise (see Fig. 12.5). Both concepts prove very useful also in more general problems of sequence alignment.

In analogy to (12.21), the probability of a set $\mathbf{r}_f$ of functional loci for given sequence data is

$$\rho(\mathbf{r}_f|a_1, \ldots, a_L) = \frac{W_L(a_1, \ldots, a_L|\mathbf{r}_f)}{W_L(a_1, \ldots, a_L)}. \quad (12.26)$$

The most likely set $\mathbf{r}_f^*$ can be obtained by the following "backward" algorithm: Given the sequence $(W_1, \ldots, W_L)$ obtained from the "forward" recursion (12.25), we can decide for every point r whether it is more likely to be a background position or the endpoint of a functional locus, ignoring all sequence information from positions $> r$. This depends on whether the leading contribution to W_r comes from the first or second term on the r.h.s. of (12.25) and defines the local optimum model $m^*(r)$. The global optimum set of functional loci respecting the no-overlap constraint is then $\mathbf{r}_f^* = \{r | b(r) = 1\}$, where $b(r)$ is given by the recursion $b(r) = \ell$ if $b(r+1) \leq 1$ & $m^*(r) = f$ and $b(r) = \max(b(r+1) - 1, 0)$ otherwise, with the initial condition $b(L+1) = 0$.

The Bayesian model can easily be extended to sequences containing several types of binding sites, which bind different transcription factors and are distinguished by their Q distributions. Dynamic programming algorithms can thus predict the likely coverage of a sequence with binding sites of known type [21]. This is the first step in extending the statistical analysis from single binding sites to entire regions of regulatory DNA. Indeed, models of this kind have been applied successfully to predict regulatory elements in eukaryotes, which typically consist of functional groups of adjacent binding sites. In the algorithms currently used, however, the scoring in (12.24) is strictly additive for groups of nonoverlapping binding sites: it does not take into account dependencies between the sites within one functional group or overlapping sites within one sequence.

12.4 Evolution of Regulatory DNA

In statistical picture developed so far, background sequences and functional sites are reduced to ensembles P_0 and Q. This picture is incomplete in two ways. On one hand, it is quite disconnected from the biophysical aspects discussed before: the specific function of binding sites hardly enters the standard formalism of position weight matrices. On the other hand, there is not yet any notion of time and dynamics. Sequences change by various mutation processes, and the observed sequence ensembles derive from this evolutionary dynamics. The evolution of functional loci is fundamentally different from that of background sequence: it is subject to *natural selection*, that is, the fitness of an organism depends on its genotype $\mathbf{a}$ at a functional locus via the effects on the regulated gene. At this point, the biophysics of binding enters the evolution of functional sequences [24–26]. Moreover, it becomes clear that the statistical framework has to be extended from individual sequences to distributions of genotypes in a population. In this section, we develop an evolutionary picture of regulatory DNA, from which we obtain expressions for the sequence ensembles P_0, Q, and the score function S. The next four paragraphs are a self-contained introduction to the underlying concepts of population genetics.

12.4.1 Deterministic Population Dynamics and Fitness

We start by describing the evolution of a large population, which contains individuals of different genotypes **a**. Each genotype is assumed to produce a specific *phenotype*, which may influence the reproductive success of the individuals carrying it. With respect to factor binding, the phenotype can be associated with the binding energy $E(\mathbf{a})$, since presumably all organismic effects of a locus depend on its genotype only via the binding energy. However, the discussion in the following paragraphs is more general. For a more detailed presentation, see e.g., [27].

We first assume that the subpopulations of a given genotype reproduce separately, i.e., there are neither transitions between genotypes through mutations nor (in a sexually reproducing population) mixing through genomic recombination. Writing the dynamics of the subpopulations in the form of simple growth laws,

$$\frac{d}{dt} N_{\mathbf{a}}(t) = F_{\mathbf{a}}(t) N_{\mathbf{a}}(t), \tag{12.27}$$

defines the (Malthusian) *fitness* $F_{\mathbf{a}}(t)$ of each genotype. For notational simplicity, we now limit ourselves to the case of just two genotypes **a** and **b**, where (12.27) can be written as growth laws for the total population size $N(t) \equiv N_{\mathbf{a}}(t) + N_{\mathbf{b}}(t)$ and for the population fraction $x(t) \equiv N_{\mathbf{b}}(t)/N(t)$ of genotype **b**,

$$\frac{d}{dt} N(t) = \bar{F}(t) N(t), \tag{12.28}$$

$$\frac{d}{dt} x(t) = \Delta F_{\mathbf{ab}}(t)\, x(t)[1 - x(t)] \tag{12.29}$$

with $\bar{F}(t) \equiv [1 - x(t)]F_{\mathbf{a}}(t) + x(t)F_{\mathbf{b}}(t)$ and $\Delta F_{\mathbf{ab}}(t) \equiv F_{\mathbf{b}}(t) - F_{\mathbf{a}}(t)$. This decomposition is useful since the overall growth rate $\bar{F}(t)$ is often strongly time-dependent because of external conditions (e.g., seasonality), while fitness differences, which reflect intrinsic properties of the phenotypes, are more stable. Different genotypes coexisting in a population frequently produce the same or very similar phenotypes and thus have equal fitness ($\Delta F_{\mathbf{ab}} = 0$).

Assuming $\Delta F_{\mathbf{ab}}$ to be constant over the time of observation, the solution of (12.29) is the evolutionary trajectory

$$x(t) = \frac{x_0 \exp[\Delta F_{\mathbf{ab}}(t - t_0)]}{1 + x_0(\exp[\Delta F_{\mathbf{ab}}(t - t_0)] - 1)} \tag{12.30}$$

with the initial condition $x(t_0) = x_0$, as shown in Fig. 12.6a. For $\Delta F_{\mathbf{ab}} \neq 0$, the fixed points of this dynamics are the monomorphic population states $x = 0$, and $x = 1$, of which $x = 1$ is stable for $\Delta F_{\mathbf{ab}} > 1$ and $x = 0$ for $\Delta F_{\mathbf{ab}} < 1$. The approach to the stationary state takes place on a characteristic time scale $\tau_d = 1/\Delta F_{\mathbf{ab}}$. In the important case of *neutral evolution* ($\Delta F_{\mathbf{ab}} = 0$), the evolutionary outcome remains indefinite. These results, which can readily

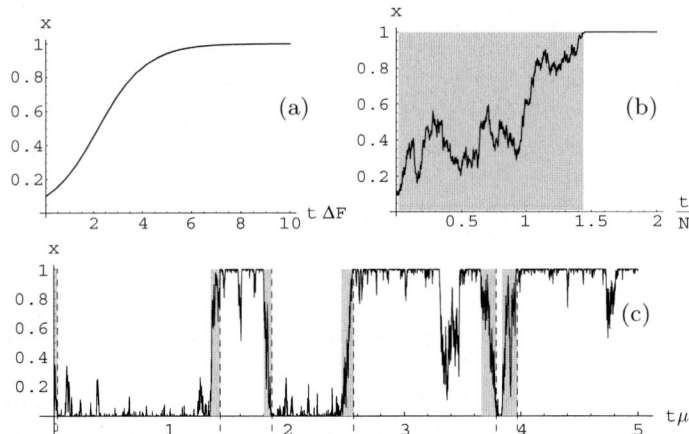

Fig. 12.6. Evolution of genotype composition $x(t)$. (**a**) Deterministic evolution with fitness difference $\Delta F_{\mathbf{ab}} > 0$, leading to certain fixation of genotype **b** (time is shown in units of $\tau_d = 1/\Delta F_{\mathbf{ab}}$). (**b**) Stochastic evolution with selection and genetic drift, leading to fixation of one of the genotypes. The time to fixation (*grey shading*) is of order τ_s ($N\Delta F_{\mathbf{ab}} = 0.5$, time is shown in units of N). (**c**) Stochastic evolution with selection, genetic drift, and mutations in the regime $N\mu \ll 1$, leading to a substitution dynamics with rates $u_{\mathbf{a}\to\mathbf{b}}$ and $u_{\mathbf{b}\to\mathbf{a}}$ given by (12.46). Substitution events are marked by *dashed lines*. The typical time between initial mutation and fixation (*grey shading*) for a given substitution, τ_s, is much shorter than the time between subsequent substitutions, $1/u_{\mathbf{a}\to\mathbf{b}}$ respectively, $1/u_{\mathbf{b}\to\mathbf{a}}$ ($N\Delta F_{\mathbf{ab}} = 0.5$, $N\mu = 0.05$, time is shown in units of $1/\mu$).

be generalized to more than two phenotypes, are a simple version of Fisher's *fundamental theorem of natural selection*: any population with initially coexisting phenotypes of different fitness will evolve toward a state where only the fittest phenotype is present.

Fisher's theorem seems to prove the popularized Darwinian notion of the "survival of the fittest." However, it rests on very restrictive assumptions that are never fulfilled in a natural population. The deterministic growth law (12.29) neglects mutations and recombinations as well as the reproductive fluctuations present in any population because of its finite number of individuals. These other evolutionary forces have to be incorporated in our theoretical picture before we can even define fitness as a measurable quantity and before the theory can address the important case of neutral evolution.

12.4.2 Stochastic Dynamics and Genetic Drift

Stochastic fluctuations of the reproduction process in a large but finite population have been studied extensively in population genetics, see [28, 29]. They are called *genetic drift*, an unfortunate name, which may falsely suggest a deterministic effect. To take these fluctuations into account, we replace (12.27) by a stochastic growth law,

$$\frac{d}{dt}N_\mathbf{a}(t) = F_\mathbf{a}(t)N_\mathbf{a}(t) + \chi_\mathbf{a}(t),\qquad(12.31)$$

where $\chi_\mathbf{a}(t)$ are Gaussian random variables with $\overline{\chi_\mathbf{a}(t)} = 0$ and

$$\overline{\chi_\mathbf{a}(t)\chi_\mathbf{b}(t')} = N_\mathbf{a}(t)\,\delta(t-t')\,\delta_{\mathbf{a},\mathbf{b}}.\qquad(12.32)$$

This form of noise is simply due to the law of large numbers, and the continuum dynamics (12.31) emerges as an effective large-N description for a plethora of discrete evolution models, which are defined at the level of individuals and have finite generation times. In the application to real populations, N has to be interpreted as the so-called *effective population size*, which can be inferred from genome data and is in general smaller than the actual population size.

In the case of two genotypes, (12.31) can again be projected onto the population fraction x,

$$\frac{d}{dt}x(t) = \Delta F_\mathbf{ab}(t)\,x(t)[1-x(t)] + \chi_x(t),\qquad(12.33)$$

where $\chi_x(t) = (\partial x/\partial N_\mathbf{a})\chi_\mathbf{a}(t) + (\partial x/\partial N_\mathbf{b})\chi_\mathbf{b}(t)$ are Gaussian random variables with zero mean and

$$\overline{\chi_x(t)\chi_x(t')} = \frac{x(1-x)}{N}\,\delta(t-t').\qquad(12.34)$$

This dynamics produces stochastic evolutionary trajectories $x(t)$ as shown in Fig. 12.6b. To capture their statistics, we convert the Langevin (12.33) into a Fokker–Planck equation for the probability distribution of the genotype composition [28, 30],

$$\frac{\partial}{\partial t}\mathcal{P}(x,t) = \frac{1}{2N}\frac{\partial^2}{\partial x^2}x(1-x)\mathcal{P}(x,t) - \Delta F_\mathbf{ab}(t)\frac{\partial}{\partial x}x(1-x)\mathcal{P}(x,t).\qquad(12.35)$$

The mathematical subtlety of this equation lies in the x-dependent diffusion "constant" $x(1-x)/2N$, which reflects the multiplicative nature of the reproduction process. As a consequence, the two monomorphic population states $x = 0$ and $x = 1$ are also fixed points of the stochastic dynamics. Any evolutionary trajectory $x(t)$ will eventually lead to one of these states with probability 1; this is called the *fixation* of the corresponding genotype in the population. In other words, the Fokker–Planck equation (12.35) describes diffusion in the interval $(0,1)$ with *absorbing boundaries*. There is a family of stationary states

$$\mathcal{P}(x) = (1-\phi)\delta(x) + \phi\delta(1-x),\qquad(12.36)$$

parametrized by the *fixation probability* ϕ of genotype **b**. The value of ϕ depends on the initial condition x_0 and can be computed by solving the backward diffusion equation

$$\frac{\partial}{\partial t}\mathcal{P}(x,t|x_0,t_0) = x_0(1-x_0)\left(\frac{1}{2N}\frac{\partial^2}{\partial x_0^2} - \Delta F_{\mathbf{ab}}(t)\frac{\partial}{\partial x_0}\right)\mathcal{P}(x,t|x_0,t_0). \tag{12.37}$$

For time-independent $\Delta F_{\mathbf{ab}}$, the stationary solution $\phi(x_0) \equiv \lim_{t\to\infty} \mathcal{P}(x=1,t|x_0,t_0)$ has the form [28, 30]

$$\phi(x_0, \Delta F_{\mathbf{ab}}, N) = \frac{1-\exp(-2N\Delta F_{\mathbf{ab}} x_0)}{1-\exp(-2N\Delta F_{\mathbf{ab}})}, \tag{12.38}$$

which for near-neutral evolution ($N\Delta F_{\mathbf{ab}} \ll 1$) reduces to

$$\phi(x_0, 0, N) = x_0 + N\Delta F_{\mathbf{ab}}\, x_0(1-x_0) + \cdots. \tag{12.39}$$

The characteristic time τ_s of the stochastic dynamics interpolates between the diffusive scale N and the deterministic scale: $\tau_s \approx \min(N, \tau_d)$. It determines the typical time of the evolution process up to fixation, shown shaded in Fig. 12.6b.

Hence, the stochastic population dynamics depends no longer only on the fitness difference of the genotypes as in the deterministic case, but also on the initial state of the population and the population size. Yet, our evolutionary picture is still incomplete. Population states with coexisting genotypes enter the dynamics as initial conditions, but since mutations are neglected, the model does not explain how this coexistence is generated and maintained.

12.4.3 Mutation Processes and Evolutionary Equilibria

At the level of an individual, mutations are rare stochastic genotype changes $\mathbf{a} \to \mathbf{b}$, which take place with rates $\mu_{\mathbf{a}\to\mathbf{b}}$, often coupled to the reproduction process. (These rates are all of the same order of magnitude, in estimates we, therefore, omit the indices.) We include mutations into the population dynamics (12.31) by their systematic effect on the genotype subpopulations,

$$\frac{d}{dt}N_{\mathbf{a}}(t) = F_{\mathbf{a}}(t)N_{\mathbf{a}}(t) + \sum_{\mathbf{b}}[\mu_{\mathbf{b}\to\mathbf{a}}N_{\mathbf{b}}(t) - \mu_{\mathbf{a}\to\mathbf{b}}N_{\mathbf{a}}(t)] + \chi_{\mathbf{a}}(t), \tag{12.40}$$

while their stochastic effect (whose variance is of order $N\mu$) is neglected since it is small against the reproductive sampling noise $\chi_{\mathbf{a}}(t)$. In the case of two different genotypes, this dynamics can again be projected onto the variable x,

$$\frac{d}{dt}x(t) = \Delta F_{\mathbf{ab}}(t)\, x(t)[1-x(t)] + \mu_{\mathbf{a}\to\mathbf{b}}[1-x(t)] - \mu_{\mathbf{b}\to\mathbf{a}}\, x(t) + \chi_x(t), \tag{12.41}$$

which leads to the Fokker–Planck equation [31]

$$\frac{\partial}{\partial t}\mathcal{P}(x,t) = \frac{1}{N}\frac{\partial^2}{\partial x^2}x(1-x)\mathcal{P}(x,t) - \Delta F_{\mathbf{ab}}(t)\frac{\partial}{\partial x}x(1-x)\mathcal{P}(x,t)$$
$$-\mu_{\mathbf{a}\to\mathbf{b}}\frac{\partial}{\partial x}(1-x)\mathcal{P}(x,t) + \mu_{\mathbf{b}\to\mathbf{a}}\frac{\partial}{\partial x}x\mathcal{P}(x,t). \quad (12.42)$$

For time-independent $\Delta F_{\mathbf{ab}}$, this equation has a single stable stationary state,

$$\mathcal{P}(x) = \frac{1}{Z}x^{-1+N\mu_{\mathbf{a}\to\mathbf{b}}}(1-x)^{-1+N\mu_{\mathbf{b}\to\mathbf{a}}}\exp(2N\Delta F_{\mathbf{ab}}\,x) \quad (12.43)$$

with a normalization constant Z that can be expressed in terms of Bessel and Gamma functions [32].

12.4.4 Substitution Dynamics

Here, we are interested in the stochastic evolution (12.42) and its equilibrium state (12.43) for $N\mu \ll 1$, which is the relevant dynamical regime for nuclear DNA in eukaryotes and in most prokaryotes (but not in viral systems). In this regime, the mutation term in (12.42) is small against the diffusion term except for values of x close to the boundaries 0 or 1. In this region, the continuum approximation of (12.42) is no longer valid, and (12.43) has to be replaced by a stationary solution $\mathcal{P}_d(N_{\mathbf{a}})$ of the underlying discrete evolution model, which gives the probability that the population contains $N_{\mathbf{a}}$ individuals of genotype $\mathbf{a}$ (with $N_{\mathbf{a}} = N - N_{\mathbf{b}} = 0, 1, \ldots, N$). The discrete solution is easily shown to have the singularity $\mathcal{P}_d(0) \simeq (N\mu_{\mathbf{a}\to\mathbf{b}})^{-1}\mathcal{P}_d(1)$. This singularity is correctly captured if we use the approximation $\mathcal{P}_d(N_{\mathbf{a}}) \simeq \int_{N_{\mathbf{a}}/N}^{(N_{\mathbf{a}}+1)/N} dx\,\mathcal{P}(x)$ for all $N_{\mathbf{a}}$ (except at the other boundary, where there is a similar singularity $\mathcal{P}_d(N) \simeq (N\mu_{\mathbf{b}\to\mathbf{a}})^{-1}\mathcal{P}_d(N-1))$ [33].

From this solution, we read off the following characteristics of the evolutionary dynamics at equilibrium, which are illustrated by the trajectory of Fig. 12.6c [32]:

1. For sufficiently small values of μ, the population remains monomorphic for most of the time. Using the shorthands $Q(\mathbf{a}) \equiv \mathcal{P}_d(N_{\mathbf{a}} = 0)$ and $Q(\mathbf{b}) \equiv \mathcal{P}_d(N_{\mathbf{a}} = N)$, we have

$$Q(\mathbf{a}) + Q(\mathbf{b}) = 1 - O(\mu N \log N). \quad (12.44)$$

2. The ratio of probabilities for the two monomorphic population states is given by the ratio of "forward" and "backward" mutation rate, the fitness difference, and the effective population size:

$$\frac{Q(\mathbf{b})}{Q(\mathbf{a})} = \frac{\mu_{\mathbf{a}\to\mathbf{b}}}{\mu_{\mathbf{b}\to\mathbf{a}}}\exp(2N\Delta F_{\mathbf{ab}}) + O(N\mu). \quad (12.45)$$

3. The monomorphic population states $x = 0$ and $x = 1$ are unstable because of mutations even at arbitrarily small values of μ, which cause occasional transitions of the entire population from genotype **a** to **b**, and vice versa. These so-called *substitutions* are marked by dashed lines in Fig. 12.6c. The substitution rate $u_{\mathbf{a}\to\mathbf{b}}$ can be evaluated as the product of creating a single mutant of genotype **b** in an initially monomorphic **a** population, $N\mu_{\mathbf{a}\to\mathbf{b}}$, and its probability of fixation, $\phi(x_0 = 1/N, \Delta F_{\mathbf{ab}}, N)$. The time between initial mutation and fixation (shown by grey shading in Fig. 12.6c) is still of order τ_s and thus much shorter than the time scale $1/\mu$, on which mutation effects become important. Hence, the fixation probability ϕ is given to leading order by (12.38), which has been derived for $\mu = 0$. Together, we have [28, 30]

$$u_{\mathbf{a}\to\mathbf{b}} = N\mu_{\mathbf{a}\to\mathbf{b}} \frac{1 - \exp(-2\Delta F_{\mathbf{ab}})}{1 - \exp(-2N\Delta F_{\mathbf{ab}})}. \tag{12.46}$$

Hence, the substitution rate $u_{\mathbf{a}\to\mathbf{b}}$ is enhanced over $\mu_{\mathbf{a}\to\mathbf{b}}$ for $\Delta F_{\mathbf{ab}} > 0$ and suppressed for $\Delta F_{\mathbf{ab}} < 0$, as shown in Fig. 12.7. For weak selection $(N|\Delta F_{\mathbf{ab}}| \ll 1)$, (12.46) becomes

$$u_{\mathbf{a}\to\mathbf{b}} = \mu_{\mathbf{a}\to\mathbf{b}}(1 + N\Delta F_{\mathbf{ab}} + \cdots). \tag{12.47}$$

This reproduces Kimura's famous original result: for neutral evolution, the substitution rate equals the mutation rate in an individual, independently of the population size. For this reason, the rates $\mu_{\mathbf{a}\to\mathbf{b}}$ are referred to as neutral mutation rates. For strong selection $(N|\Delta F_{\mathbf{ab}}| \gg 1 \gg |\Delta F_{\mathbf{ab}}|)$, (12.46) takes the asymptotic forms

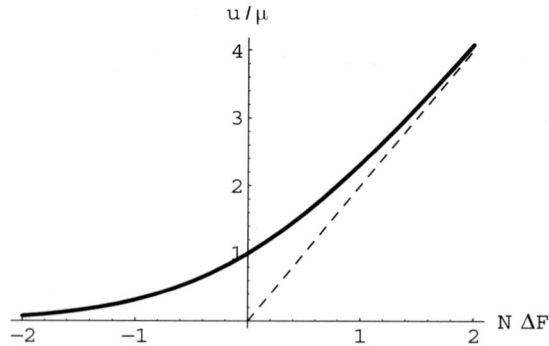

Fig. 12.7. Substitution rate in a population versus mutation rate in an individual. The ratio of these rates, $u_{\mathbf{a}\to\mathbf{b}}/\mu_{\mathbf{a}\to\mathbf{b}}$, depends on the product $N\Delta F_{\mathbf{ab}}$ of effective population size and fitness difference between the genotypes (in the relevant regime $N \gg 1$, $\Delta F_{\mathbf{ab}} \ll 1$, $N\Delta F_{\mathbf{ab}}$ finite). The substitution rate $u_{\mathbf{a}\to\mathbf{b}}$ is equal to $\mu_{\mathbf{ab}}$ for neutral mutations ($\Delta F_{\mathbf{ab}} = 0$), reduced for deleterious mutations ($\Delta F_{\mathbf{ab}} < 0$), and enhanced for advantageous mutations ($\Delta F_{\mathbf{ab}} > 0$)

$$u_{\mathbf{a}\to\mathbf{b}} = \mu_{\mathbf{a}\to\mathbf{b}} \begin{cases} 2N|\Delta F_{\mathbf{ab}}|\exp(2N\Delta F_{\mathbf{ab}}) & (2N\Delta F_{\mathbf{ab}} \ll 1), \\ 2N\Delta F_{\mathbf{ab}} & (2N\Delta F_{\mathbf{ab}} \gg 1). \end{cases} \quad (12.48)$$

The backward substitution rate $u_{\mathbf{b}\to\mathbf{a}}$ is given by a formula similar to (12.46) with $\Delta F_{\mathbf{ba}} = -\Delta F_{\mathbf{ab}}$. Forward and backward substitution rate have the simple ratio

$$\frac{u_{\mathbf{a}\to\mathbf{b}}}{u_{\mathbf{b}\to\mathbf{a}}} = \frac{\mu_{\mathbf{a}\to\mathbf{b}}}{\mu_{\mathbf{b}\to\mathbf{a}}} \exp(2N\Delta F_{\mathbf{ab}}) \quad (12.49)$$

for $N \gg 1$. Comparing with (12.45), we obtain the consistency condition

$$\frac{u_{\mathbf{a}\to\mathbf{b}}}{u_{\mathbf{b}\to\mathbf{a}}} = \frac{Q(\mathbf{b})}{Q(\mathbf{a})}. \quad (12.50)$$

Hence, for sufficiently small mutation rates ($\mu N \log N \ll 1$), a simple picture emerges. The evolution of a population can be described as a sequence of transitions between monomorphic genotype states (substitutions). The substitution rate u is determined by the corresponding mutation rate in an individual, the fitness difference between the genotypes, and the effective population size.

12.4.5 Neutral Dynamics in Sequence Space, Sequence Entropy

This evolutionary picture can be generalized to multiple genotypes, for example, the 4^ℓ-dimensional sequence space of genomic loci $\mathbf{a} = (a_1, \ldots, a_\ell)$. Transitions between different sequence states are point mutations $\mathbf{a} \to \mathbf{b}$, which change exactly one nucleotide. (We neglect here insertion and deletion processes, which change the length of the sequence). We first discuss neutral evolution, where the substitution rate $u_{\mathbf{a}\to\mathbf{b}}$ equals the mutation rate in an individual, $\mu_{\mathbf{a}\to\mathbf{b}}$, for all elementary transitions $\mathbf{a} \to \mathbf{b}$. Bona fide neutral mutation rates can be inferred from DNA sequence alignments of sufficiently close species, recent insights have also come from studying repeat elements.

We assume that the neutral dynamics has an equilibrium distribution $P_0(\mathbf{a})$, which obeys *detailed balance*, i.e., the relation

$$\frac{\mu_{\mathbf{a}\to\mathbf{b}}}{\mu_{\mathbf{b}\to\mathbf{a}}} = \frac{P_0(\mathbf{b})}{P_0(\mathbf{a})} \quad (12.51)$$

holds for each pair of sequence states linked by an elementary transition process $\mathbf{a} \to \mathbf{b}$. This says that the probability current at equilibrium, $\mu_{\mathbf{a}\to\mathbf{b}} P_0(\mathbf{a}) - \mu_{\mathbf{b}\to\mathbf{a}} P_0(\mathbf{b})$, vanishes for each elementary transition. Clearly, any distribution $P_0(\mathbf{a})$ satisfying the conditions (12.51) is stationary under the dynamics with rates $\mu_{\mathbf{a}\to\mathbf{b}}$, but not every such dynamics has a stationary distribution, which satisfies (12.51) (the simplest counterexample involving three states and a circular probability current $\mathbf{a} \to \mathbf{b} \to \mathbf{c}$ at stationarity). However, as will be verified below, detailed balance is a good approximation for the genomic substitution dynamics at least in prokaryotes. (There are known violations at CpG islands in eukaryotes [34]). In the simplest type of models, every nucleotide a mutates independently of all other positions with uniform rates $\mu_{a \to b}$ (i.e., $\mu_{\mathbf{a} \to \mathbf{b}} = \mu_{a \to b}$ for any two sequences $\mathbf{a} = (\ldots, a, \ldots)$ and $\mathbf{b} = (\ldots, b, \ldots)$ differing by exactly one nucleotide). This produces a factorized equilibrium distribution $P_0(\mathbf{a})$ of the form (12.12).

We can project the equilibrium distribution onto a measurable quantity as independent variable. For binding site sequences, a convenient choice is the binding energy E, and the projected distribution $P_0(E)$ has the form (12.13). Hence, we can define the *sequence entropy* [35]

$$S_0(E) = \log P_0(E), \tag{12.52}$$

which counts the log density of sequence states $\mathbf{a}$ at energy E, weighted by the distribution $P_0(\mathbf{a})$.

12.4.6 Dynamics Under Selection, the Score-Fitness Relation

The dynamics of substitutions can be studied in the same way for evolution under selection, which is specified at the level of genotypes by an arbitrary fitness function $F(\mathbf{a})$ [17, 36]. This generalizes the results of [37] for a model with selection acting independently at different nucleotide positions, i.e., $F(\mathbf{a}) = \sum_{i=1}^{\ell} f_i(a_i)$. For each elementary transition $\mathbf{a} \to \mathbf{b}$, the substitution rate $u_{\mathbf{a} \to \mathbf{b}}$ is determined by the neutral rate $\mu_{\mathbf{a} \to \mathbf{b}}$, the fitness difference $\Delta F_{\mathbf{ab}}$, and the effective population size N according to (12.46). Given the detailed balance (12.51) of neutral evolution and the relation (12.49) between forward and backward rates, it then follows immediately that the evolutionary dynamics under selection also obeys detailed balance, as given by (12.50) with an equilibrium distribution $Q(\mathbf{a})$ of the form (12.45). Thus, we have [17, 36]:

The equilibrium distribution $Q(\mathbf{a})$ of fixed genotypes generated by a substitution dynamics (12.46) with fitness function $F(\mathbf{a})$ is related to its neutral counterpart $P_0(\mathbf{a})$ by

$$Q(\mathbf{a}) = P_0(\mathbf{a}) \exp[2NF(\mathbf{a}) + \text{const.}], \tag{12.53}$$

with the constant given by normalization.

We can project (12.53) onto the fitness as independent variable. Defining the distribution $Q(F) \equiv \sum_{\mathbf{a}} Q(\mathbf{a}) \delta(F(\mathbf{a}) - F)$, similarly $P_0(F)$, and the sequence entropy $S_0(F) \equiv \log P_0(F)$, the projected identity takes the form

$$Q(F) = \exp[2NF + S_0(F) + \text{const.}] \tag{12.54}$$

For binding site sequences, we have a similar projection on the binding energy, $Q(E) = \exp[2NF(E)+S_0(E)+\text{const.}]$, since all genotypes with the same "phenotype" E have the same fitness, i.e., the same score S. The projected identities express the equilibrium distribution under selection in terms of fitness and sequence entropy, reflecting the balance between stochasticity (genetic drift) and selection [17]. For strong selection, the exponent $2NF - S_0$ is dominated by the fitness term, and $Q(F)$ takes appreciable values only at points of near-maximal fitness, i.e., where $F_{\max} - F \lesssim 1/2N$. For moderate selection, there is a nontrivial balance between both terms, and for weak selection, the Q distribution can be approximated by its neutral counterpart $P_0 = \exp(S_0)$. Clearly, the roles of fitness and sequence entropy are formally analogous to those of energy and entropy in statistical physics of thermodynamic systems, if $2N$ is identified with the inverse temperature $1/k_\text{B}T$. Some consequences of this analogy are discussed in [38].

The dynamics of substitutions establishes a rather general evolutionary grounding of genome statistics, if we identify the equilibrium distributions $P_0(\mathbf{a})$ and $Q(\mathbf{a})$ with the genomic distributions discussed in Sect. 12.3 as already anticipated by our notation. Comparing (12.53) and (12.14) gives a relation between fitness and score [15, 17]:

The log-likelihood score $S(\mathbf{a}) = \log[Q(\mathbf{a})/P_0(\mathbf{a})]$ equals the fitness function multiplied by twice the effective population size up to a constant,

$$S(\mathbf{a}) = 2NF(\mathbf{a}) + \text{const.}. \tag{12.55}$$

This relation allows us to use sequence data of a given genome to infer quantitative patterns of its evolution. We now discuss specific consequences for the evolution of regulatory DNA; an application to protein evolution can be found in [37].

12.4.7 Measuring Selection for Binding Sites

We first give a precise definition of functionality for regulatory (and other) elements: A binding *locus* is functional if the genotype at that locus is under selection (for binding of the corresponding factor). Nonfunctional loci have evolutionarily neutral genotypes. This definition asks whether binding at a given locus makes a difference to the organism or not. It is weaker than that of a functional *binding site*, which is a functional locus with a sequence $\mathbf{a}$ that is likely to actually bind the factor. A functional locus can lose its binding sequence because of deleterious mutations, leading to suboptimal fitness of the organism. Conversely, a nonfunctional locus can have by chance a sequence, which does bind the factor: this is a spurious binding site without consequences for the organism.

To measure the selection on functional sites in silico, we apply the identity (12.55) to the genomic distributions $P_0(\mathbf{a})$ and $Q(\mathbf{a})$. (Assuming equilibrium for most loci seems to be justified for our example of CRP binding sites in *E. coli* since we find very similar distributions in the distant bacterial species *Salmonella typhimurium*, and the factor protein itself is highly conserved between these species.) After projection onto the energy, the fitness landscape $2NF(E)$ for CRP binding sites is thus given by Fig. 12.4b [15]. The fitness is constant in the no-binding region ($E \gtrsim E_s \approx 13$) since the evolution is always neutral in that region. This constant is set to 0 in our normalization, i.e., $F(E)$ measures the fitness gain of functional sites due to factor binding. Loci with strong binding are also under strong selection, with effective fitness values $2NF$ of order 10. Genetic drift counteracts selection, producing also loci with weaker binding and reduced effective fitness. This fitness "landscape" is thus qualitatively of the form predicted from the underlying biophysics [17, 24]. Of course, it should be kept in mind that this landscape results from averaging over a family of binding sites, which may have a spectrum of individual selection coefficients and selected binding strengths.

12.4.8 Nucleotide Frequency Correlations

A further consequence of (12.54) is the generic occurrence of nucleotide frequency correlations within functional loci [17]. If the fitness function $F(\mathbf{a})$ is not additive in the nucleotide positions, nucleotide frequencies are correlated in selected genotypes even if they are independent under neutral evolution. This happens quite generically since selection acts on the entire genotype $\mathbf{a}$ as a functional unit and not on its single nucleotides. For binding sites, fitness effects follow from the expression level of the regulated gene, which depends on the sequence $\mathbf{a}$ via the binding probability of the corresponding transcription factor. While the binding energy is often approximately additive in the nucleotide positions as given by (12.1), the binding probability (12.10) is a strongly nonlinear function of the energy. This introduces correlations between nucleotide frequencies at *any two* positions within functional loci, preventing factorization of the distribution $Q(\mathbf{a})$.

12.4.9 Stationary Evolution of Binding Sites

Functional loci with a substantial level of selection (as found for the CRP binding sites in *E. coli*) evolve in a way quite different from background sequence. This is quantified in Fig. 12.8a, which shows pairs of binding energies (E_1, E_2) for experimentally verified CRP binding sites in *E. coli* and the corresponding sites regulating orthologous genes in *S. typhimurium* [15, 26]. The evolutionary distance t between the two species and characteristics of the neutral mutation process can be inferred from alignments of background sequence. The "phenotypic" evolution of CRP binding is quantified by the *energy transition probabilities* $G_0(E_2|E_1)$ under neutral evolution and $G_f(E_2|E_1)$ under stationary selection [15]. These are readily obtained by simulating the substitution dynamics over a time interval t for given initial value E_1, both with neutral rates $\mu_{\mathbf{a} \to \mathbf{b}}$ and with rates $u_{\mathbf{a} \to \mathbf{b}}$ given by (12.46) and the fitness function

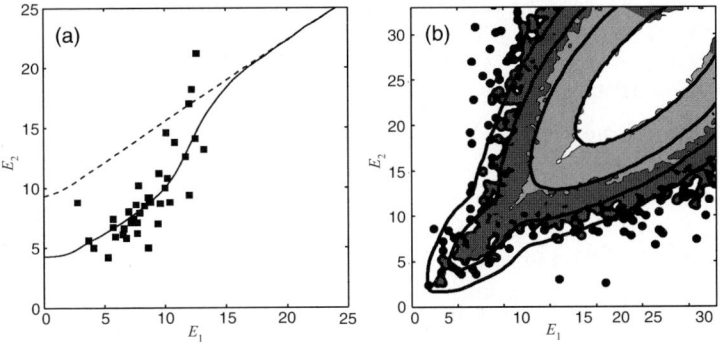

Fig. 12.8. Evolution of binding sites. (**a**) Binding energy pairs (E_1, E_2) for 32 experimentally verified CRP binding sites in *E. coli* from the DPInteract database [39] and their aligned orthologs in *S. typhimurium* (*dots*). Conditional expectation value for the binding energy in *S. typhimurium* under neutral evolution, $\langle G_0(E_2|E_1)\rangle$ (*dashed line*), and under selection, $\langle G_f(E_2|E_1)\rangle$ (*solid line*). (**b**) Distribution of energy pair counts $W_{\text{dat}}(E_1, E_2)$ (*filled contours*), compared to the distribution $W(E_1, E_2)$ given by the Bayesian model (12.59). The symmetry of these distributions under exchange of E_1 and E_2 reflects detailed balance of the substitution dynamics (adapted from [15, 40])

$2NF(E)$ measured in *E. coli*. The resulting conditional expectation values $\langle G_0(E_2|E_1)\rangle$ and $\langle G_f(E_2|E_1)\rangle$ for the binding energy in *S. typhimurium* are also shown in Fig. 12.8a. The data conform to the selection model, showing a substantially stronger conservation of binding energy than expected for neutral evolution [15, 26, 40].

We can now build a probabilistic model for cross-species comparisons [15]. It is based on the joint distributions of energy pairs

$$P_0(E_1, E_2) = G_0(E_2|E_1) \, P_0(E_1) \tag{12.56}$$

under neutral evolution and

$$Q(E_1, E_2) = G_f(E_2|E_1) \, Q(E_1) \tag{12.57}$$

under stationary selection, which are determined by the corresponding distributions in one species and the energy transition probabilities. Detailed balance of the substitution dynamics implies

$$\frac{P_0(E_2)}{P_0(E_1)} = \frac{G_0(E_2|E_1)}{G_0(E_1|E_2)} \quad \text{and} \quad \frac{Q(E_2)}{Q(E_1)} = \frac{G_f(E_2|E_1)}{G_f(E_1|E_2)}, \tag{12.58}$$

i.e., the joint distributions $P_0(E_1, E_2)$ and $Q(E_1, E_2)$ must be symmetric functions of their arguments. These distributions combine into a model for pairs of aligned loci, which generalizes the single-species model (12.22) and takes the form

$$W(E_1, E_2) = (1 - \lambda)P_0(E_1, E_2) + \lambda Q(E_1, E_2). \qquad (12.59)$$

(This model can be extended further to include nonstationary selection.) The distribution $W(E_1, E_2)$ with a fraction of functionality $\lambda = 0.0018$ is in excellent agreement with the count distribution $W_{\mathrm{dat}}(E_1, E_2)$ obtained from *E. coli* and *S. typhimurium*, as shown in Fig. 12.8b. The symmetry of W_{dat} thus corroborates the underlying assumption of detailed balance. Analogous Bayesian models can be defined for more than two species related by a phylogeny. This approach has been applied to binding site prediction in bacteria [15]; a related study of several species of fungi has been reported in [41].

12.4.10 Adaptive Evolution of Binding Sites

What does this picture say about the adaptive evolution of transcriptional regulation in response to a newly arising selection pressure? The evolution from a genotype with marginal binding ($E(\mathbf{a}) \approx E_s$) to strong binding requires only about three uphill point mutations in the fitness landscape of Fig. 12.4b, i.e., there is an effective fitness gain $2N\Delta F \approx 3$ per mutation. Hence, according to (12.48), the rate of uphill substitutions per locus is enhanced by a factor $2N\Delta F \cdot d(\mathbf{a}, \mathbf{a}^*)$ at least of order 10 over the neutral point mutation rate per nucleotide. At the same time, the downhill rate is strongly suppressed. This shows that the adaptive formation of a binding site from background sequence can indeed be a rapid mode of regulatory evolution, because of the substantial level of selection [17].

However, this mode is only efficient if adaptation can set in immediately after the selection pressure is established. In larger regulatory regions, the exact position of a binding site is often not important. We assume that the initial genome contains a set of $\tilde{L}$ *shadow sites*, i.e., positions $r_1, \ldots, r_{\tilde{L}}$ where a given sequence $\mathbf{a}$ would have the same regulatory effect. If one of these shadow sites has already a genotype with marginal binding, it acts as a "seed" for the onset of adaptation [42]. On the other hand, if all shadow sites of the initial genome have energy $E > E_s$, there is typically a substantial waiting time of neutral evolution before one of them reaches the threshold energy E_s. Assuming the initial genome to be entirely background sequence, it will contain at least one such seed if $\int_{E<E_s} P_0(E)\mathrm{d}E \gtrsim 1/\tilde{L}$, which is a joint condition on $\tilde{L}$ and the site length ℓ: the shadow regulatory region must be long enough and binding sites must be short enough. The example shows that the evolvability of regulation imposes constraints on genome architecture [17].

12.5 Toward a Dynamical Picture of the Genome

The relationship $S = 2NF + \mathrm{const.}$ between score and fitness is a cornerstone of the theoretical picture developed so far, which links its population genetic, bioinformatic, and biophysical arches. It relates a key evolutionary variable with the statistics of genomic frequency counts. The physical binding energy is an appropriate phenotypic variable on which fitness and score depend, because molecular function is determined by binding interactions.

We have discussed this picture for transcription factor binding sites, but it can be applied more generally to functional elements in genomes. It relates the statistics of these elements in one genome with their evolutionary dynamics, which is observed in cross-species comparisons. This dynamics is shaped by selection: The components of functional elements are coupled by a common fitness function. Hence, *functional correlations lead to evolutionary correlations*. These can be traced in the Q distribution over fixed genomes of a functional element; other methods use the statistics of polymorphisms within a population.

Thus, the picture of the genome as a system with multiple interactions has a fundamental dynamical significance. This is important since it allows us to trace functional modules from evolutionary patterns. We conclude the chapter with a brief outlook on various levels of functional integration for regulatory sequences.

12.5.1 Evolutionary Interactions Between Sites

Regulatory function is often determined not by single binding sites, but jointly by a group of sites in the same regulatory region [43]. An important mechanism is *binding cooperativity*, i.e., the formation of a protein complex between two (or more) factors bound to their corresponding DNA sites. The binding energy of this complex has the form $E = E_1 + E_2 + \Delta E_{12}$, where E_1 and E_2 are the energies of the factors bound individually and $\Delta E_{12} < 0$ is the energy gain because of the protein–protein interaction, which is of the order of a few $k_B T$. Cooperative binding has a number of functional effects [1]:

1. It increases the signal-to-noise ratio for the targeting of regulatory input to a specific gene, which is important in larger eukaryotic genomes, where single spurious binding sites are abundant in background sequence.
2. It sharpens the response of the binding probability to variations in the factor concentrations around their threshold value. This follows from the thermodynamics of two factors, which is a straightforward generalization of the case of a single factor discussed in Sect. 12.2.
3. It implements logical connections between regulatory input signals to a given gene. The simplest example is an AND connection between two factors, where the regulated gene is affected only if both factors are simultaneously present. This happens if the binding energies and factor concentrations are such that individual binding is weak but joint binding is strong. Larger groups of binding sites can encode a whole repertoire of more complicated logical functions [44].

Regulatory modules with several jointly acting binding sites are frequently found in eukaryotes. The functional coupling of sites in a module translates into interactions between these sites in their sequence evolution. The genomic functional element, i.e., the subset of the regulatory region on which selection acts, is the module as a whole. Its fitness $F(E_1, E_2, \Delta E_{12}, \dots)$ is a joint function of the binding energies as the relevant phenotypic variables [17, 24]. The evolutionary dynamics under this selection allows for a large number of *compensatory changes*, i.e., pairs of correlated substitutions changing two binding energies such that the fitness remains constant. These lead to nucleotide frequency correlations between different sites. Such compensatory changes have indeed been observed in experiments on *Drosophila* promoters [45].

12.5.2 Site–Shadow Interactions

In larger regulatory regions, there is a number of shadow sites where a binding sequence **a** would have a similar regulatory effect as at the functional sites present. In that case, the genomic functional element contains not only the functional binding sites but also the shadow sites. Once a functional site has disappeared due to deleterious mutations, a shadow site can turn functional by adaptive evolution as described in Sect. 12.4. The resulting evolutionary dynamics leads to sequence turnover with the actual binding sites present at different but functionally equivalent positions [36]. Substantial sequence turnover has been observed in a number of case studies [45–50]. Also the number of actual sites is subject to evolutionary variation since the same regulatory effect, i.e., the same fitness, can be distributed over fewer stronger or more weaker sites. With increasing number $\tilde{L}$ of shadow positions, one expects that the number of actual sites grows while individual sites get weaker [36].

12.5.3 Gene Interactions

Evolutionary interactions are not limited to regulatory elements for the same gene. An example are gene duplications and the subsequent evolution of the daughter genes. Selection acts jointly on this pair of genes [51], which have initially identical functions, eventually leading to either loss of one of them or to *subfunctionalization*, which has been argued to be an important mode of genome evolution in eukaryotes [52, 53]. This process can take place by regulation, i.e., via a correlated distribution of the regulatory elements on the daughter genes. More generally, the evolution of genes in a regulatory network is correlated if their functions are coupled either in series (i.e., one gene acts on the other) or in parallel (i.e., they are part of alternative pathways for the same function). Although some regulatory networks in model organisms – e.g., the embryonic development in the sea urchin [54] – have been studied in detail, we lack a coherent view of their functional evolution to date.

12.5.4 Evolutionary Innovations

Under stationary selection, functional elements are more conserved than background sequence, and the score-fitness relation quantifies the amount of conservation. But evolution is, of course, not limited to conservation. On one hand, there is typically a multitude of different genotypes yielding the same molecular function, and the evolutionary dynamics continuously plays with these alternatives. On the other hand, organisms face long-term changes of their environment, which lead to new selection pressures and a response by adaptive evolution of *new functions*. If regulation is to account for a large part of the diversification in higher eukaryotes, loss or gain of regulatory function should be an important mode of molecular evolution. Changes in regulatory DNA leading to new functions of gene networks have been observed [55], and it is possible to extend the statistical models described in Sect. 12.4 to include evolutionary gain or loss of function of individual binding sites [15]. On a broader scale, understanding the molecular basis of evolutionary innovations is a major challenge for theory and experiment in the coming years. It will profoundly change our dynamical view of the genome.

References

1. M. Ptashne, A. Gann, in *Genes and Signals* (Cold Spring Harbour Laboratory Press, Cold Spring Habor, 2002)
2. D. Tautz, Curr. Opin. Genet. Dev. **10**, 575 (2000)
3. G.A. Wray, M.W. Hahn, H. Abouheif, J.P. Balhoff, M. Pizer, M.V. Rockman, R.A. Romano, Mol. Biol. Evol. **20**, 1377 (2003)
4. O.G. Berg, R.B. Winter, P.H. von Hippel, Biochemistry **20**, 6929 (1981)
5. R.B. Winter, P.H. von Hippel, Biochemistry **20**, 6948 (1981)
6. R.B. Winter, O.G. Berg, P.H. von Hippel, Biochemistry **20**, 6961 (1981)
7. P.H. von Hippel, O.G. Berg, Proc. Natl. Acad. Sci. USA **83**, 1608 (1986)
8. A. Sarai, Y. Takeda, Proc. Natl. Acad. Sci. USA **86**, 6513 (1989)
9. D. Fields, Y. He, A. Al-Uzri, G. Stormo, J. Mol. Biol. **271**, 178 (1997)
10. G.D. Stormo, D.S. Fields, Trends Biochem. Sci. **23**, 109 (1998)
11. M. Oda, K. Furukawa, K. Ogata, A. Sarai, N. Nakamura, J. Mol. Biol. **276**, 571 (1998)
12. K. Omagari et al., FEBS Lett. **563**, 55 (2004)
13. U. Gerland, D. Moroz, T. Hwa, Proc. Natl. Acad. Sci. USA **99**, 12015 (2002)
14. M. Djordjevic, A.M. Sengupta, B.I. Shraiman, Genome Res. **13**, 2381 (2003)
15. V. Mustonen, M. Lässig, Proc. Natl. Acad. Sci. USA **15**, 15936 (2005)
16. M. Slutsky, L.A. Mirny, Bioinformatics **15**, 2539 (2002)
17. J. Berg, S. Willmann, M. Lässig, BMC Evol. Biol. **4**, 42 (2004)
18. R. Durbin, S.R. Eddy, A. Krogh, G. Mitchison, in *Biological Sequence Analysis* (Cambridge University Press, Cambridge, 1998)
19. G. Stormo, G.W. Hartzell, Proc. Natl. Acad. Sci. USA **86**, 1183 (1989)
20. G. Hertz, G. Stormo, Bioinformatics **15**, 563 (1999)
21. N. Rajewsky, N.D. Socci, M. Zapotocky, E.D. Siggia, Genome Res. **12**, 298 (2002)

22. E. van Nimwegen, M. Zavolan, N. Rajewsky, E.D. Siggia, Proc. Natl. Acad. Sci. USA **99**, 7323 (2002)
23. B. Lenhard, A. Sandelin, L. Mendoza, P. Engström, N. Jareborg, W.W. Wasserman, J. Biol. **2**, 13 (2003)
24. U. Gerland, T. Hwa, J. Mol. Evol. **55**, 386 (2002)
25. A.M. Moses, D.Y. Chiang, M. Kellis, E.S. Lander, M.B. Eisen, BMC Evol. Biol. **3**, 19 (2003)
26. C.T. Brown, C.G. Callan, Proc. Natl. Acad. Sci. USA **101**, 2404 (2003)
27. J. Hofbauer, K. Sigmund, in *The Theory of Evolution and Dynamical Systems* (Cambridge University Press, Cambridge, 1988)
28. M. Kimura, J.F. Crow, in *An Introduction to Population Genetics Theory* (Harper & Row, New York, 1973)
29. M. Kimura, in *The Neutral Theory of Molecular Evolution* (Cambridge University Press, Cambridge, 1983)
30. M. Kimura, Genetics **47**, 713 (1962)
31. M. Kimura, T. Ohta, Genetics **61**, 763 (1969)
32. M. Rouzine, A. Rodrigo, J.M. Coffin, Microbiol. Mol. Biol. Rev. **65**, 151 (2001)
33. D. Grün, M. Lässig, to be published.
34. P. Arndt, T. Hwa, Bioinformatics **21**, 2322 (2005)
35. L. Peliti, Europhys. Lett. **57**, 745 (2002)
36. J. Berg, M. Lässig, Biophysics (Moscow) **48**(Suppl. 1), 36 (2003)
37. A.L. Halpern, W.J. Bruno, Mol. Biol. Evol. **15**, 910 (1998)
38. G. Sella, A. Hirsh, Proc. Natl. Acad. Sci. USA **102**, 9541 (2005)
39. K. Robison, A.M. McGuire, G.M. Church, J. Mol. Biol. **284**, 241 (1998)
40. V. Mustonen, M. Lässig, Proc. Natl. Acad. Sci. USA **104**, 2277 (2007)
41. A.M. Moses, D.Y. Chiang, A.P. Pollard, N.I. Iyer, M.B. Eisen, Genome Biol. **5**, R98 (2004)
42. S. MacArthur, J. Brookfield, Mol. Biol. Evol. **21**, 1064 (2004)
43. D. Arnosti, Annu. Rev. Entomol. **48**, 579 (2003)
44. N. Buchler, U. Gerland, T. Hwa, Proc. Natl. Acad. Sci. USA **100**, 5136 (2003)
45. M. Ludwig, C. Bergman, N. Patel, M. Kreitman, Nature **403**, 564 (2000)
46. M. Ludwig, N. Patel, M. Kreitman, Development **125**, 949 (1998)
47. A. McGregor, P. Shaw, J. Hancock, D. Bopp, M. Hediger, N. Wratten, G. Dover, Evol. Dev. **3**, 397 (2001)
48. E. Dermitzakis, C. Bergman, A. Clark, Mol. Biol. Evol. **20**, 703 (2002)
49. J. Scemama, M. Hunter, J. McCallum, V. Prince, E. Stellwag, J. Exp. Zool. **294**, 285 (2002)
50. J. Costas, F. Casares, J. Vieira, Gene **310**, 215 (2003)
51. A. Wagner, Genome Biol. **3**, 1012 (2002)
52. M. Lynch, J.S. Conery, J. Struct. Funct. Genomics **3**, 35 (2003)
53. M. Lynch, J.S. Conery, Science **302**, 1401 (2003)
54. E. Davidson, Curr. Opin. Genet. Dev. **9**, 530 (1999)
55. A.P Gasch, A.M. Moses, D.Y. Chiang, H.B. Fraser, M. Berardini, M.B. Eisen, PLoS Biol. **2**, E398 (2004)

Part V

Populations

13

Drift and Selection in Evolving Interacting Systems

T. Ohta

There are various levels of interacting networks, within a protein or nucleic acid molecule, among proteins and nucleic acids, and genetic regulatory networks. At all these levels, interacting systems are dynamically evolving and repeated appearance of modular structures is noted. These modular structures repeatedly appear at different levels from proteins to genetic regulatory networks. Evolution of primary structure of proteins and nucleic acids depends on all these systems. For the evolution of such interacting systems, both drift and selection play important roles. Therefore, the near neutrality holds. The nearly neutral theory argues that slightly deleterious mutant substitutions disturb such systems mildly and occur by drift. Compensatory mutant substitutions are expected to follow. Drift and selection often become inseparable, such that formation and maintenance of networks depends on the interaction of drift and selection. By considering the effects of drift and selection on evolution of genetic regulatory elements that determine gene expression, the nearly neutral theory may be extended to morphological evolution, because organismal development is mainly controlled by regulation of gene expression.

It is now evident that random genetic drift is a main force of evolution of the primary structure of DNA and proteins. However, genetic information is steadily inherited, and for this, natural selection is responsible for the maintenance. Also genes and gene regulations (regulation systems) are dynamically evolving so that they are promptly responsive to environmental changes. Positive selection works in such occasions. The most interesting issue for me is how drift and selection interact each other. It is related to evolution of highly complex interaction systems at various levels. In this chapter, in connection to the nearly neutral theory, series of ideas and hypotheses are reviewed.

13.1 Hierarchy of Networks

The most fundamental network is within a protein or nucleic acid folding. To properly fold for functional proteins or RNA molecules, there are certain requirements on amino acid or RNA sequences. For example, Bastolla et al. [1] investigated this problem and estimated that the average fraction of "neutral connectivity" for myoglobin fold is about 0.6. In other words, the fraction, 0.4, of all amino acid substitutions results in disruption of the fold. This fraction is not constant but varies according to amino acid sequences [1].

The neutral connectivity is an advanced version of the neutral space [2,3], and is directly related to the rate of protein evolution. More than three decades ago, Dickerson [4] presented very clearly the relationship between the rate of evolution and structural constraint of proteins. He compared the evolutionary rates among fibrinopeptides A and B, globins, cytochrome c, and histone IV, and concluded that the more rigid the specifications for the protein, the slower is the accumulation of amino acid changes.

The next level of network is physical interaction among proteins and nucleic acids. In fact, various studies are being done on proteomes. Such interactions would impose constraints on amino acid or nucleic acid sequence changes. Fraser et al. [5] examined the correlation between the number of interactions and evolutionary rate of proteins, and found a significant negative correlation between the two. In other words, proteins with more interactions evolve slowly, in agreement with the prediction on constraints coming from "among-proteins" interactions, i.e., proteins that interact with many other proteins evolve slowly as compared with those with less interaction. However, it is noted that the Pearson's rank correlation ($r = -0.24$, with $P = 0.002$) is not high. This is thought to reflect interactions at other levels, such as protein folding mentioned before.

The most interesting network for evolutionary biologists is the genetic regulatory network. This is because gene regulation is fundamental to organismal development and therefore to evolution of form and shape of organisms. An interesting finding is the "turnover" of regulatory elements of transcription factor genes. The expression pattern of the gene, *even-skipped*, is fine tuned at a larval stage of *Drosophila melanogaster*. This pattern has been conserved among *Drosophila* species examined [6]. Nevertheless, an enhancer element that locates upstream of the coding region is turning over. Ludwig et al. [7] constructed a chimera of the element between *D. melanogaster* and *D. pseudoobscura*, and found that the function of the chimeric enhancer is not quite normal even if the two native ones work perfectly in *D. melanogaster*. From this finding, they suggested that the stabilizing selection is operating on the expression pattern. In other words, slightly deleterious mutations that mildly disturb the pattern are followed by compensatory mutant substitutions. Here, drift and selection interact for such a turnover process.

At all these levels of interactions, there exist repetitive structures. It is well known that proteins often consist of domains, which are made from modules.

Proteins owe their structures and functions to compact assembly of modules. At the "among-protein interaction level," again characteristic repetitive motifs are found such that motif constituents are conserved in "among-protein interaction networks" [8,9] (see also the collection of papers in [10]).

Enhancers of genes often form modular structures [11]. The modular structure must be useful for evolution of new regulatory networks via cooption of gene expression [6,11]. Once a new network becomes useful for the organisms, it will be fixed and established in evolution. Repetitive structures are also found in genetic regulatory networks themselves. Some structures, often called motifs, repeatedly appear in development and evolution, such that many combinations of transcription factors and their target genes occur much more frequently than random expectation [10,12]. Interaction of drift and selection is related to the formation and maintenance of all these interaction systems.

13.2 Drift and Selection, a Historical Perspective

The late 1960s were a remarkable time for the emergence of the field of molecular evolution. At that time, biochemical data were beginning to accumulate and they needed to be looked at from the point of view of evolution. Motoo Kimura was looking for data that he could connect with population genetics theory. The relevant raw data that were available pertained amino acid sequences. Zuckerkandl and Pauling [13] compared amino acid sequences of hemoglobin α and cytochrome c among several mammals and discovered the molecular clock that was totally unexpected to evolutionary biologists at that time.

It was Kimura [14] who first paid attention to the number of mutant substitutions per generation in a whole genome. His interest came from the concept of the genetic load, particularly the load due to advantageous mutant substitutions [15]. The load is called the cost of natural selection. Haldane [15] estimated the amount of genetic death needed for an advantageous mutation to be fixed in the population, and argued that the number of advantageous mutant substitutions in a whole genome would be one in 300 generations. This is because too much genetic death would cause extinction of the species. Kimura extrapolated from the number of mutant substitutions per protein locus to the same number per total genome, and obtained one nucleotide substitution per two years in mammalian lineage. He argued that this figure was much too large as compared to the Haldane's estimate. This led him to propose the neutral theory of molecular evolution.

In the next year, King and Jukes [16] published the article under the title, "Non-Darwinian Evolution," where they argued that most mammalian DNA does not code for proteins and the load argument was not appropriate. In the same year, Kimura [17] published another article where he argued that the molecular clock was a strong evidence for the neutral theory. He further predicted that genes in living fossils might be expected to have changed by the same amount as corresponding genes in more ordinary species.

During the 1970s, electrophoresis was used to measure protein polymorphisms. A most notable observation was a narrow range of heterozygosity among various species. Lewontin [18] observed that the heterozygosity among diverse species ranged from 5.6% in the house mouse to 18.4% in the fly, *Drosophila willistoni*. He argued that such a small range of heterozygosity could not be explained by the neutral theory and was strong evidence against the theory.

Almost ever since the neutral theory was proposed, I have had several questions about the theory. The first one was whether or not natural selection was so simple as to distinguish neutral mutations from selected ones. It seemed to me that there might be numerous mutations whose effects were so small that both random drift and selection influenced their behavior. The second question was concerned with the molecular clock: the clock looked as if it depended on chronological time rather than on generation number. This was contrary to the belief that generation length provided the appropriate time scale over which mutation rate should be measured. The third question concerned with narrow range of heterozygosity in many species as measured by electrophoresis. Remember that, under the neutral theory, the heterozygosity should depend on the population size, since neutral mutations may accumulate indefinitely in large populations.

I had been much puzzled by these three questions and realized that by bringing very slightly deleterious mutations into the neutral theory, these puzzles could be explained. The slightly deleterious mutation theory was first published in [19], and then in [20]. Later, it has been named the nearly neutral theory.

13.3 Molecular Clock and Near-Neutrality

Evolutionary rate of primary structure of proteins and RNAs, such as ribosomal and transfer RNA, is roughly constant, a phenomenon called molecular clock. However, it is now known that some irregularities exist in molecular clock. It is related to the networks at all levels considered above, since they are maintained or modified in evolution. Interactions coming from higher order structures have the strongest effects. It is well known that histone IV is very conservative because of the structural constraint, whereas fibrinopeptides A and B change freely. Interactions at other levels also have significant effects, and molecules with similar structures evolve by different rates. Examples are functional domains of ftz and bcd, which are believed to have arisen from those of Hox genes and are evolving much faster than the latter [21], and different evolutionary rates among gene members belonging to the growth hormone-prolactin family [22].

Effect of protein interaction on evolutionary rate is seen in the following example. The $\alpha_1\beta_2$ domain contact area of hemoglobin α and β has been relatively conserved in vertebrates as compared with the changes in the area of nascent α and β genes [23]. In many cases, however, effect of protein interactions may not be easily found even if they exist.

As mentioned already, genetic regulatory interactions are most interesting for molecular evolutionary biologists. If a gene is recruited and beginning to perform a new function, amino acid sequence of the gene product would need some modifications. The recruitment may result in an addition of a function to already existing ones. The gene product cannot be freely modified, and the elevation of evolutionary rate may not be revealed.

This process is sometimes coupled with gene duplication. Since duplicated gene copies are free to change, acceleration of amino acid substitutions is easily noted. Lynch and Force [24] formulated a simple case of gene duplication with differential expression. They call it "subfunctionalization" of duplicate genes. Their interesting finding is that the probability of subfunctionalization is higher in small populations than in large ones. This process is intrinsically stochastic. In any case, gene duplication often accompanies acceleration of amino acid substitution rate [25, 26].

Genes that have fixed functions for a long time usually show molecular clock. However, note that even if the functions do not change, nearly neutral mutant substitutions are characterized by rapid evolution in small populations and by slow change in large ones [20, 27, 28]. Another confounding factor is variation in mutation rate among lineages, i.e., mutation rate depends on the number of cell generations, mutation–repair systems, and others. I have examined the mean and the variance of amino acid substitution rates of 49 genes of three orders of mammals, primates, artiodactyls, and rodent [29]. The mean number of substitutions per site of the three branches is given in Fig. 13.1. In the figure, the rodent branch is relatively long while the primate branch is short, reflecting the generation time effect. However, the pattern is not so conspicuous in nonsynonymous tree as compared with synonymous tree. This is because weak selection (mostly slightly deleterious) had stronger effect in rodent lineage than in primate lineage. Remember that the former has had larger population size than the latter. However, when divergence pattern of mouse–rat sequences is compared with that of human–mouse or human–rat sequences, such a difference is not found [30]. Estimation of synonymous substitution number may not be quite accurate for calculating divergence of human–rodent sequences, and the pattern was not found. Also mutation–repair system differs between rodent and primate [31] and causes the discrepancy. More recently, Jorgensen et al. [32] performed a much larger-scale analyses of mammalian sequences than earlier studies [29], and showed the highest average ratio of nonsynonymous to synonymous substitutions in the human lineage, followed by the pig and then the mouse lineages. So the previous results were verified.

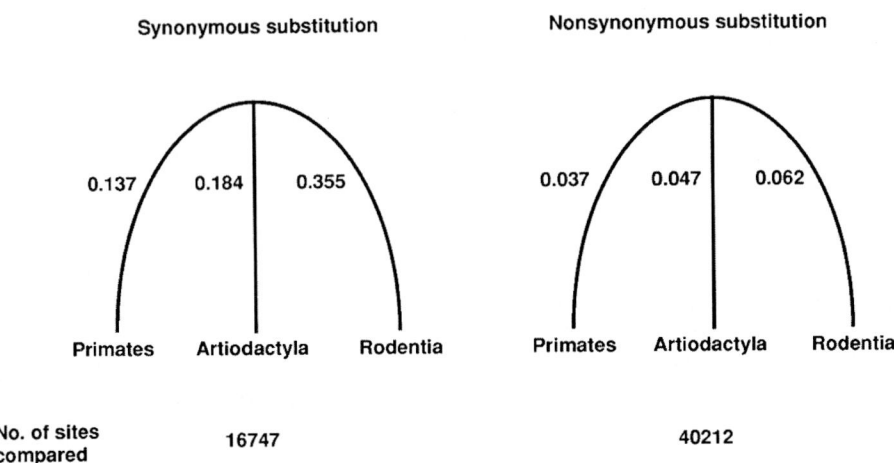

Fig. 13.1. Star phylogenies of 49 genes. Figures beside each branch are the estimated number of substitutions per site (adapted from [29])

Genome sequence data have become available, and large-scale analyses are being carried out. Many of such analyses reveal abundance of slightly deleterious substitutions. Lu and Wu [33] found that synonymous substitutions on the X chromosome of human and chimpanzee are less frequent than those on the autosomes. Based on the consideration that genes on X are more effectively selected via hemizygous males than those on autosomes, these authors conclude the abundance of slightly deleterious synonymous changes. They further observed that X-linked nonsynonymous substitutions are more frequent than autosomal ones. This would suggest occurrence of recessive advantageous amino acid substitutions that are weakly selected [33]. Gu et al. [34] carried out a genome-wide comparison of the evolutionary rates between a laboratory strain and a wild strain of *Saccharomyces cerevisiae*. They found that genes in the former strain tend to evolve faster than that in the latter, and attributed this result to the increase of drift in the laboratory strain. Keightley et al. [35] paid attention to the conservation of noncoding regions of mammalian genomes that are thought to be regulatory regions, and found that hominid genomes are less conserved than rodent genomes. They argue that this is caused by the reduction of the effectiveness of weak selection in hominids. Hughes [36] carried out statistical analyses on polymorphisms of coding regions of bacterial populations and showed that various statistics indicate abundance of slightly deleterious amino acid substitutions.

The variance of the evolutionary rate is often larger than that of the simple Poisson process and indicates some irregularities of molecular clock [37]. Such substitution patterns are thought to reflect influences coming from various interacting systems.

13.4 Mutants' Effects on Fitness

A most critical quantity on near neutrality is the fixation probability of mutant genes in a population. Let us examine the simplest case of semidominance with selective advantage, s. The fixation probability, u, is a monotonically increasing function of the product of the population size N and s, Ns,

$$u = \frac{1 - \exp(-4Nsp)}{1 - \exp(-4Ns)}, \qquad (13.1)$$

where p is the initial frequency in the population [38]. It is clear that, when discussing evolution by nearly neutral mutants, one has to consider all mutants around $Ns = 0$. So it is important to examine the frequency distribution of occurrence of new mutations around this point.

There have been various attempts to estimate mutants' effects on fitness. Recent three papers [39–41] emphasize adaptive amino acid substitutions but present evidence that most of amino acid changes are weakly selected. Sanjuan et al. [42] estimated the distribution of fitness effects of mutations (in terms of single nucleotide substitution) of an RNA virus by direct measure, and found that distribution is skewed but continuous at the neutrality point. Nielsen and Yang [43], by using the codon-based maximum likelihood method, estimated the distribution of selection coefficient for viral data and for mitochondrial gene data of primates. It was found that the normal and gamma distribution fit the data. Piganeau and Eyre-Walker [44], on the other hand, used the McDonald–Kreitman (MK) test [45], relative numbers of nonsynonymous to synonymous divergences within species and between closely related species for estimating the frequency distribution. They showed that the model of constant selection coefficient does not fit the observed pattern of polymorphism versus between-population divergence, by analyzing mitochondrial genes of several species. Their results indicate that partially reflected gamma distribution fits most data sets.

It is convenient to assume that the distribution of selection coefficient for nearly neutral mutations is normal. Let u_s and σ^2 be the mean and the variance of the distribution. The most important factor is twice the product of the population size N and the standard deviation of the distribution, $2N\sigma$. Note that the factor 2 comes because we are considering the diploid population. This product is analogous to $2Ns$ of the ordinary selection model where s is the selection coefficient. The present model is called the fixed model of selection coefficient [46, 47], in that the distribution is fixed irrespective of mutant substitutions. Note that in the shift model, on the other hand, the fitness distribution shifts according to the substituted mutants' effects, since the population shifts back to the original state after mutant substitution [2, 48].

Results of our simulation studies on the fixed model show that, when $4N\sigma > 3$, several advantageous mutants are quickly fixed depending upon the starting condition and thereafter almost all new mutations are deleterious [47]. When $4N\sigma$ is in the range, 0.2–3.0, both random drift and selection affect the population fitness, which reaches equilibrium among mutation, selection, and drift. Nielsen and Yang's analyses on viral and mitochondrial data suggest that this value is small and nearly neutral model is applicable. Piganeau and Eyre-Walker's analysis indicates that the distribution is of high leptokurtosis. Because McDonald–Kreitman test depends on segregating polymorphisms, strongly deleterious mutants are included and may have influenced. Furthermore, bottlenecks at the speciation that is not incorporated in the analyses may have some effects.

One should recognize that the distribution would differ depending on function and environment. Thatcher at al. [49] attempted to measure fitness effects of nonessential genes in yeast. Competition experiment of disruption strains against wild type was performed, and it was found that the distribution of selection coefficient is continuous, from almost no effect to considerable decrease. It was also noted that some disruption strains showed marginal fitness effect, whereas fitness of others depends on environments. In other words, they are contingent. Remold and Lenski [50] examined the effects on fitness of 18 insertion mutations of *E. coli*, in five different genetic backgrounds (two genomes isolated from different populations evolved on glucose, two genomes from different populations evolved on maltose, and their common ancestral genome), and in two different environments (glucose and maltose environments). They found that half the mutations had no effects of genetic backgrounds and of environments. The remaining half had some effects of genetic backgrounds and/or of environments. One-third had the effects of genetic backgrounds in an environment-dependent way. One-sixth of mutations had different effects in different genetic backgrounds in an environment-independent way. They conclude that fitness effects of many mutations depend on both genetic backgrounds and environments.

As for mitochondrial genes, their functions belong to the housekeeping type, and most of them have been fixed long time ago. In terms of the long term selection effects in evolution, the distribution of fitness effect would be narrow with mean at highly negative side as compared with genes of contingent function. For genes that are more responsive to environmental factors, the fitness distribution would be of more flat shape. This is because mutants' effects of contingent types such as those of gene regulation and signal transduction, the mutants would act on phenotypes and morphological characters that are targets of natural selection.

How does environment affect the fitness distribution, particularly in relation to simple versus diverse environments? Let us consider the influence of a bottleneck on the distribution of effects on fitness. When the population size is small and the population occupies a simple environment, the variance of the distribution would be large as compared with the value of a large population occupying diverse environment. This is because the fitness effect is averaged over many environmental conditions for diverse environment [27]. On the other

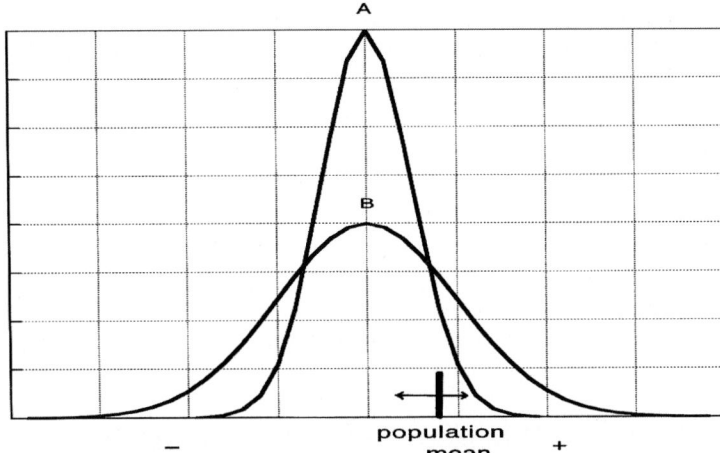

Fig. 13.2. Schematic diagram showing the frequency distribution of the fitness effects of new mutations. *Curve A* is for a large population that occupies a heterogeneous environment, and *curve B* is for a small population that occupies a simple environment (adapted from [27])

hand, the mean of the distribution would be unaltered. Figure 13.2 shows the comparison of the distribution in a simple environment with that in a diverse environment.

The situation may be compared with the case of artificial selection. Innan and Kim [51] showed that artificial selection differs from natural selection such that beneficial alleles may be fairly common in the founding population, resulting in fewer hitchhiking effects. As pointed out by the authors, even in nature, for isolated small populations, environmental simplicity together with bottleneck effects may provide similar conditions as observed in the artificially selected lines.

The most responsive class of mutations to environmental factors is the one that changes gene expression. In many genes, regulation of their expression is controlled by the interaction between transcription factors and regulatory elements. As presented before, the turnover of a regulatory element of a transcription factor gene of *D. melanogaster* is governed by stabilizing selection for keeping the pattern [7], i.e., slightly deleterious mutant substitutions are followed by compensatory ones, and the nearly neutral theory is applicable. The distribution of mutants' effects at regulatory elements would have larger variance than that of amino acid changes.

As already argued, the fitness distribution of mutants' effects is narrow with deep negative peak for mitochondrial genes that have had fixed functions for a long period of time. This is because the past selection had brought the system to a well-adapted state. On the other hand, the distribution of relatively new genes like hormones is flat and the mean may be close to zero [52]. The distribution is also flat for regulatory elements that are being coopted. Population size effect that reflects environmental diversity is significant in providing opportunities of finding out favorable combinations of cooptions.

Lynch and Cornery [53] proposed that genome complexity has increased in eukaryotes because of its small population size as compared to prokaryotes. In other words, drift enabled various nonadaptive processes to occur and thus provided novel substrates for evolution. In terms of distribution of mutants' effects on fitness, many of such nonadaptive processes are of a contingent type, and transition from nonadaptive to adaptive ones would be dependent on genetic as well as on environmental factors.

13.5 Evolution of Form and Shape: Cooption

Organismal development is controlled by spatial and temporal regulation of gene expression. Evolutionary processes of new forms and shapes are accomplished by cooption or recruitment of genes expressed. It is a process of forming a new regulatory network. It appears that associations of enhancer and transcription factor involve loose interactions among available transcription factors and their binding sites. New recognition sites are being created by random mutations. According to Hahn et al. [54], spurious transcription binding sites are very weakly selected against $Ns \approx 0.1$. Hence they are almost neutral and spread in populations by random drift. While drifting in populations, certain favorable combinations of binding sites would appear and cooption would take place.

Let us consider this process in a little more detail. In analogy with ordinary mutant substitutions, cooption would consist of an occurrence of a favorable or at least nondeleterious gene expression in one genome (phase 1) and increase of the frequency of this locus in the population (phase 2). The question is under what conditions does this rapid change in cooption occur? In other words, which are among a subdivided large population as in Wright's shifting balance theory, a small population as in the nearly neutral model, and a large population as in Fisher's model, gives the best condition for rapid evolution?

Wray et al. [55] review various aspects on transcriptional regulations in eukaryotes, and summarize the studies on correlation between gene expression and anatomy: induced mutations, comparison of expressions, and quantitative genetics. Through such investigations, it is now clear that gene expression is highly polymorphic in natural populations, and many important phenotypic variations may be attributed to such polymorphisms of gene expression. It is also possible that variations coming from gene regulation are mostly quantitative. Furthermore, activities of transcription factors are redundant with overlapping functions, i.e., more than one coexisting transcription factors may recognize the identical binding site [55]. Therefore, quantitative regulation of gene expression must be dependent on complicated networks among transcription factor-binding site interplay. Such a network should be robust, and with high variability of expression level at each locus.

In fact, the phases 1 and 2 become inseparable. Consider the situation where many newly arisen enhancers are drifting in a population. Transcription factors often interact synergistically with other transcription factors, and cooption takes place if two nearby recognition sites are available. According to [6], the average cell contains roughly 60 different transcription factors in case of yeast. They estimate that three recognition sites of six base pairs exist in every 200 base pairs if 60 transcription factors recognize 60 different sequences of six base pairs. This abundance of recognition sites implies that two nearby sites for two kinds of transcription factors are readily available, and many genes are expressed awaiting cooption. It is also remembered that not all new binding sites are functional, since expression of near-by genes is context dependent such as chromatin structure and the relationship with other binding sites [55].

Some of such new enhancers increase their frequencies by drift, and cooption would start with them. There must be numerous trial and errors before real cooption takes place. It means reformation of self-organizing regulatory networks different from the previous one. It would be likely that frequency of a new enhancer is high or fixed in a population via random drift before cooption. Then the phase 1 of cooption may be different from the case of amino acid substitutions. If a transcription factor were newly expressed, it would alter the downstream pathway of gene expression. Since gene transcription is redundant and an element of complex network, the fitness effect of a new binding site would be positive or negative depending much on genetic as well as on outside factors. So the selection coefficient of new gene expression would be more responsive to environment than amino acid changes. The mean of the distribution of selection coefficient is close to zero, and the variance is larger than the case of amino acid substitutions.

For amino acid changes of mitochondria, the value of $2N\sigma$ is estimated around one [43]. For a new enhancer, this value would be larger, and numerous nearly neutral variations are expected that are slightly deleterious as well as slightly advantageous. In fact, it has been reported that regulatory elements are characterized by very high polymorphisms [56].

According to Davidson [11], evolution of genetic regulatory network is characterized by "bottom-up" evolution of genetic regulatory network pattern. In this process, a set of terminally expressed (battery) genes may be kept during evolution, however the regulatory network changes and becomes more complex by cooption of additional transcription factors. Such modification of network enables differential expression of genes in space and time of developing organisms with more complicated forms than their ancestor.

Now consider effects of genetic background and environmental factors on the fitness distribution for new enhancers. Wilkins [57] reviewed changing patterns of mutants' effects on gene expression depending on genetic, as well as environmental conditions. Genetic background effects come from variations in interacting pathways and networks. Any mutations at a gene participating in a pathway may influence other genes in this pathway because of their connection through pathway (see [57, pp. 356–360]). Mutants' expression also depends on environmental factors such as temperature. It is well known that, in heterozygotes of Hsp90 mutations, monstrous character appears only at high temperature [58]. Therefore, as argued before, the variance of the distribution is smaller for diverse environment than for plain environment (see Fig. 13.2). The difference between curve A and curve B would be larger for regulatory mutations than for amino acid changes, since the former is more responsive to environment.

So it is likely that most cooption processes are nonadaptive, nearly neutral spreading, and it is expected that rapid cooption occurs in small populations in simple environment. However, at some rare occasions, advantageous spreading of new expression pattern would take place. It may be that, after a series of cooptions, another new gene recruitment becomes truly advantageous as in the Fisher model, filling a gap of an interaction network. It is also possible that two or more new enhancers taking place in different geographic regions of a species meet to acquire a new useful interaction network in a subdivided population as in the Wright model. In terms of sheer numbers, however, nearly neutral spreading of new enhancers is thought to be most prevalent.

The process may be compared with the stepwise increase of fitness of replicating RNA population. Schuster [59] reviews on RNA evolution, and points out that stepwise increase of fitness is compatible with the neutral or nearly neutral theory. The phenotype of RNA is considered to depend on secondary structure, and the formation of secondary structure occurs in a stepwise pattern. In other words, some neutral and nearly neutral variations that are silent accumulated before a better secondary structure is formed.

Acknowledgement

I thank Dr. Tomoko Steen for her many valuable suggestions to improve the presentation.

References

1. U. Bastolla, M. Porto, H.E. Roman, M. Vendruscolo, J. Mol. Evol. **57**, S103 (2003)
2. M. Kimura, *The Neutral Theory of Molecular Evolution* (Cambridge University Press, Cambridge, 1983)
3. N. Takahata, Genetics **116**, 169 (1987)
4. R.E. Dickerson, J. Mol. Evol. **1**, 26 (1971)
5. H.B. Fraser, A.E. Hirsh, L.M. Steinmetz, C. Scharfe, M.W. Feldman, Science **296**, 750 (2002)

6. S.B. Carroll, J.K. Grenier, S.D. Weatherbee, *From DNA to Diversity, Molecular Genetics and the Evolution of Animal Design* (Blackwell Science, Massachusetts, 2001)
7. M.Z. Ludwig, C. Bergman, N.H. Patel, M. Kreitman, Nature **403**, 564 (2000)
8. E. Ravasz, A.L. Somera, D.A. Mongru, Z.N. Oltvai, A.-L. Barabasi, Science **297**, 1551 (2002)
9. S. Wuchty, Z.N. Oltvai, A.-L. Barabasi, Nat. Genet. **35**, 176 (2003)
10. G. Schlosser, G.P. Wagner, *Modularity in Development and Evolution* (University of Chicago Press, Chicago, 2004)
11. E.H. Davidson, *Genomic Regulatory Systems* (Academic Press, San Diego, CA, 2001)
12. Z.N. Oltvai, A.-L. Barabasi, Science **298**, 763 (2002)
13. E. Zuckerkandl, L. Pauling, in *Evolving Genes and Proteins*, ed. by V. Bryson, H.J. Vogel (Academic, New York, 1965) pp. 97–166
14. M. Kimura, Nature **217**, 624 (1968)
15. J.B.S. Haldane, J. Genet. **55**, 511 (1957)
16. J.L. King, T.H. Jukes, Science **164**, 788 (1969)
17. M. Kimura, Proc. Nat. Acad. Sci. USA **63**, 1181 (1969)
18. R.C. Lewontin, *The Genetic Basis of Evolutionary Change* (Columbia University Press, New York, 1974)
19. T. Ohta, M. Kimura, J. Mol. Evol. **1**, 18 (1971)
20. T. Ohta, Nature **246**, 96 (1973)
21. M. Akam, Cell **57**, 347 (1989)
22. T. Ohta, Genetics **134**, 1271 (1993)
23. B. Tiplody, M. Goodman, J. Mol. Evol. **9**, 343 (1977)
24. M. Lynch, A. Force, Genetics **154**, 459 (2000)
25. W.-H. Li, in *Population Genetics and Molecular Evolution*, ed. by T. Ohta, K. Aoki (Japan Scientific Society, Tokyo, 1985) pp. 333–352
26. T. Ohta, Genetics **138**, 1331 (1994)
27. T. Ohta, J. Mol. Evol. **1**, 305 (1972)
28. T. Ohta, Annu. Rev. Ecol. Syst. **23**, 263 (1992)
29. T. Ohta, J. Mol. Evol. **41**, 115 (1995)
30. Rat Genome Sequencing Project Consortium, Nature **428**, 493 (2004)
31. G. Bernardi, *Structural and Evolutionary Genomics, Natural Selection in Genome Evolution* (Elsevier, Amsterdam, 2004)
32. F.G. Jorgensen, A. Hobolth, H. Hornshoj, C. Bendixen, M. Fredholm, M.H. Schierup, BMC Biol. **3**, 2 (2005)
33. J. Lu, C.-I. Wu, Proc. Natl. Acad. Sci. USA **102**, 4063 (2005)
34. Z. Gu, L. David, D. Petrov, T. Jones, R.W. Davis, L.M. Steinmetz, Proc. Natl. Acad. Sci. USA **102**, 1092 (2005)
35. P.D. Keightley, M.J. Lercher, A. Eyre-Walker, PLOS Biol. **3**(2), e42 (2005)
36. A.L. Hughes, Genetics **169**, 533 (2005)
37. J.H. Gillespie, *The Causes of Molecular Evolution* (Oxford University Press, Oxford, 1991)
38. M. Kimura, Genetics **47**, 713 (1962)
39. C.D. Bustamante, R. Nielsen, S.A. Sawyer, K.M. Olsen, M.D. Purugganan, L.D. Hartl, Nature **416**, 531 (2002)
40. J.C. Fay, G.J. Wyckoff, C.-I. Wu, Nature **415**, 1024 (2002)
41. N.G.C. Smith, A. Eyre-Walker, Nature **415**, 1022 (2002)

42. R. Sanjuan, A. Moya, S.F. Elena, Proc. Natl. Acad. Sci. USA **101**, 8396 (2004)
43. R. Nielsen, Z. Yang, Mol. Biol. Evol. **20**, 1231 (2003)
44. G. Piganeau, A. Eyre-Walker, Proc. Nat. Acad. Sci. USA **100**, 10335 (2003)
45. J.H. McDonald, M. Kreitman, Nature **351**, 652 (1991)
46. T. Ohta, H. Tachida, Genetics **126**, 219 (1990)
47. H. Tachida, Genetics **128**, 183 (1991)
48. T. Ohta, Theor. Pop. Biol. **10**, 254 (1976)
49. J.W. Thatcher, J.M. Shaw, W.J. Dickinson, Proc. Nat. Acad. Sci. USA **95**, 253 (1998)
50. S.K. Remold, R.E. Lenski, Nat. Genet. **36**, 423 (2004)
51. H. Innan, Y. Kim, Proc. Nat. Acad. Sci. USA **101**, 10667 (2004)
52. T. Ohta, Proc. Nat. Acad. Sci. USA **99**, 16134 (2002)
53. M. Lynch, J.S. Cornery, Science **302**, 1401 (2003)
54. M.W. Hahn, J.E. Stajich, G.A. Wray, Mol. Biol. Evol. **20**, 901 (2003)
55. G.A. Wray, M.W. Hahn, E. Abounheif, J.P. Balhoff, M. Pizer, M.W. Rockman, L.A. Romono, Mol. Biol. Evol. **20**, 1377 (2003)
56. M.V. Rockman, G.A. Wray, Mol. Biol. Evol. **19**, 1991 (2002)
57. A.S. Wilkins, *The Evolution of Developmental Pathways* (Sinauer Associates, Sunderland, MA, 2001)
58. S.L. Rutherford, S. Lindquist, Nature **396**, 336 (1998)
59. P. Schuster, in *Evolutionary Dynamics*, ed. by J.P. Crutchfield, P. Schuster (Oxford University Press, Oxford, 2003) pp. 163–215

14
Adaptation in Simple and Complex Fitness Landscapes

K. Jain and J. Krug

The notion that evolution can be viewed as a hill-climbing process in an adaptive landscape was introduced in 1932 by Wright [1], and remains one of the most powerful images in evolutionary biology [2]. Since the discovery of the molecular structure of genes, it has been clear that the substrate over which the adaptive landscape should be properly defined is the space of genetic sequences [3]. Nevertheless, apart from a few landmark papers [4, 5], adaptation has not been in the focus of the theory of molecular evolution, which instead has concentrated on the effects of stochastic drift in a neutral (*flat*) fitness landscape [6]. This situation is presently changing [7, 8]. Long-term evolution experiments on microbial populations [9] are beginning to produce a wealth of data, on the phenotypic as well as on the genotypic level, which make it meaningful to ask precise questions about the timing and size of adaptive events, and what they can tell us about the structure of the underlying adaptive landscape.

In this chapter, we introduce a class of sequence-based models of adaptation, which have been the subject of much recent interest in theoretical population genetics as well as in biologically inspired statistical physics. These models describe the behavior of a population of haploid, asexual individuals, each characterized by a genetic sequence of fixed length, in an adaptive landscape, which assigns a fitness value to each genotype. The population is exposed to the competing influences of *mutations*, which tend to increase the genetic variability, and *selection*, which focuses the population in regions of high fitness. The dynamics is deterministic, which implies that the genetic drift induced by the stochastic sampling noise in finite populations is neglected, and the adaptive landscape is generally taken to be time-independent. In view of the vastness of the field, the selection of topics is unavoidably biased by the interests and preferences of the authors. For a more comprehensive coverage we refer the reader to several recent review articles [10–14].

The chapter is organized as follows. In Sect. 14.1 the key concepts and their mathematical representation are introduced, and several types of mutation–selection dynamics are described, leaving the form of the adaptive landscape unspecified. In Sect. 14.2, we consider simple fitness landscape comprising a single adaptive peak or possibly two competing peaks. Here the central theme is the *error threshold* phenomenon, which refers to the sudden delocalization of the population from the fitness peak as the mutation rate increases beyond a critical value. As is described in this book in the chapter by Ester Lázaro, the error threshold and the related concept of a *quasispecies* play an important role in the population dynamics of RNA viruses and in the development of antiviral strategies. Because of its similarity to a phase transition, the error threshold has been thoroughly analyzed using a range of methods from statistical physics. We give an elementary derivation of the critical mutation rate, and describe several modifications of the basic model, including fitness peaks with a variable amount of epistasis, diploid populations, semiconservative replication, and time-dependent landscapes.

Section 14.3 is devoted to complex fitness landscapes consisting of many peaks and valleys. Such landscapes can be modeled by ensembles of random functions, which links this subject to the statistical physics of disordered systems. Whereas so far the discussion has been restricted to static or steady-state properties, time-dependent aspects of mutation–selection dynamics are discussed in Sect. 14.4. Finally, experimental realizations (in vitro as well as in vivo) of the models are described in Sect. 14.5, and some concluding remarks are presented in Sect. 14.6.

14.1 Basic Concepts and Models

In the following discussion, the constituents of a population carry a string $\sigma \equiv \{\sigma_1, \ldots, \sigma_N\}$ where each of the N letters σ_i is taken from an alphabet of size $\ell \geq 2$. In classical population genetics, σ represents the configuration of alleles (variants of a gene) σ_i located at gene loci i. Typically, one-locus, ℓ allele models where ℓ can take values between two (wild type and mutant) to infinity (continuum of alleles) have been considered [15]. In the language of population genetics, we are here concerned with multilocus models with complete linkage [11].

At the molecular level, σ represents the genetic sequence of an individual. For DNA(RNA)-based organisms, $\ell = 4$ corresponding to the nucleotide bases A, T(U), C, and G and the sequence length N varies from a few thousands for viruses to about 10^9 for humans. Thus, the total number 4^N of sequences available is hyperastronomically large. The minimum value of $\ell = 2$ can be obtained by lumping A and G together in purins and C, T, and U in pyrimidines. The sequences may also represent proteins composed of a few hundred amino acids taken from an alphabet of size $\ell = 20$ [3].

14.1.1 Fitness, Mutations, and Sequence Space

The essence of natural selection is that the relative reproductive success of an individual determines whether the corresponding genotype becomes more or less abundant in the population. The *fitness* of an individual is a quantitative measure of its reproductive success; depending on the context, it may be defined as the *viability* of an organism, i.e. the probability to survive to the age of reproduction [2], the replication rate of a microbe, the binding affinity of regulatory proteins to DNA [16] or of antibodies produced by B-cells to pathogens [17], the program execution speed for digital organisms [18], or the cost function in an optimization problem [19].

In principle, one should assign a fitness to the phenotype, which then should be related to the genotype; unfortunately, the genotype–phenotype map is complicated and largely unknown except for a few cases [20] (see also Sect. 14.3.1). This problem is usually outflanked by associating fitness $W(\sigma)$ with the genotype itself and define it to be *the expected number of offspring produced by an individual with sequence σ* [14]. This definition applies to the case of discrete generations, and is known as *Wrightian* fitness. To pass to continuous time dynamics we write

$$W(\sigma) = \exp[w(\sigma)\Delta t] \approx 1 + w(\sigma)\Delta t, \quad \Delta t \to 0, \qquad (14.1)$$

where Δt is the generation time and $w(\sigma)$ is referred to as the *Malthusian* fitness [10]. For future reference, we note that multiplication of the Wrightian fitnesses by a common factor implies a constant additive shift of the Malthusian fitnesses.

In Sect. 14.1.2, we will discuss models in which *mutations* occur either as copying errors in the genetic material during cell division or induced by some external influences. In Table 14.1, the spontaneous mutation rates for some organisms are shown. They differ by orders of magnitude between RNA-based viruses whose mutation rate per genome exceeds unity, and DNA-based organisms, which can afford the complex replication machinery needed to reduce the mutation rate to a much lower level. It has been suggested that the high mutation rate of RNA viruses, rather than being due to the lack of correction mechanisms, may constitute an adaptation to the rapidly fluctuating environments that these organisms encounter (see the chapter by E. Lázaro in this book). Within the DNA organisms, the mutation rate per base is seen to

Table 14.1. Spontaneous mutation rates for various organisms taken from [21]

Organism	Genome size	Rate per base	Rate per genome
Bacteriophage $Q\beta$	4.5×10^3	1.4×10^{-3}	6.5
Vesicular stomatitis virus	1.4×10^4	2.5×10^{-4}	3.5
Bacteriophage λ	4.9×10^4	7.7×10^{-8}	0.0038
E. coli	4.6×10^6	5.4×10^{-10}	0.0025
C. elegans	8.0×10^7	2.3×10^{-10}	0.018
Mouse	2.7×10^9	1.8×10^{-10}	0.49
Human	3.2×10^9	5.0×10^{-11}	0.16

The first two organisms have RNA as genetic material and the rest are DNA-based decrease with increasing sequence length, and the mutation rate per genome is roughly constant for similar organisms. However, mutation rates per genome in higher eukaryotes become comparable to those of DNA-based microbes if referred to the *effective* genome size, which excludes noncoding regions [21].

Before we describe the mutation–selection models, we need to specify the space of sequences on which the evolutionary dynamics operate. The structure and geometry of the sequence space depends on the nature of the allowed moves that change one sequence into another. In the simplest case of a genome of fixed length N subject only to point mutations (which we will restrict ourselves to throughout this chapter), the natural choice for the sequence space is the Hamming space with ℓ^N points. Two sequences σ and σ' are separated by the Hamming distance $d(\sigma, \sigma')$ which is given by

$$d(\sigma, \sigma') = \sum_{i=1}^{N} (1 - \delta_{\sigma_i, \sigma'_i}). \qquad (14.2)$$

The Hamming distance simply counts the number of letters in which the two sequences differ, that is, the number of point mutations needed to mutate σ into σ' (and vice versa). The Hamming space for $N = 3$ and $\ell = 2$ is shown in Fig. 14.1(left). The sequences are located at the corners of a cube, which for general N becomes the N-dimensional *hypercube*.

To give an example of a sequence space with a somewhat different geometry, we consider the graph bipartitioning problem (GBP) [22] (see also Sect. 14.3.3). In the GBP, as the name suggests, the problem is to partition a graph with given connections into two sets A and B with equal number of vertices, such that the number of connections between A and B is minimized. A bipartitioning configuration is mapped onto a binary sequence by setting $\sigma_i = 1$ if the vertex i belongs to set A, and $\sigma_i = -1$ else. Thus the sequence space consists of those $\binom{N}{N/2}$ configurations σ for which $\sum_i \sigma_i = 0$,

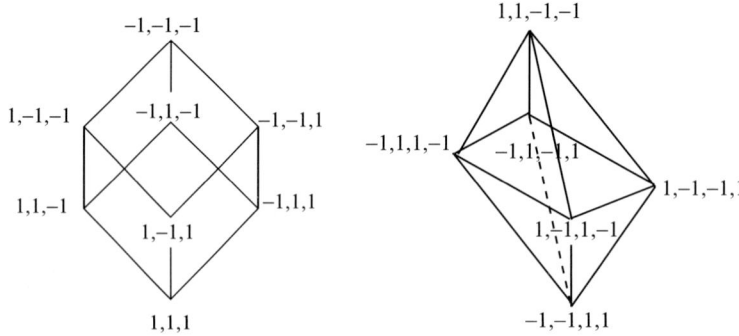

Fig. 14.1. Examples of sequence spaces. *Left* panel: Hamming space of binary sequences of length $N = 3$. *Right* panel: Graph bipartitioning problem space for $N = 4$. In both cases $\sigma_i = \pm 1$ and nearest neighbors are connected by *lines*

a subset of the Hamming space. An elementary move exchanges a pair of vertices between the sets A and B. Two configurations are said to be at a distance $d_{\text{GBP}}(\sigma, \sigma') = d$ if they can be related by d exchange moves, so that d_{GBP} is half of the Hamming distance defined above. The GBP sequence space for $N = 4$ is shown in Fig. 14.1(right).

The Hamming space as well as the GBP sequence space are symmetric and regular graphs, in the sense that each vertex has the same number of neighbors and all vertices are equivalent. This is no longer true if mutations that change the sequence length through deletions, insertions, or gene duplications are taken into account. Genetic recombination, which is of crucial importance for sexual reproduction, leads to additional complications, because it introduces moves, which involve pairs of sequences [19].

We return to the case of point mutations acting on sequences of fixed length N, and proceed to derive an expression for the mutation probabilities taking one sequence to another. If the mutations change a letter σ_i to any one of the other $\ell - 1$ values with a probability μ, independent of the identity of the letter and the other letters in the sequence, then the *probability* to mutate a sequence σ' to σ can be written as

$$p(\sigma' \to \sigma) = \left(\frac{\mu}{(\ell-1)(1-\mu)}\right)^{d(\sigma,\sigma')} (1-\mu)^N. \tag{14.3}$$

Obviously, this probability is the same for all α_d sequences, which are at a constant Hamming distance d from sequence σ, where α_d is given by

$$\alpha_d = \binom{N}{d}(\ell-1)^d. \tag{14.4}$$

This can be seen by noting that there are $\binom{N}{d}$ ways of choosing d letters at which a sequence differs from σ and each of these d letters can take $\ell - 1$ values. For large N, most of the sequences are located in a belt of width $\sim \sqrt{N}$ around the distance $d_{\max} = N(\ell-1)/\ell$ away from σ. Using (14.4), it is easily checked that $\sum_\sigma p(\sigma' \to \sigma) = 1$.

Similar to the transition from Wrightian to Malthusian fitness, in the continuous time limit the mutation probability (14.3) has to be replaced by the *mutation rate* $\gamma(\sigma' \to \sigma)$, such that for generation time $\Delta t \to 0$

$$p(\sigma' \to \sigma) \approx \delta_{\sigma',\sigma} + \Delta t\, \gamma(\sigma' \to \sigma). \tag{14.5}$$

Denoting the mutation rate per letter by $\tilde{\mu}$ and setting $\mu = \tilde{\mu}\Delta t$ in (14.3) yields

$$\gamma(\sigma' \to \sigma) = \begin{cases} 0 & d(\sigma',\sigma) > 1 \\ \tilde{\mu}/(\ell-1) & d(\sigma',\sigma) = 1 \\ -N\tilde{\mu} & d(\sigma',\sigma) = 0. \end{cases} \tag{14.6}$$

The normalization condition for mutation rates reads $\sum_\sigma \gamma(\sigma' \to \sigma) = 0$.

14.1.2 Mutation–Selection Models

We now discuss models of adaptation that incorporate the two competing processes discussed above, namely, mutation and selection. Mutation increases genetic diversity, while selection tend to contain the population at fit sequences. In case selection wins out, one obtains a population in which individuals are genetically closely related else a heterogeneous population distributed over the entire sequence space results. In this chapter, we will mainly discuss the so-called *coupled models* in which the mutations occur only during replication. In the *paramuse models*, on the other hand, mutation and selection occurs in parallel, and they will be discussed here briefly. We refer the reader for more details to the reviews [10,11] and references therein. Although one may expect both types of mutation mechanisms to be relevant in describing evolution, the jury is still out on their relative importance. For this reason, both classes of models have been analyzed in detail and the relationship between them has been explored, with regard to both static [23] and dynamic [24] properties.

The models discussed below work under the following two assumptions:

(1) *Infinite population*, i.e., the total population size $M \gg \ell^N$, the total number of genotypes available. Under this assumption, a deterministic description suffices and we can write down the time evolution equation for the average population fraction $X(\sigma, t)$ of sequence σ at time t. Although this is often unrealistic, the analysis is simpler in this limit, which in many cases can be adapted to the finite population case to provide quantitative agreement with experiments [25–27]. The infinite population limit can be justified if the population is known to be localized in a small region of sequence space around a fitness peak, if one is interested in a short piece of the genome such as a single regulatory binding site [16] (see also Sect. 14.2.4) or if one works in the population genetics setting, where the letters in the sequence are alleles of a gene, rather than single nucleotides.
(2) *Asexual reproduction* that dominates in the lower forms of life such as virus and bacteria, and digital organisms. We will mainly consider haploid organisms but diploids are briefly discussed in Sect. 14.2.5. However, we do not consider the case of sexual reproduction; a comparison between sexual and asexual reproduction modes in the context of sequence space models can be found in [28].

Paramuse Models

In the paramuse models, introduced by Crow and Kimura [6], one assumes error-free replication and mutations are induced by the environment through radiation, thermal fluctuations, etc. [10]. The equation for the rate of change $\dot{X}(\sigma, t) = \partial X(\sigma, t)/\partial t$ of the fraction $X(\sigma, t)$ of the population with sequence σ is given by

$$\dot{X}(\sigma,t) = \left[w(\sigma) - \sum_{\sigma'} w(\sigma')X(\sigma',t)\right] X(\sigma,t) + \sum_{\sigma'} \gamma(\sigma' \to \sigma)X(\sigma',t)\,. \tag{14.7}$$

The first term is the selection term while the contribution from the mutations is contained in the last term. The evolution (14.7) is nonlinear in $X(\sigma,t)$ because of the second term on the right-hand side, which is required to ensure the normalization $\sum_\sigma X(\sigma,t) = 1$. This nonlinearity can be eliminated by passing to un-normalized population variables $Z(\sigma,t)$ defined by

$$Z(\sigma,t) = X(\sigma,t) \exp\left[\sum_{\sigma'} w(\sigma') \int_0^t d\tau X(\sigma',\tau)\right], \tag{14.8}$$

which satisfy the linear equation [29]

$$\frac{\partial Z(\sigma,t)}{\partial t} = w(\sigma)Z(\sigma,t) + \sum_{\sigma'} \gamma(\sigma' \to \sigma)Z(\sigma',t)\,. \tag{14.9}$$

Equation (14.7) follows from (14.9) using the relation

$$X(\sigma,t) = \frac{Z(\sigma,t)}{\sum_{\sigma'} Z(\sigma',t)}\,. \tag{14.10}$$

Inserting the explicit form (14.6) for the mutation rates, it can be shown that the vector $\mathbf{Z}(t) = (Z(\sigma^1,t),\ldots,Z(\sigma^S,t))$, where the index labels the $S = \ell^N$ points in sequence space, obeys a Schrödinger equation in imaginary time

$$\frac{\partial \mathbf{Z}(t)}{\partial t} = H\mathbf{Z}(t) \tag{14.11}$$

with quantum spin Hamiltonian H in one dimension. Specifically, for $\ell = 2$, one obtains the Hamiltonian of an Ising chain in the presence of a transverse magnetic field (mutations) with general interactions (specified by the fitness landscape) [29]; for an explicit example see (14.39). This model has been solved exactly for a variety of fitness landscapes using methods of quantum statistical physics [24, 29, 30]. A similar analysis has also been carried out for the biologically relevant case of $\ell = 4$ [31].

Coupled (Quasispecies) Dynamics

In the quasispecies model introduced by Eigen in the context of prebiotic evolution [4, 32, 33], the mutations are copying errors that occur during the reproduction process. This implies that the population fraction $X(\sigma,t)$ evolves according to

$$\dot{X}(\sigma,t) = \sum_{\sigma'} p(\sigma' \to \sigma)W(\sigma')X(\sigma',t) - \left(\sum_{\sigma'} W(\sigma')X(\sigma',t)\right) X(\sigma,t)\,, \tag{14.12}$$

which can be linearized by a transformation analogous to (14.8) to yield the linear equation

$$\dot{Z}(\sigma,t) = \sum_{\sigma'} p(\sigma' \to \sigma) W(\sigma') Z(\sigma',t). \qquad (14.13)$$

In discrete time, this model takes the form

$$X(\sigma,t+1) = \frac{\sum_{\sigma'} W(\sigma') p(\sigma' \to \sigma) X(\sigma',t)}{\sum_{\sigma'} W(\sigma') X(\sigma',t)} \qquad (14.14)$$

where the denominator arises due to the normalization. The discrete time analog of the transformation (14.8) is given by

$$Z(\sigma,t) = X(\sigma,t) \prod_{\tau=0}^{t-1} \sum_{\sigma'} W(\sigma') X(\sigma',\tau). \qquad (14.15)$$

As before, the un-normalized variables obey a linear equation given by

$$Z(\sigma,t+1) = \sum_{\sigma'} p(\sigma' \to \sigma) W(\sigma') Z(\sigma',t). \qquad (14.16)$$

The use of the Wrightian (discrete time) fitness $W(\sigma)$ in the continuous time (14.12) requires some explanation. First, it ensures that the stationary solutions of (14.12) and (14.14) are identical. Second, it reflects the fact that (14.12) is invariant (up to a rescaling of time) under multiplication of the fitnesses by a constant factor, $W(\sigma) \to CW(\sigma)$, which is an exact symmetry of the discrete time equation (14.14), whereas the continuous time paramuse dynamics (14.7) is invariant under additive shifts $w(\sigma) \to w(\sigma) + C$ [10, 23]. In fact, (14.12) is *not* the continuous time limit of (14.14). Instead, inserting (14.1) and (14.5) in (14.14) and taking $\Delta t \to 0$, one obtains the paramuse dynamics (14.7). In this sense (14.12) is somewhat intermediate between the discrete time model (14.14) and the continuous time dynamics (14.7).

For the discrete time model (14.14), one can represent the evolutionary histories as configurations on a two-dimensional lattice with the two axes directed along the sequence and along time, with a spin variable $\sigma_i(t)$ at each site. Writing the evolution (14.16) for the vector $\mathbf{Z}(t)$ in the form

$$\mathbf{Z}(t+1) = T_{t+1,t}\, \mathbf{Z}(t) \qquad (14.17)$$

then suggests to interpret $T_{t+1,t}$ as the transfer matrix of a two-dimensional classical spin model, which relates the probability of a configuration in one row of the lattice to the next one [34]. For $\ell = 2$, this $2^N \times 2^N$ matrix can be written (up to a multiplicative constant) as

$$T_{t+1,t}[\{\sigma_i(t+1)\},\{\sigma_i(t)\}] = \exp\left[\ln W(\{\sigma_i(t)\}) + J \sum_{i=1}^{N} \sigma_i(t+1)\sigma_i(t)\right] \qquad (14.18)$$

where
$$J = \frac{1}{2}\ln(\mu^{-1} - 1). \qquad (14.19)$$

Thus $T_{t+1,t} = \exp[-\tilde{H}]$, where $\tilde{H}$ is the Hamiltonian of a two-dimensional Ising model[1] with nearest neighbor interactions of strength J along the time direction and general interactions [determined by the fitness landscape $W(\sigma)$] along the sequence direction [35]. The expression (14.19) shows that the interactions along the time direction are ferromagnetic (antiferromagnetic) whenever $\mu < 1/2$ ($\mu > 1/2$), while for $\mu = 1/2$ the sequence is completely randomized in each time step and the interaction vanishes.

Clearly, to obtain the distribution of sequences at time slice t, one needs to solve iteratively for all the $t-1$ preceding layers. In the steady-state for which $t \to \infty$ one requires the properties of the last "surface" layer coupled to a semi-infinite "bulk." Since the transfer matrix (14.18) does not contain any couplings along the sequence direction in the last layer $t+1$, the boundary condition for this semi-infinite spin model corresponds to a free surface [36].

14.2 Simple Fitness Landscapes

So far we have discussed the general equations governing the evolution of a population with mutations, but the fitness landscape was not specified. We do so now and begin with landscapes that are "simple" in that the fitness depends only on the distance from a given (*master*) sequence, which is usually the genotype of highest fitness.[2] Such landscapes are called *permutation invariant*, because the fitness depends only on the number of mismatches relative to the master sequence, but not on their position. Using this symmetry, the ℓ^N population variables can be grouped into $N+1$ *error classes*, which greatly facilitates both numerical and analytic work [37].

14.2.1 The Error Threshold: Preliminary Considerations

Much of this section is devoted to a discussion of the error threshold phenomenon, which refers to the loss of genetic integrity when mutations are increased beyond a certain threshold. We consider only the stationary population distribution, which is established after a long time. The linearity of both the continuous and discrete time evolution equations ((14.9), (14.14) and (14.16)) implies that the stationary distribution is identical to the *principal eigenvector* of the matrix multiplying the population vector on the right-hand side, i.e., the eigenvector with the largest eigenvalue. The principal eigenvalue

---

[1] The Ising Hamiltonian $\tilde{H}$ should not be confused with the Hamiltonian H of the quantum spin chain in (14.11).
[2] In the context of population genetics, the master genotype is often referred to as the *wild type*.

is related to the mean population fitness in the stationary state. In this sense, the analysis of different fitness landscapes and mutation schemes is reduced to the investigation of the spectral properties of the corresponding evolution matrices [38].

The error threshold separates two regimes of mutation–selection balance characterized by a qualitatively different structure of the principal eigenvector. For small mutation rates, the eigenvector is localized around the master sequence, i.e., only the entries corresponding to the dominant genotype and a few of its nearby mutants carry appreciable weight. Following Eigen and Schuster [32], such a localized population distribution is referred to as a *quasispecies*. When the mutation rate is increased beyond the error threshold, the principal eigenvector becomes delocalized and the population spreads uniformly throughout the sequence space. In this regime, finite population effects, which are neglected in the models considered here, become extremely important; rather than covering the entire sequence space, which is impossible given the vast number of sequences, a finite population forms a localized cloud, which wanders about randomly [39].

Since both the eigenvectors and eigenvalues of any finite matrix depend smoothly on its entries, the error threshold can become *sharp*, in the sense of being associated with some nonanalytic behavior of the population distribution or the mean population fitness, only in the limit $N \to \infty$. We shall see below that in order to maintain the localized quasispecies in this limit, it is usually necessary to either reduce the single site mutation probability μ, such that the mutation probability per genome μN remains constant, or to increase the selective advantage of the master sequence with increasing N.

14.2.2 Error Threshold in the Sharp Peak Landscape

We demonstrate the error threshold in the case of a single sharp peak landscape, which is defined as

$$W(\sigma) = W_0 \delta_{\sigma,\sigma_0} + (1 - \delta_{\sigma,\sigma_0}), \quad W_0 > 1. \qquad (14.20)$$

Here σ_0 denotes the master sequence, and W_0 is the selective advantage of the master sequence relative to the other sequences, whose Wrightian fitness has been normalized to unity. We anticipate the error threshold to occur for $\mu \to 0$, $N \to \infty$, keeping the mutation rate per genome μN finite. Let us consider the coupled model in discrete time[3] defined by (14.14) with the choice (14.20). In the limit $\mu \to 0$, the mutations taking the mutants back into the master sequence can be neglected,[4] and the only nonzero contribution to $X(\sigma_0)$ on the right-hand side of (14.14) is that for $\sigma' = \sigma$. This yields

---

[3] Recall that in the steady-state, both versions of the coupled model are identical.
[4] Neglecting back mutations toward the master sequence is common in population genetics, where it is referred to as a *unidirectional* mutation scheme [11]. It simplifies the analytic treatment [28, 40], and will be used repeatedly in this chapter as an approximation, which is expected to become exact for $\mu \to 0$, $N \to \infty$.

$$X(\sigma_0) = \frac{W_0 e^{-\mu N} - 1}{W_0 - 1}, \qquad (14.21)$$

which is an acceptable solution provided $\mu N \leq \ln W_0$. Thus, a phase transition occurs at the critical mutation probability

$$\mu_c = \frac{\ln W_0}{N} \qquad (14.22)$$

beyond which the population cannot be maintained at the peak of the landscape. Close to μ_c, the fraction of population at the master sequence behaves as

$$X(\sigma_0) \approx \frac{N}{W_0 - 1}(\mu_c - \mu) \qquad (14.23)$$

thus approaching zero continuously at μ_c. The above results are also confirmed by a detailed numerical analysis for finite μ and N, in which the population was grouped into error classes at constant Hamming distance from the master sequence and the population in the error classes as well as the eigenvalues of the evolution matrix were followed as a function of μ [37].

The way in which the error threshold condition (14.22) combines mutation rate, sequence length, and selective advantage is the central result of quasispecies theory. In particular, it shows that in order to maintain a localized quasispecies at finite single site mutation rate in the limit $N \to \infty$, the selective advantage has to increase *exponentially* with N [41]. Under the assumption that typical selective advantages do not depend strongly on sequence length, (14.22) also provides some rationalization for the observation that the product μN is roughly constant within classes of similar organisms (see Sect. 14.1.1). On the other hand, at given achievable values of the replication accuracy and the selective advantage, the condition $\mu < \mu_c$ place an upper bound $N_{\max}$ on the sequence length, beyond which genetic integrity is lost. Elsewhere in this book, Ester Lázaro presents substantial evidence that RNA viruses have evolved to reside close to this threshold, possibly because this allows them to maintain a maximal genetic variability, which is needed to rapidly adapt to changing environments (see also Sect. 14.5.2).

Neglecting back mutations to the master sequence allows to derive an expression for the mean Hamming distance to the master sequence, which reads [40]

$$\langle d(\sigma, \sigma_0) \rangle = \frac{W_0 N \mu}{W_0 e^{-N\mu} - 1}. \qquad (14.24)$$

The mean Hamming distance is finite for $\mu < \mu_c$ and diverges as $(\mu_c - \mu)^{-1}$ as the error threshold is approached. This provides an alternative characterization of the threshold. A related quantity, which has been proposed as an order parameter for the transition, is the *mean overlap*

$$m = 1 - \frac{2\langle d(\sigma, \sigma_0) \rangle}{N} \qquad (14.25)$$

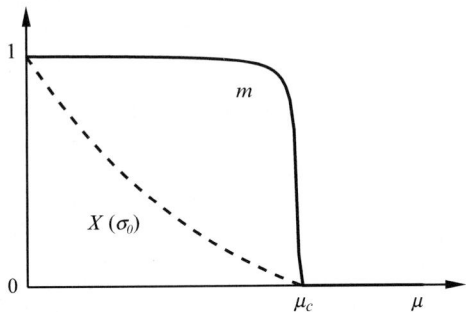

Fig. 14.2. Continuous transition in the fraction $X(\sigma_0)$ of the master sequence and the (almost) discontinuous one in the overlap m as a function of mutation rate μ, for $N = 1,000$ and $W_0 = 4$

between the master sequence and a randomly chosen sequence [41]. Since $\langle d(\sigma, \sigma_0) \rangle$ remains finite for $N \to \infty$ in the localized phase, the overlap is $m = 1$ in this limit and jumps discontinuously to $m = 0$ at the threshold. Figure 14.2 displaying the two order parameters considered in the above discussion illustrates that the nature of the transition – continuous or discontinuous – depends to some extent on the quantity under consideration.[5]

Yet another characterization of the error threshold relies on the notion of the *consensus sequence* σ^c, which carries at each site i that letter σ_i^c, which is most frequently represented in the population. It is easy to see that, for symmetry reasons, the consensus sequence in the sharp peak landscape (14.20) coincides with the master sequence, $\sigma^c = \sigma_0$, throughout the localized phase; this is true for general permutation-invariant single peak landscapes. In the delocalized phase, where the population is uniformly spread throughout sequence space for $N \to \infty$, all letters appear with equal probability and the consensus sequence cannot be defined. This is an artifact of the assumption of infinite population size: a finite population retains some genetic structure even in a flat fitness landscape and diffuses through sequence space as a cloud centered around a moving consensus sequence $\sigma^c(t)$ [39]. Thus, at the error threshold, the consensus sequence ceases to be pinned to the master sequence and becomes time-dependent. This criterion to locate the transition is particularly useful in complex fitness landscapes, where the most-fit master sequence is not known [42] (see Sect. 14.3). Similarly, in experimental studies of microbial populations such as RNA viruses, the consensus sequence is taken to represent the (unknown) wild type genome, and the genetic spread of the population

---

[5] In contradiction to the discussion above, a numerical study based on the mapping to a two-dimensional Ising model described in Sect. 14.1.2 deduced that both m and $X(\sigma_0)$ change smoothly at the transition [36]. However, in this study, a scaling analysis with genome length (akin to finite size scaling analysis in statistical mechanics) was not carried out to obtain the behavior in the limit $N \to \infty$.

around σ^c is interpreted as a measure of the balance between mutational and selective forces (see the chapter by E. Lázaro).

14.2.3 Exact Solution of a Sharp Peak Model

A variant of Eigen's model was solved exactly for any N in [43]. The model is defined in discrete time but the mutations are restricted to mutants within Hamming distance equal to one, as for the continuous time mutation rates (14.6). In addition, mutations are assumed to occur in the whole population *before* the reproduction process. With the fitness landscape (14.20) this leads to the linear evolution equation

$$Z(\sigma, t+1) = [1 + (W_0 - 1)\delta_{\sigma,\sigma_0}]$$
$$\times \left[(1 - N\mu)Z(\sigma, t) + \mu \sum_{\sigma'} Z(\sigma', t)\, \delta_{d(\sigma',\sigma),1} \right]$$

for the un-normalized population variables. Note that the model is well defined only for $N\mu < 1$.

At large times, $Z(\sigma, t+1) \approx \Lambda Z(\sigma, t)$ where Λ is the largest eigenvalue of the evolution matrix on the right-hand side of (14.26). In the delocalized phase, the population is spread over the entire sequence space with mean fitness $W = 1$, so that $\Lambda = 1$ whereas in the localized phase, a finite fraction has fitness $W_0 > 1$ and hence $\Lambda > 1$. For any N, the eigenvalue Λ is determined by the exact equation

$$\frac{W_0}{W_0 - 1} = \frac{1}{2^N} \sum_{k=0}^{N} \binom{N}{k} \frac{\Lambda}{\Lambda - 1 + 2k\mu}. \qquad (14.26)$$

Because of the $k = 0$ term on the RHS of the above equation, it is evident that Λ can take a value equal to 1 only in the $N \to \infty$ limit. Thus, there is no phase transition for any finite N.

In the limit $N \to \infty$, $\mu \to 0$ with $N\mu < 1$ fixed the eigenvalue is given by the expression

$$\Lambda = \max\{1, W_0(1 - N\mu)\}, \qquad (14.27)$$

which sticks to unity beyond the critical mutation strength

$$\mu_c = \frac{W_0 - 1}{W_0 N}. \qquad (14.28)$$

Incidentally, the above expression for μ_c can be obtained using (14.21) by expanding the exponential to first order in μN. This is required to ensure that $\mu N < 1$ is satisfied for any $W_0 > 1$. In both cases, the selective advantage needed to localize the quasispecies is the inverse of the *copying fidelity*, i.e., the probability of creating an error-free offspring.

The behavior of other quantities at the threshold follows from that of Λ. For example, the fraction of the population residing at the master sequence is given by

$$X(\sigma_0) = \frac{W_0(\Lambda - 1)}{(W_0 - 1)\Lambda}, \qquad (14.29)$$

which vanishes linearly in $\mu_c - \mu$ at the threshold, and the mean Hamming distance from the master sequence is

$$\langle d(\sigma, \sigma_0)\rangle = \frac{N\mu}{\Lambda - 1}, \qquad (14.30)$$

which diverges as $(\mu_c - \mu)^{-1}$. The expressions (14.28)–(14.30) are valid in the asymptotic limit $N \to \infty$, but systematic expansions of these quantities in powers of $1/N$ are also available [43].

Comparing the expressions (14.23) and (14.24) to (14.29) and (14.30), respectively, we see that $X(\sigma_0)$ and $\langle d(\sigma, \sigma_0)\rangle$ behave qualitatively similar in the two models as the error threshold is approached. This is a simple example of the principle of *universality* commonly encountered at physical phase transitions, which states that the way in which singular quantities vanish or diverge at the transition is independent of detailed properties of the model.

14.2.4 Modifying the Shape of the Fitness Peak

Since the sharp peak landscape (14.20) was chosen for its simplicity, and not because it is expected to be biologically realistic, it is important to investigate how the error threshold phenomenology depends on the shape of the fitness peak. In this section, we discuss some illustrative examples. A method for solving the stationary quasispecies equation for general peak shapes has been developed by Peliti [44]. It employs a strong selection limit, in which the fitness is written as $W(\sigma) = \exp[N\phi(\sigma)]$ and the limit $N \to \infty$ is carried out at fixed mutation probability μ.

Peak Height vs. Peak Width

We first consider a landscape with one sharp global maximum and a broad peak of lower fitness separated by a flat landscape. This is defined as

$$W(\sigma) = W_0 \delta_{\sigma,\sigma_0} + W_N \delta_{\sigma,\sigma^N} + W_{N-1} \delta_{d(\sigma,\sigma^N),1} + \sum_{j \neq 0, N-1, N} \delta_{d(\sigma,\sigma_0),j}, \qquad (14.31)$$

where σ^N is the sequence at maximal Hamming distance N from σ_0 and $W_0 > W_N > W_{N-1} > 1$. By placing the two fitness peaks at the two poles σ_0

and σ^N of the sequence space, the permutation symmetry of the landscape is preserved and the population can be subdivided into error classes. The coupled model with the landscape (14.31) has been studied in both continuous [45] and discrete time [36]. Interestingly, with increasing mutation rate, the quasispecies shifts abruptly from the sequence σ_0 to the broader peak around σ^N finally delocalizing over the whole sequence space. For large mutation rates, the quasispecies is more comfortable at the lower peak surrounded by an extended region of elevated fitness than at the (globally optimal) isolated master sequence.

Mesa Landscapes

Broad fitness peaks arise naturally in the evolution of regulatory binding sites [16, 46, 47]. In this context, the fitness of a given regulatory sequence can be plausibly related to the binding probability of the corresponding transcription factor. Simple thermodynamic models predict that the binding probability depends on the number of mismatches $d(\sigma, \sigma_0)$ with respect to the regulatory master sequence σ_0 through a Fermi function,

$$p_b(d) = \frac{1}{1 + \exp[\epsilon(d - d_0)/k_B T]}, \quad (14.32)$$

where ϵ is the binding energy per mismatch, ϵd_0 is the chemical potential corresponding to the concentration of the transcription factor, and $k_B T$ is the thermal energy at temperature T. For $\epsilon/k_B T \gg 1$, the binding probability drops abruptly from $p_b = 1$ to $p_b = 0$ when d exceeds the number d_0 of tolerable mismatches; a typical value of this ratio is $\epsilon/k_B T \approx 2$.

In the simplest scenario, the selective advantage of a regulatory sequence is assumed to be proportional to the binding probability. This leads to a *mesa-shaped* fitness landscape, with a plateau of constant fitness and radius d_0 around the master sequence. In [16], a detailed study of the error threshold in this landscape was presented for continuous time paramuse dynamics with fitness landscape $w(d) = w_0 p_b(d)$. An exact solution is possible in a limit where d becomes a continuous variable and the Fermi function (14.32) is replaced by a step function. Provided $d_0 \ll N$, the error threshold is found to take place at a critical mutation strength μ_c given by

$$\mu_c = \frac{2w_0}{N(1 + \eta^2/d_0^2)}, \quad (14.33)$$

where η is a constant of order unity. The critical mutation strength is seen to increase with increasing d_0, illustrating the enhanced stability of the quasispecies with increasing width of the fitness peak. In the localized phase, the majority of the population is located near the mesa edge at $d = d_0$, reflecting the exponential increase of the number (14.4) of available genotypes with distance d. This is a purely entropic effect, which leads to a maximal *fuzziness* of regulatory motifs.

Somewhat more realistically, one expects that the fitness depends not only on the ability of the sequence to bind the transcription factor in a certain cellular state, but also on its ability to *avoid* binding in other states. This can be modeled by a fitness function, which is proportional to the difference between two Fermi functions (14.32) with different values of d_0, leading to a *crater* landscape with a rim of high fitness around a fitness minimum at $d = 0$ [47].

Epistasis: Coupled Dynamics

Not all landscapes display the error threshold phenomenon. We illustrate this point using the multiplicative (or Fujiyama) landscape as an example. In this case

$$W(\sigma) = \prod_{i=1}^{N} e^{\lambda \sigma_i} = \exp[\lambda(N - 2d(\sigma_0, \sigma))], \qquad (14.34)$$

where for simplicity we choose $\ell = 2$ and let σ_i take values ± 1. For $\lambda > 0$, the master sequence is $\sigma_0 = (1, 1, 1, \ldots, 1)$ and the Hamiltonian $\tilde{H}$ obtained from (14.18) is

$$\tilde{H} = \sum_{i=1}^{N} [-J\sigma_i(t+1)\sigma_i(t) - \lambda \sigma_i(t)]. \qquad (14.35)$$

Because of the absence of interactions along the sequence space direction, one obtains, for each position i, an one-dimensional Ising model in the presence of magnetic field λ. This model is well known to lack a phase transition and due to the λ term, the spins tend to align in the direction of the field. Correspondingly, a finite fraction of the population is maintained at the master sequence for any value of the mutation rate. The full population distribution has been worked out in [48].

In genetic terms, the multiplicative form (14.34) implies that the different gene loci contribute independently to the fitness, which is referred to as the absence of *epistatic interactions*. In general, one must distinguish between *synergistic* or *negative* epistasis, in which the (deleterious) effect of an additional mutation *increases* with increasing distance from the wild type (master sequence), and *diminishing returns* or *positive* epistasis, when the effects of mutations decreases with increasing distance.[6] The sharp peak landscape (14.20) is an extreme case of positive epistasis, because after the first mutation away from the master sequence, any additional mutation does not affect the fitness at all. An extreme limit of negative epistasis is represented by the case of truncation selection, where the Wrightian fitness vanishes beyond a critical Hamming distance d_c [49]. As we discuss below, whether or not an error

---

[6] This nomenclature is based on [11,40], but it does not appear to be unambiguous; in [49] a definition of positive and negative epistasis is used, which is opposite to the present one.

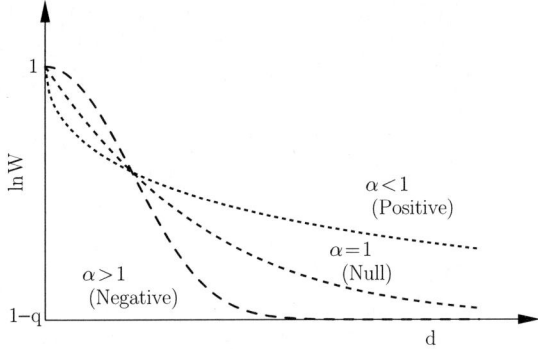

Fig. 14.3. Illustration of the fitness landscape (14.36) with $s = q = 0.5$ and three different values of α

threshold occurs depends on the behavior of the landscape at large Hamming distance from the master sequence.

Consider a general fitness landscape defined by [40]

$$W(\sigma) = q(1-s)^{d(\sigma,\sigma_0)^\alpha} + 1 - q, \qquad (14.36)$$

where $0 \leq q, s \leq 1$, and $\alpha > 0$. Two cases need to be distinguished: for $q < 1$, the lower bound on the fitness is nonzero and when $s \to 1$ it becomes of sharp peak type (14.20) with the (relative) selective advantage $1/(1-q)$ for the master sequence, while for $q = 1$, the multiplicative form (14.34) with $\lambda = -\ln(1-s)$ is recovered for $\alpha = 1$, and $\alpha > 1$ ($\alpha < 1$) describes a situation with negative (positive) epistasis (Fig. 14.3).

The error threshold can be computed in the unidirectional approximation (no back mutations toward the master sequence), and in the limit $N \to \infty$, $\mu \to 0$, for $\alpha = 1, q < 1$, it has been shown that the critical mutation strength $\mu_c = N^{-1} \ln[1/(1-q)]$, which is of exactly the same form as the sharp peak result (14.22). For $q = 1$, a similar analysis shows that (14.36) displays an error threshold only when $\alpha < 1$, with a critical mutation strength given by $\mu_c = N^{\alpha-1}\lambda$ [40]. Note that in this case, the correct scaling is obtained in the limit $N \to \infty$, $\mu \to 0$ keeping $\mu N^{1-\alpha}$ fixed.

The above behavior can be understood using the following result for general *bounded* Wrightian fitness landscapes with $0 < W_{\min} \leq W(\sigma) \leq W_{\max} < \infty$. For such landscapes, the master sequence is lost from the population at a critical mutation probability, which satisfies (in the unidirectional approximation and for $N \to \infty$) [40]

$$\mu_c \leq \frac{1}{N} \ln(W_{\max}/W_{\min}); \qquad (14.37)$$

a similar result is proved in [50]. If the right-hand side of (14.37) diverges as $N \to \infty$, this would imply that there is no finite error threshold and the master sequence is maintained at any mutation rate while its vanishing would be consistent with the existence of a sharp transition for $\mu \to 0$, $N \to \infty$. For $q = 1$, the ratio between the largest and the smallest fitness is $W_{\max}/W_{\min} = e^{\lambda N^\alpha}$, so that the right-hand side of (14.37) vanishes for $N \to \infty$ only when $\alpha < 1$ whereas it goes to zero for any $\alpha > 0$ for $q < 1$, in agreement with the results cited above. The case of the multiplicative landscape (14.34) is special; here $W_{\max}/W_{\min} = e^{2\lambda N}$ and (14.37) would suggest a finite error threshold.[7] However, as discussed earlier, the master sequence is maintained at any mutation rate for $\alpha = q = 1$.

The general conclusion from these considerations is that the existence of an error threshold requires positive epistasis. This can be understood from the following qualitative argument [40]: For the case of positive epistasis, the selection force toward the fitness peak that has to be overcome by mutations is largest close to the peak; once this initial barrier has been surpassed, the population delocalises completely. In contrast, for negative epistasis, each additional step away from the fitness peak requires a larger mutation pressure than the previous step, and hence the population remains localized.

Epistasis: Paramuse Models

Since Malthusian fitness is essentially the logarithm of Wrightian fitness, the absence of epistatic interactions in continuous time models implies a *linear* dependence of the fitness $w(\sigma)$ on $d(\sigma, \sigma_0)$. To investigate the effects of epistasis, a quadratic fitness landscape of the form

$$w(\sigma) = a[1 - 2d(\sigma, \sigma_0)/N] + \frac{1}{2}b[1 - 2d(\sigma, \sigma_0)/N]^2 \qquad (14.38)$$

has been considered [11], with $a > 0$ and $b > 0$ ($b < 0$) for positive (negative) epistasis. This choice of parameters leads, through the mapping described in Sect. 14.1.2, to the quantum spin Hamiltonian

$$H = \tilde{\mu} \sum_{i=1}^{N}(\sigma_i^x - 1) + a \sum_{i=1}^{N} \sigma_i^z + \frac{b}{2N} \sum_{ij} \sigma_i^z \sigma_j^z, \qquad (14.39)$$

where σ_i^x and σ_i^z denote the x- and z-components of the quantum mechanical spin operator. As in the discrete time case, in the absence of epistasis ($b = 0$) the spins at different sites i are independent. Epistasis introduces a coupling between any pair i, j of spins, independent of their position in the sequence. In the language of statistical mechanics, this is an interaction of *mean field* type; it is ferromagnetic for $b < 0$ and antiferromagnetic for $b > 0$.

---

[7] The unidirectional approximation erroneously predicts a transition at $\mu_c = s$ [40].

An explicit solution of the model has been presented for the case $a = 0$, $b > 0$ [11]. In the limit $N \to \infty$, the mean overlap (14.25) is given by the expression
$$m = \max[1 - \tilde{\mu}/b, 0], \tag{14.40}$$
which, in contrast to the case of the sharp peak landscape, vanishes *continuously* at $\tilde{\mu} = b$. In general, an error threshold exists only if $-b \leq a < 0$. This implies that the fitness displays a minimum at a distance $0 < d_{\min} \leq N/2$ from the master sequence.

14.2.5 Beyond the Standard Model

In this section, we discuss a few biologically motivated generalizations of the mutation–selection models described so far, while however maintaining the basic simplicity of the fitness landscape.

Diploid Models

The evolution equations for diploid organisms are similar to those for the haploid case,[8] except that the fitness $W(\sigma)$ is replaced by the marginal fitness
$$\tilde{W}(\sigma, t) = \sum_{\sigma'} W(\sigma, \sigma') X(\sigma', t), \tag{14.41}$$
where $W(\sigma, \sigma')$ is the fitness of an individual with diploid genotype (σ, σ'), and $X(\sigma, t)$ is the fraction of individuals carrying sequence σ in either one of their two sets of genes [23, 51]. The analog of the sharp peak landscape (14.20) is given by
$$W(\sigma, \sigma') = \begin{cases} W_0 & \sigma = \sigma' = \sigma_0 \\ W_1 & \text{either } \sigma = \sigma_0 \text{ or } \sigma' = \sigma_0 \\ W_2 & \text{both } \sigma, \sigma' \neq \sigma_0 \end{cases} \tag{14.42}$$
with $W_0 \geq W_1 \geq W_2$. In the absence of dominance effects ($W_1 = \sqrt{W_0 W_2}$ for Wrightian fitness or $w_1 = (w_0 + w_2)/2$ for Malthusian fitness), the problem can be reduced to the haploid case. However, in general, a transformation to a linear equation, as described in Sect. 14.1.2, is unknown for the diploid case; the equations are inherently nonlinear because of the dependence of the marginal fitness (14.41) on the population distribution. As a consequence, there are multiple solutions for the fraction $X(\sigma_0)$ of wild type individuals. Nevertheless, error threshold phenomena occur whose locations depend on the relative values of W_0, W_1, and W_2. For instance, the critical mutation rate is roughly doubled when compared to with haploid model in the case of complete dominance of the wild type ($W_0 = W_1 > W_2$).

---

[8] In general, for diploid organisms the mode of reproduction – e.g., sexual, parthenogenetic, or selfing – has to be taken into account [28]. The model discussed here and in [23, 51] is a special case of parthenogenesis.

Semiconservative Replication

Although the quasispecies model described in Sect. 14.1.2 is appropriate for organisms with RNA as genetic material, it needs to be amended for DNA-based organisms. The genotype corresponding to a double stranded DNA molecule can be represented by $\{\sigma, \bar{\sigma}\}$ where $\bar{\sigma}$ is the complementary strand of σ. The replication process involves splitting the DNA and pairing each strand with the complementary bases to produce two daughter DNAs. Thus, only one strand of the original DNA is conserved in the daughter DNA. However, copying errors and subsequent (imperfect) repair result in a different DNA genotype $\{\sigma', \bar{\sigma}'\}$. Thus, the (un-normalized) number of individuals of genotype $\{\sigma, \bar{\sigma}\}$ evolves in time as [52]

$$\dot{Z}(\{\sigma,\bar{\sigma}\},t) = -W(\{\sigma,\bar{\sigma}\})Z(\{\sigma,\bar{\sigma}\},t) + \sum_{\{\sigma',\bar{\sigma}'\}} (p(\sigma' \to \{\sigma,\bar{\sigma}\})$$
$$+ p(\bar{\sigma}' \to \{\sigma,\bar{\sigma}\})) \, W(\{\sigma',\bar{\sigma}'\})Z(\{\sigma',\bar{\sigma}'\},t) \,, \quad (14.43)$$

where $p(\sigma' \to \{\sigma, \bar{\sigma}\})$ is the probability that parent strand σ' produces $\{\sigma, \bar{\sigma}\}$ and the first-term represents the loss of the original genome. For the sharp peak landscape, the error threshold occurs at

$$\mu_c = \frac{2}{N} \ln\left(\frac{2W_0}{1+W_0}\right), \quad (14.44)$$

which saturates for $W_0 \to \infty$ unlike (14.22), so that the loss of the master sequence cannot be avoided by increasing its selective advantage. This can be traced back to the destruction of the parent genome in the semiconservative case, which implies that, at sufficiently high mutation probability per genome, increasing the reproduction rate of the master sequence actually accelerates its extinction.

Dynamic Landscapes

The assumption of a static fitness landscape is good when evolution occurs on short time scales or in long-term, controlled experiments in the laboratory. However, natural populations are usually subjected to dynamic environments such as that of pathogens living in a host with a dynamic immune system. For the problem of formation of quasispecies in the presence of a dynamic sharp peak landscape, two cases need to be distinguished – one when the fitness W_0 of the master sequence σ_0 is fixed but its location shifts at periodic time intervals of length τ to a nearest neighbor [53], and the other in which the location is kept fixed but the height of the peak changes with time [54,55].

In the former case, besides the usual upper limit on the mutation rate, an analytical approximation of the model shows the existence of a lower limit also [53]. The latter arises because when the peak shift occurs, at least one individual should be present at the new location so that it can replicate and form the quasispecies. For too low mutation rates, this may not happen and this effect is quite likely to be more pronounced for finite populations.

In the case of a time-dependent peak height $W_0(t)$ of the master sequence, the characteristic timescale τ of variation of the fitness landscape must be compared with the response time of the population, which is the inverse of the relative growth rate of the master sequence compared with its mutants. When τ is large compared with the response time, the population fraction at the master sequence follows the landscape quasistatically. For rapidly changing landscapes, the time-averaged population undergoes an error threshold transition at the mutation strength μ_c given by [54, 55]

$$(1-\mu_c)^N = \left(\frac{\int_0^T W_0(t)\mathrm{d}t}{T}\right)^{-1}, \qquad (14.45)$$

which generalizes (14.22) by replacing the static fitness by an average over a time interval of length $T \gg \tau$. For periodic $W_0(t)$ with period τ the fraction $X(\sigma_0, t)$ also changes periodically with the same period but with a phase shift that increases with decreasing τ [55]. Because of this time lag, the master sequence achieves maximum population when its fitness has already dropped from the maximum amplitude.

Parental Effects

Digital organisms are computer programs with a set of instructions (genome) including copy commands due to which they can be replicated. During the copying process, some instructions can get deleted, repeated, or replaced. An evolved program can perform complex logic operations by using a simple logic operator available to it. Such complex organisms are selected by allotting them more CPU time thus increasing their replication rate defined as the ratio of the number of logical instructions that they can execute to the number of instructions that they have to perform in order to produce a new program [18]. While the latter depends on the individual's own genome, the CPU time available to it is a parental influence.

The situation is analogous to the case of biological organisms, which obtain proteins etc. from the parent besides the genome. In such a case, the fraction $X(\sigma', \sigma, t)$ of population at sequence σ with ancestor σ' evolves as [56]

$$\dot{X}(\sigma', \sigma, t) = \sum_{\sigma''} A(\sigma'')W(\sigma')p(\sigma' \to \sigma)X(\sigma'', \sigma', t) - f(t)X(\sigma', \sigma, t), \quad (14.46)$$

where $f(t) = \sum_{\sigma'',\sigma'} A(\sigma'')W(\sigma')X(\sigma'',\sigma',t)$ and $A(\sigma')$ is the contribution to the fitness from the ancestor. In the absence of parental effects, $A(\sigma) = 1$ for all sequences and the original (14.12) is obtained for $X(\sigma, t) = \sum_{\sigma'} X(\sigma', \sigma, t)$.

This can be generalized by weighting the population variable by the parental contribution and defining the normalized variable

$$X(\sigma, t) = \sum_{\sigma'} A(\sigma')X(\sigma', \sigma, t) / \sum_{\sigma', \sigma''} A(\sigma')X(\sigma', \sigma'', t), \qquad (14.47)$$

which reduces the ℓ^{2N} variables in (14.46) to ℓ^N. Interestingly, in the steady state the population $X(\sigma, t)$ obeys the quasispecies (14.12) with fitness $A(\sigma)W(\sigma)$. Thus the available results for the standard quasispecies model can be directly applied to this case. In particular, for the sharp peak landscape the fraction $X(\sigma_0, \sigma_0)$ at the master sequence increases (relative to the null case when there are no parental effects) if the ancestral fitness $A(\sigma_0) > A(\sigma)$ for $\sigma \neq \sigma_0$. It is also possible to obtain the opposite trend if the ancestral effect is deleterious and has to be compensated by the fitness of the individual itself, such as when $A(\sigma_0) < A(\sigma)$ and $W(\sigma_0) > W(\sigma)$.

Heterogeneous Mutations

The accuracy of replication depends on enzymes called polymerases, which can be present in different types with their respective accuracies. For example, as discussed by E. Lázaro in Chap. 15 of this book, RNA virus strains that show resistance to certain mutagens may possess polymerases with a particularly high copying fidelity. In the presence of p polymerases with concentrations c_k and replication error μ_k, $k = 1, \ldots, p$, the mutation probability (14.3) generalizes to

$$p(\sigma' \to \sigma) = \sum_{k=1}^{p} c_k \left(\frac{\mu_k}{(\ell - 1)(1 - \mu_k)} \right)^{d(\sigma, \sigma')} (1 - \mu_k)^N. \qquad (14.48)$$

One may expect that by increasing the concentration of the polymerase with low error rate, the error threshold can be increased (even to infinity). That this indeed is the case was demonstrated in [57] for $p = 2$ with concentration c of an error-free polymerase with replication error probability $\mu_1 = 0$ and $1 - c$ of an error-prone polymerase with $\mu_2 = \mu > 0$. For the sharp peak landscape, one can find the fraction $X(\sigma_0) = W_0 p(\sigma_0 \to \sigma_0)/(W_0 - 1)$ of the master sequence by neglecting the back mutations as before where $p(\sigma_0 \to \sigma_0) \approx c + (1 - c)e^{-\mu N}$ for $\mu \to 0$ and $N \to \infty$. Then the master sequence can localize the population if

$$\mu > \mu_c = \frac{1}{N} \ln \left(\frac{1 - c}{W_0^{-1} - c} \right), \qquad (14.49)$$

which reduces to (14.22) for $c = 0$ as expected and increases with increasing c. Since the argument of the logarithm should be positive for real μ, it follows that $c < c' = 1/W_0$ and on exceeding c', the master sequence continues to localize population for any mutation rate.

14.3 Complex Fitness Landscapes

We now turn our attention to "complex" landscapes, which do not possess the symmetries of the simple ones discussed in the last section. Realistic landscapes are expected to have hills, valleys, basins, and ridges [19]. A pictorial representation of such a *rugged* fitness landscape drawn over a two-dimensional plane is shown in Fig. 14.4. Despite the intuitive appeal of such pictures, however, it should be kept in mind that they are metaphors rather than models of biological reality. Real fitness landscapes extend over the very high dimensional, discrete space of genotype sequences, and there are indications that the intuition gained in our experience with low-dimensional landscapes fails when applied to such abstract objects [2].

Researchers trying to construct realistic fitness landscapes have followed one of two basic approaches. One approach is to study simple model systems for which the mapping from genotype to phenotype can be carried out explicitly. This has been pursued in great detail for the case of RNA sequences, which will be briefly described in Sect. 14.3.1, as well as for proteins; for a detailed discussion we refer to the chapters by P. Schuster and P. Stadler, and by U. Bastolla, M. Porto, H.E. Roman, and M. Vendruscolo in this book. The second approach, which was conceptually inspired by the statistical physics of disordered systems [58, 59], is to regard a given fitness landscape as the realization of an *ensemble of random functions* with prescribed statistical properties. In this case, an important quantity characterizing the ruggedness of the landscape is the correlation coefficient $\rho(d, N)$ between the fitnesses of two genotypes at Hamming distance d, which is defined as

$$\rho(d, N) = \frac{\langle w(\sigma)w(\sigma')\rangle - \langle w(\sigma)\rangle^2}{\langle w(\sigma)^2\rangle - \langle w(\sigma)\rangle^2}, \quad d = d(\sigma, \sigma'). \tag{14.50}$$

Here, the angular brackets stand for an average over the ensemble of landscape configurations and the denominator ensures that $\rho(0, N)$ is scaled to unity. We have defined (14.50) in terms of Malthusian fitness, but the Wrightian case can be treated in the same way. Examples of random fitness landscapes will be discussed in Sects. 14.3.2–14.3.4.

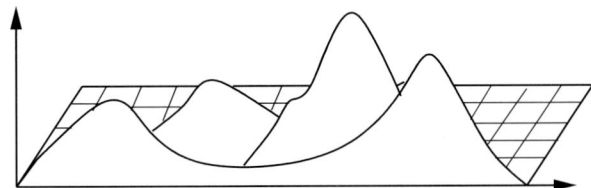

Fig. 14.4. Schematic representation of a rugged fitness landscape defined over a two-dimensional genotype space

14.3.1 An Explicit Genotype–Phenotype Map for RNA Sequences

For the description of evolution experiments with self-replicating RNA molecules (see Sect. 14.5.1), it is natural to assume that the fitness of a given RNA sequence depends only on the three-dimensional shape that the molecule folds into in the solution. As an approximation to the full three-dimensional shape (the *tertiary* structure of the molecule), its *secondary* structure, defined as the set of allowed base pairings that satisfies the no-knot constraint and minimizes the free energy, can be used. In contrast to tertiary structure, the secondary structure can be computed from the sequence by efficient algorithms. Although this does not yet solve the problem of how to assign a fitness to the genotype, it allows to study in great detail the mapping from the genotype (the sequence) to the phenotype (the secondary structure) [20, 60, 61].

The most important feature of this mapping is that it is *many-to-one*. Indeed, the number of secondary structures of random RNA sequences of length N behaves asymptotically as [61]

$$\mathcal{N}_{\text{RNA}} \approx 1.4848 \times N^{-3/2} \times (1.8488)^N, \tag{14.51}$$

whereas the number of sequences is 4^N. Thus, exponentially many sequences fold into the same secondary structure for large N. Since sequences with the same secondary structure must be assigned the same fitness, it follows that the fitness landscape contains large regions of constant fitness, which are therefore selectively *neutral*. Typically, there are a few common structures (which are represented by many sequences) and many more rare ones, with the distribution of the number of sequences mapping to a given structure following a power law. The most common structures form *neutral networks* extending throughout sequence space, such that any randomly chosen sequence is close to a sequence on this network. Similar networks have also been found in the sequence space of proteins [62–64], see the chapter by U. Bastolla, M. Porto, H.E. Roman, and M. Vendruscolo in this book. Some aspects of the evolutionary process on such neutral networks will be discussed in Sects. 14.3.4 and 14.4.1.

14.3.2 Uncorrelated Random Landscapes

The simplest kind of random fitness landscape is the uncorrelated landscape where the fitnesses are independent random variables drawn from some common probability distribution [59]. In this case, the correlation function (14.50) reduces to $\rho(d, N) = \delta_{d,0}$. An example from this class is the Random Energy Model (REM) of spin glass theory [65–67], for which the (Wrightian) fitness is given by

$$W(\sigma) = \exp[\kappa E(\sigma)], \tag{14.52}$$

where the "energies" E are independent Gaussian random variables with distribution
$$P(E) = \frac{1}{\sqrt{\pi N}} \exp(-E^2/N), \qquad (14.53)$$
and κ is an "inverse selective temperature."

This model displays a phase transition, which is quite similar to the error threshold in the single peak landscape. At high mutation rates the population is delocalized, while at low mutation rates it is frozen into the master sequence, which in this case is simply the sequence $\sigma_{\max}$ with the largest value $E_{\max}$ of $E(\sigma)$ in the particular realization (the "ground state" configuration of the REM). The scaling with N in (14.53) is chosen such that this maximal value is proportional to N, $E_{\max} = N\sqrt{\ln 2}$ to leading order. At the transition, the mean overlap (14.25) jumps discontinuously from one to zero [41].

The critical mutation probability required for delocalization can be computed along the lines used in Sect. 14.2.2 for the sharp peak landscape. Neglecting back mutations to $\sigma_{\max}$, a nonzero population fraction $X(\sigma_{\max})$ is maintained if the product of $W_{\max} = \exp[\kappa E_{\max}]$ with the probability $(1-\mu)^N$ of producing an error-free offspring is greater than the mean population fitness $\bar{W}$ in the delocalized phase [12]. The latter is obtained by averaging (14.52) with respect to the distribution (14.53), which yields $\bar{W} = \exp[\kappa^2 N/4]$. Comparing the two expressions, one finds [12, 41, 67]
$$\mu_c = 1 - \exp[\kappa^2/4 - \kappa\sqrt{\ln 2}]. \qquad (14.54)$$
The critical mutation probability reaches its maximal value $\mu_c = 1/2$ at the value $\kappa_c = 2\sqrt{\ln 2}$ of the inverse selective temperature, which coincides with the glass transition of the REM [65]. For $\kappa > \kappa_c$, the selective advantage of the most fit sequence is so great that it dominates the population even in the limiting case $\mu = 1/2$, when a complete reshuffling of genotypes occurs in each generation.

We note that, in contrast to most examples discussed in Sect. 14.2, the expression (14.54) is independent of the sequence length N. This is a consequence of the scaling of the random energies in (14.53). Indeed, this scaling implies that the ratio $W_{\max}/W_{\min} = \exp[2E_{\max}]$ grows exponentially in N, and hence the right-hand side of (14.37) is independent of N.

14.3.3 Correlated Landscapes

An example of a random fitness landscape with correlations can be constructed from the Sherrington–Kirkpatrick (SK) spin glass model, which is defined by the energy function
$$E_{\text{SK}}(\sigma) = \frac{1}{N} \sum_{i<j} J_{ij}\sigma_i\sigma_j. \qquad (14.55)$$
Here $\sigma_i = \pm 1$ and the J_{ij} are independent Gaussian random variables with zero mean and unit variance. A similar energy function arises for the graph bipartitioning problem (GBP) discussed in Sect. 14.1.1,

$$E_{\text{GBP}}(\sigma) = -\sum_{i<j} J_{ij}\sigma_i\sigma_j, \tag{14.56}$$

where the spins satisfy the vanishing total spin constraint. In this case $J_{ij} = J > 0$ if the sites i and j are connected by an edge of the graph, and $J_{ij} = 0$ else [22,42]. Through (14.52), energy functions (14.55) and (14.56) can be directly interpreted as Malthusian fitness landscapes [42,58,66]. They belong to a large class of random landscapes for which the correlation function behaves as [60]

$$\rho(d, N) \approx 1 - a_1 \frac{d}{N} + \mathcal{O}\left(\left(\frac{d}{N}\right)^2\right) \tag{14.57}$$

for $N, d \to \infty$ but $d/N \ll 1$, with a constant a_1 which is independent of N. The significance of this behavior becomes clear if we interpret d/N as a continuous variable. For random functions of a real variable, the linear dependence of the correlation function for small arguments is typical of a nondifferentiable process with independent increments (such as Brownian motion), whereas for a differentiable random process the correlation function varies quadratically at small distances. In this sense the linear behavior in (14.57) is indicative of the ruggedness of the landscape.

A simple modification of the argument leading to (14.54) gives some insight into how the fitness correlations affect the location of the error threshold [42]. We assume that in the localized phase the bulk of the population is located at some distance $d^* = \mathcal{O}(1)$ from the most fit genotype $\sigma_{\max}$, with corresponding energy values $\bar{E} \approx \rho(d^*)E_{\max}$. Equating the resulting mean population fitness $\bar{W} = \exp[\kappa \bar{E}]$ to the product $(1-\mu)^N W_{\max}$ and using (14.57) then yields, for large N, the estimate

$$\mu_c \approx \frac{\kappa a_1 d^* E_{\max}}{N^2}. \tag{14.58}$$

Together with the scaling of the ground state energy as $E_{\max}^{\text{SK}} \sim N^{1/2}$ and $E_{\max}^{\text{GBP}} \sim N^{3/2}$ for the SK-model and the GBP, it follows that $\mu_c^{\text{SK}} \sim N^{-3/2}$ and $\mu_c^{\text{GBP}} \sim N^{-1/2}$, respectively, in agreement with simulations [42].

A family of random landscapes in which the ruggedness can be tuned are the NK landscapes[9] introduced by Kauffman and Levin [5,68]. In this model, the Malthusian fitness[10] of a genotype is written as a sum of contributions from the N loci,

$$w(\sigma) = \frac{1}{N} \sum_{i=1}^{N} w_i, \tag{14.59}$$

where each w_i is a function of σ_i and K other loci chosen at random[11]. The number of possible states of σ_i and its K chosen neighbors is then ℓ^{K+1}, and

---

[9] A related family was defined in [66] in analogy to Derrida's p-spin model of spin glasses [65].
[10] A Wrightian version of the model is discussed in [69].
[11] Other schemes for choosing the interacting loci are described in [17,60,69].

each of these states is assigned a random fitness drawn from some continuous probability distribution. For large N the additive form of (14.59) ensures that the $w(\sigma)$ become Gaussian by virtue of the central limit theorem.

For $K = 0$ the loci are independent, and the model becomes equivalent to the multiplicative fitness landscape without epistasis discussed in Sect. 14.2.4; in particular, there is a unique fitness peak. At the other extreme $K = N - 1$, the w_i are independent random variables and the model reduces to the uncorrelated landscape of Sect. 14.3.2. With increasing K the number of fitness maxima increases and their height decreases [68], and the correlation function is given by[12] [71]

$$\rho(d, N) = \begin{cases} \dfrac{(N - K - 1)! \, (N - d)!}{N! \, (N - K - d - 1)!} & : d \leq N - K - 1 \\ 0 & : \text{else.} \end{cases} \quad (14.60)$$

This shows how the correlations decay more rapidly with increasing epistasis (increasing K), and reduces to $\rho(d, N) = 1 - d/N$ for $K = 0$ and $\rho(d, N) = \delta_{d,0}$ for $K = N - 1$, respectively (see Fig. 14.5).

Another model with tunable correlations was introduced in a study of evolutionary dynamics in the limit of infinite genome size but with a finite population [72]. For $N \to \infty$ every mutation creates a genotype that has not been previously represented in the population. The fitnesses can then be created "on the fly" according to the transition probability

$$\text{Prob}[w(\sigma)|w(\sigma')] \sim \exp[-(w(\sigma) - \lambda^d w(\sigma'))^2] \quad (14.61)$$

where $d = d(\sigma, \sigma')$ and the parameter $0 \leq \lambda \leq 1$ determines the decay of the correlations as $\rho(d, N) \sim \lambda^d$.

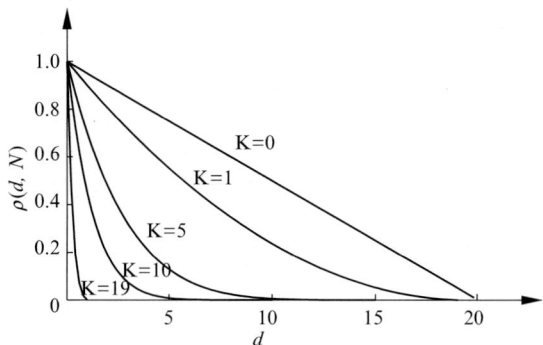

Fig. 14.5. Correlation function (14.60) for the NK-model with $N = 20$

---

[12] The expressions for the correlation function given in [60, 70] are incorrect, because it is not taken into account that the d mutations separating the two genotypes in (14.50) must affect different sites in the sequence.

14.3.4 Neutrality

We have seen in Sect. 14.3.1 that realistic fitness landscapes obtained from mapping sequences to structures contain extended regions that are selectively neutral. It has been argued that this is a general feature of high-dimensional fitness landscapes, which has important consequences for the way in which evolutionary dynamics should be visualized [2]. Rather than consisting of valleys and hilltops, as suggested by the low-dimensional rendition in Fig. 14.4, such a *holey landscape* would display a network of ridges of approximately constant fitness, along which a population can travel large genetic distances without ever having to cross an unfavorable low-fitness region.[13]

Several properties of the stationary population distribution for the quasispecies model on a neutral network can be inferred without specifying the precise structure of the network [73]. It is only assumed that the viable genotypes make up a connected graph $\mathcal{G}$ of constant fitness, which is surrounded by genotypes that are lethal or at least of very low fitness. Mutations are restricted to nearest neighbor sequences. Then the key observation is that the stationary population distribution $X(\sigma)$ on the network[14] is the principal eigenvector of the *adjacency matrix* of the graph, which is a matrix that has unit entries for pairs of viable sequences that are connected by a single point mutation, and zero entries otherwise. The corresponding eigenvalue Λ is equal to the *population neutrality* $\langle \nu \rangle$,

$$\Lambda = \langle \nu \rangle = \sum_{\sigma \in \mathcal{G}} \nu(\sigma) X(\sigma), \qquad (14.62)$$

where $\nu(\sigma)$ is the number of viable neighbors of sequence σ (the *degree* of the corresponding node of $\mathcal{G}$). The weighting by the population fraction $X(\sigma)$ in (14.62) is significant. For any graph $\mathcal{G}$, the principal eigenvalue of the adjacency matrix satisfies the bounds [74]

$$\bar{\nu} \leq \Lambda \leq \nu_{\max}, \qquad (14.63)$$

where $\bar{\nu}$ and $\nu_{\max}$ denote the average and maximal degrees of the graph. For a random graph with a range of degrees, the relations (14.62) and (14.63) imply

---

[13] The evolutionary importance of paths of viable genotypes that connect distant points in sequence space was emphasized by Smith [3]. He illustrates the issue with a game where the goal is to transform one word into another by changing one letter at a time, with the requirement that all intermediate words are meaningful (i.e., "viable"). An example is the path WORD → WORE → GORE → GONE → GENE.

[14] The population on the network is normalized to unity, $\sum_{\sigma \in \mathcal{G}} X(\sigma) = 1$, which does not include the individuals in the lethal region. Although these individuals do not reproduce, they constitute a finite fraction of the population, which is replenished by mutations from viable genotypes.

that generally $\langle \nu \rangle > \bar{\nu}$, which shows that the population preferentially resides at nodes where the number of viable neighbors is larger than on average. This has been referred to as the evolution of *mutational robustness* [73]. The heterogeneity of the node degree along the neutral network has important consequences also for the evolutionary dynamics, because it induces strong fluctuations in the rate of neutral substitutions [63, 64, 75].

Neutral networks can be modeled as random subgraphs in sequence space. Such subgraphs are defined through a simple modification of the uncorrelated landscape model of Sect. 14.3.2, where each sequence σ is randomly assigned fitness $W(\sigma) = 1$ (*viable*) with probability P and $W(\sigma) = 0$ (*lethal*) with probability $1 - P$. Each connected region of viable genotypes then constitutes a random subgraph. For small P these regions are small and isolated, but at the *percolation threshold* $P = P_c$ given by

$$P_c = \frac{1}{(\ell - 1)N} \qquad (14.64)$$

a giant network appears, which spans the sequence space and which, for $P > P_c$, contains a finite fraction of all sequences [2, 76]. Since N is a large number, the fraction of viable genotypes needed to create such a spanning network is remarkably small [3].

For subgraphs of the binary hypercube ($\ell = 2$) with random assignment of links (rather than sites), it has been shown that the principal eigenvalue of the adjacency matrix is asymptotically given by [74]

$$\Lambda \approx \max[NP, \sqrt{\nu_{\max}}]. \qquad (14.65)$$

Taking $N \to \infty$ at fixed P one finds that $\nu_{\max} \sim N$, so that $\Lambda \to NP = \bar{\nu}$. In this limit, the neutral network behaves like a regular graph, and no significant mutational robustness develops. On the other hand, if $P \to 0$ as $N \to \infty$ with NP fixed, one obtains $\nu_{\max} \sim N/\ln N \gg \bar{\nu}$, and the mutational robustness effect is significant.

14.4 Dynamics of Adaptation

In this section, we turn our attention to time-dependent aspects of the adaptive process. In rugged fitness landscapes, the population is faced with the task of reaching ever higher fitness peaks by traversing fitness valleys or neutral networks, which typically gives rise to a pattern of episodic or *punctuated* evolution. This phenomenon will be discussed in general terms in Sect. 14.4.1, and a specific model study [77] will be summarized in Sect. 14.4.2. In Sect. 14.5.3, we describe an approach to evolutionary dynamics that is suited for landscapes that are *smooth*, in the sense that a simple (linear) relation between fitness and genetic distance can be assumed.

14.4.1 Peak Shifts and Punctuated Evolution

The existence of multiple fitness peaks of different height, as illustrated in Fig. 14.4, immediately suggests that evolutionary histories should generally display two distinct regimes: periods of stabilizing selection, where the population resides near a local fitness maximum, and *peak shifts* in which the population moves quickly from one fitness peak to another of greater height. The stationary distributions in a single peak landscape that were discussed at length in Sect. 14.2 can be viewed as an approximate description of the first regime. The necessity of peak shifts for explaining the succession of biological forms in the paleontological data has been recognized for a long time,[15] but the underlying mechanisms (and even the relevance of the concept itself) remain controversial.

Mathematical analysis of peak shifts driven by stochastic fluctuations in finite populations (genetic drift) generally show that the waiting time for the shift is vastly larger than the time required for the transition itself [79, 80]. This can be argued to support the scenario of *punctuated equilibrium* in macroevolution [81], which states that evolutionary changes (including both speciation and phenotypic changes within a lineage) occur during relatively short time intervals, which are separated by long periods of no discernible change (*stasis*).

However, for realistic population sizes the stochastically driven peak shifts may be far too rare to be relevant, and in fact they may not be needed at all, if the picture of a holey landscape spanned by a network of neutral ridges described in Sect. 14.3.4 is generally applicable [2]. Evolution in such a landscape will nevertheless be punctuated, because a population moving by genetic drift across a neutral network can increase its fitness only by finding a path to another network of higher fitness. If these paths are rare, a natural separation of time scales between (phenotypic) stasis and sudden fitness jumps arises. This scenario is well established for simulations of in vitro evolution of RNA sequences [20, 82]. Borrowing a concept from statistical physics, it can be said that in this case the population is confined by *entropic* barriers rather than by fitness barriers [83].

14.4.2 Evolutionary Trajectories for the Quasispecies Model

In the deterministic mutation–selection models of interest in this chapter, stochastic fluctuations cannot be invoked to drive peak shifts. Nevertheless, a population initially placed near one fitness peak in a multipeaked landscape is able to relocate to a higher peak, by developing tails of mutants, which (since the number of individuals is formally infinite) with time explore the entire sequence space. Once a small mutant population has been established at the

---

[15] A famous example is the transition from browsing to grazing behavior in equids [78].

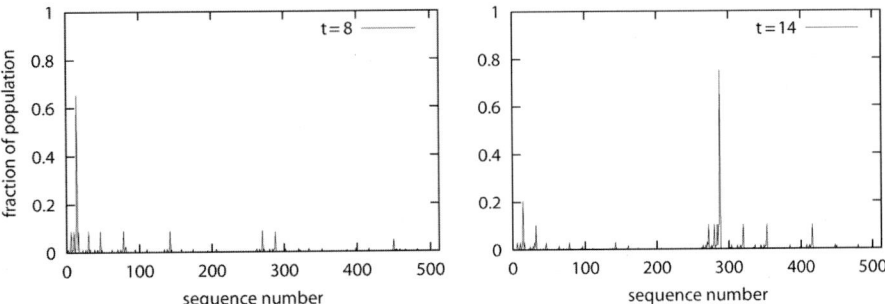

Fig. 14.6. Example of a peak shift event for quasispecies dynamics with binary sequences of length $N = 9$ in an uncorrelated random fitness landscape. At time $t = 8$, the most populated sequence is near the origin (sequence number 1), but at time $t = 14$ it has moved to a sequence number close to 300. The peaks of lower height represent the first and second neighbor mutants of the most populated sequence. They are not adjacent because of the linear arrangement of the sequences

distant fitness peak, it starts to compete with the majority at the original peak and, if the newly populated peak is higher, it will eventually come to dominate the population. In this way, the majority of the population can shift between peaks without ever actually having to traverse a fitness valley (Fig. 14.6).

The time $t_\times$ required for a single peak shift in the discrete time quasispecies model has been estimated numerically for a simple degenerate two-peak landscape, given by (14.31) with $W_N = W_0$ and $W_{N-1} = 1$ [84]. The population was first allowed to equilibrate in a single peak landscape and then the second peak was turned on. The result is

$$t_\times \sim \left(\frac{\ln W_0}{N\mu}\right)^N \sim \left(\frac{\mu_c}{\mu}\right)^N \qquad (14.66)$$

which, somewhat surprisingly, has the same form as the time required for a finite population to cross a fitness valley [83]; of course in the latter case there is an additional dependence on the population size.

The evolutionary trajectories that result from multiple peak shifts in an uncorrelated rugged fitness landscape have been studied in detail in a strong selection limit motivated by the zero temperature limit of the statistical physics of disordered systems [77, 85–87]. Writing

$$Z(\sigma,t) = e^{\kappa F(\sigma,t)}, \quad W(\sigma) = e^{\kappa E(\sigma)}, \quad \mu = e^{-\kappa}, \qquad (14.67)$$

with κ denoting the inverse selective temperature (see Sect. 14.3.2), and starting with an initial condition $Z(\sigma, 0) = \delta_{\sigma,\sigma^{(0)}}$ where $\sigma^{(0)}$ is a randomly chosen sequence, the dynamics takes the following form in the $\kappa \to \infty$ limit:

$$F(\sigma, t+1) = \max_{\sigma'} \left[F(\sigma',t) + E(\sigma') - d(\sigma,\sigma')\right], \quad t \geq 2 \qquad (14.68)$$
$$F(\sigma, 1) = E(\sigma^{(0)}) - d(\sigma, \sigma^{(0)}). \qquad (14.69)$$

Here, the logarithmic fitnesses $E(\sigma)$ are independent random variables chosen from a common distribution $p(E)$. As already discussed in Sect. 14.3.2, one expects the whole population to be localized at the fittest genotype in the large time limit. At any finite time, in the strong selection limit, the population can be identified with the most populated genotype. The behavior of this genotype is essentially unaffected by dropping the mutation term for times $t > 1$, so that the dynamics reduces to [86]

$$F(\sigma, t) = F(\sigma, 1) + (t-1)E(\sigma), \quad t \geq 2. \qquad (14.70)$$

This illustrates the fact that, after the entire sequence space has been "seeded" by mutants of the original genotype $\sigma^{(0)}$ at time $t = 1$, the subsequent evolution consists in the competition of independent populations located at the fitness peaks. Distant peaks of high fitness are disadvantaged by a small initial population but may come to dominate at later times.

Since the seeding population $F(\sigma, 1)$ of a sequence only depends on its distance from the initial genotype $\sigma^{(0)}$, within each *shell* of constant $k = d(\sigma, \sigma^{(0)})$ only the most fit genotype is a contender for global leadership. Thus, the dynamics of the ℓ^N variables (14.70) can be reduced to $N + 1$ shell population variables $F(k, t)$ whose fitnesses $E(k)$ are chosen from the distribution

$$p_k(E) = \alpha_k \, p(E) \left(\int_{E_{\min}}^{E} p(x) \mathrm{d}x \right)^{\alpha_k - 1}. \qquad (14.71)$$

This is the distribution of the maximum among α_k independent random variables with distribution $p(E)$, and α_k is the number of sequences in shell k, as defined in (14.4).

The representation of the "evolutionary race" as a problem of crossing straight lines[16] is illustrated in Fig. 14.7. At a given time t, the most populated sequence located in shell k^* leads until it is overtaken by a shell $k^{*'}$ with $E(k^{*'}) > E(k^*)$ and so on, until the global fitness maximum takes over. A natural question of interest is to identify the sequences that take part in this evolutionary trajectory, and to determine their number. It is clear that for a sequence to participate in the trajectory, it is necessary that it constitutes a *fitness record*, in the sense that its fitness exceeds the fitnesses of all sequences that are closer to $\sigma^{(0)}$. An analytical treatment of the statistics of these independent but nonidentically distributed records shows that the average number of records encountered on the way to the global maximum is [77]

$$\mathcal{R} \approx \frac{(\ell - \ln \ell - 1)}{\ell - 1} N \qquad (14.72)$$

for large N, and that essentially all records are located within the distance $d_{\max} = N(\ell - 1)/\ell$ near which most of the sequences (including the most

---

[16] The problem is related to models of highway traffic, where each vehicle is equipped with a fixed random speed and overtaking is forbidden [88].

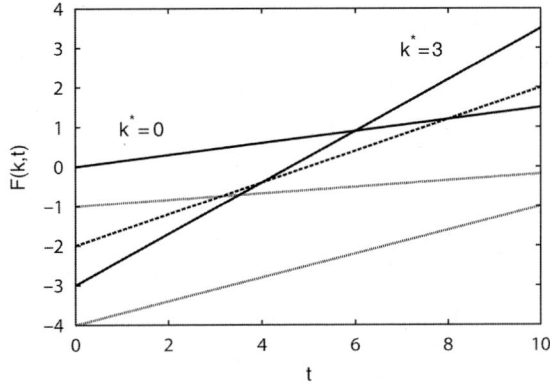

Fig. 14.7. Illustration of the linear dynamics (14.70). For each shell of constant Hamming distance k from $\sigma^{(0)}$ only the *line* with the largest slope is drawn. *Dashed lines* are fitness records, *dotted lines* are nonrecords, and *solid lines* are records that are not bypassed

fit sequence) reside. For $\ell = 2$, the inter-record spacing between the jth and $j+1$th record is of the order $\sqrt{N/j}$ where $j = 1$ labels the last record (the global maximum). Thus, a few records separated by distances of order $\sqrt{N}$ occur near $d_{\max}$ and the rest are clustered away from it.

However, many records are *bypassed* by fitter sequences that arise further away from $\sigma^{(0)}$ but manage to catch up with the current leader at an earlier time. For unbounded fitness distributions with Gaussian or exponential tails, the number of nonbypassed records (which is the number of sequences that take part in a trajectory) is found to be only of order $\sqrt{N}$ with a uniform spacing $\sim \sqrt{N}$, which suggests that the competition among the contenders is strong when the average fitness of the population is still low. For fat-tailed power law distributions the average number of records that are not bypassed is asymptotically equal to unity, which implies that the population relocates to the global fitness maximum in a single step.

Several statistical properties of the *timing* of peak shifts turn out to be independent of the fitness distribution $p(E)$ [77, 85, 86]. Specifically, denoting by T_j the time at which the jth peak shift occurs, with $j = 1$ denoting the last shift (which reaches the global fitness maximum), $j = 2$ the penultimate peak shift and so on (Fig. 14.8), the corresponding distributions display universal power law tails

$$P_j(T_j) \sim (T_j)^{-(j+1)}. \qquad (14.73)$$

In particular, the expected value of T_1 is infinite. The prefactors of these power laws depend, however, on the fitness distribution and the sequence length, in such a way that, e.g., the typical value of T_1 tends to unity for fitness distributions with a power law tail.

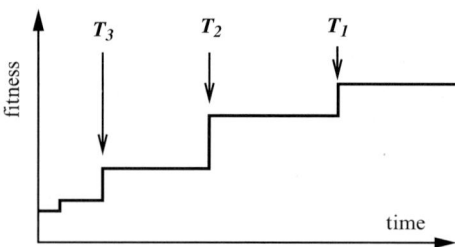

Fig. 14.8. Timing of evolutionary jumps

14.4.3 Dynamics in Smooth Fitness Landscapes

So far we have discussed landscapes in which the fitnesses can be very different from each other and as described above, the evolutionary trajectory can change in a stepwise manner if the landscape has local maxima. Smoothly varying landscapes for which the system does not get trapped in such metastable states are the subject of the following discussion. *Smoothness* will be taken to imply here that there is a simple (linear) relationship between the fitness of a genotype and its genetic distance from the master sequence. Individuals can then be characterized by their fitness alone, and the description can be based on a one-dimensional *fitness space* [89].

The prototypical case in which this reasoning applies is that of the multiplicative fitness landscape discussed in Sect. 14.2.4. We work in the Malthusian setting and assume that the fitness $w(\sigma)$ is simply equal to the number of matches with the master sequence, $w = 0, \ldots, N$. Then, the fraction $Y(w,t)$ of individuals with fitness w at time t evolves as [89]

$$\dot{Y}(w,t) = (w - \bar{w})Y(w,t) + \tilde{\mu}[(w+1)Y(w+1,t) \\ + (N-w+1)Y(w-1,t) - NY(w,t)], \quad (14.74)$$

which is just the paramuse (14.7) evaluated for the present fitness landscape, with $\bar{w}$ denoting the mean fitness of the population. For large N the fitness w can be treated as a continuous variable. Setting $r = (w - N/2)/\sqrt{N}$, $\bar{\mu} = \tilde{\mu}/\sqrt{N}$, and $\tau = \sqrt{N}t$, (14.74) reduces for $N \to \infty$ to the drift-diffusion equation

$$\frac{\partial Y}{\partial \tau} = (r - \bar{r})Y + \frac{\bar{\mu}}{2}\frac{\partial^2 Y}{\partial r^2} + \frac{\partial}{\partial r}(2\bar{\mu}rY). \quad (14.75)$$

Analysis of (14.74) and (14.75) and related equations [90] shows that the mean fitness diverges in finite time, since the equations ignore the fact that at least one individual is required to initiate the reproduction process. This can be circumvented by imposing a cutoff Y_c inversely proportional to the population size, below which the selection term does not operate.

With this modification, one finds that at short times, the population which was initially spread over a fitness range gets localized about the maximum available fitness leading to a fast growth of average fitness. This is followed by the collective motion of the localized "species" as a traveling wave with constant speed and width (as long as the population is far from the boundaries $w = 0$ and N of the fitness space). A finite population size analysis of discrete models (described in the infinite population limit by the above continuum equations) shows that both speed and variance of the wave diverge linearly with increasing population size, which is consistent with the finite time singularity that appears in the absence of a cutoff [90].

Quantitative agreement with finite population simulations requires a more careful treatment in which the most fit nonempty mutant class is treated stochastically, while keeping deterministic differential-difference equations of the type (14.74) for the remainder of the population. In addition, the continuum limit of (14.74) should be carried out on the level of $\ln Y$ rather than for Y itself, which leads to a nonlinear drift-diffusion equation replacing (14.75) [27]. Recent applications of fitness space models that go beyond the present discussion include studies of the in vitro evolution of DNA sequences selected for protein binding [46], viral populations undergoing serial population transfers [91], and the effects of recombination in asexual populations [92].

14.5 Evolution in the Laboratory

Viruses and bacteria are suitable candidates for testing the theory of asexual evolution because of their simple genomes and high replication rates. For instance, RNA viruses that are characterized by high mutation rates and small genome (see Table 14.1 and Chap. 15 by E. Lázaro in this book) can produce about 10^4 copies in an hour. Their typical population numbers are of the order of 10^{11}, thus getting close to the infinite population condition for the applicability of quasispecies theory. Interestingly, evolution can also occur in nonliving systems such as RNA extracted from a bacteriophage, which we now proceed to discuss in 14.5.1.

14.5.1 RNA Evolution In Vitro

Early in vitro studies of adaptation to a given environment were carried out on a simple system comprising RNA molecules and the enzyme RNA replicase, which is required to catalyze the RNA replication reaction. In the first of a series of experiments, the time interval during which the reaction is allowed to proceed was gradually reduced with the number of generations, thus selecting the rapidly growing molecules [93]. By the 74th generation, the initial baseline strain with a genome length of a few thousand bases evolved to a 15 times faster replicating (but no longer pathogenic) chain of merely a few hundred bases, by casting off the parts of the genome, which do not participate in the in vitro replication process. Subsequently, experiments using such short RNA were performed under different conditions and selection pressure [94, 95]. In particular, the formation of a quasispecies consisting of only 40% of the master sequence and many mutants has been demonstrated [96].

14.5.2 Quasispecies Formation in RNA Viruses

Inside a cell, a virus is subjected to the constantly changing environment of the host, whereas the quasispecies concept described in earlier sections assumes an infinite population evolving toward a stationary state in a static landscape. Nevertheless, evidence for quasispecies formation has been obtained in in vivo experiments on RNA viruses by examining their genetic heterogeneity [97], and the quasispecies concept now plays an important role in virology [98–100]; for a detailed discussion, we refer to the chapter by Ester Lázaro in this book.

The first such experiment was performed on a Qβ phage population derived from the wild type [25]. On sampling about 10% of its sequence, it was found that on an average, the genome of the derived phage differs from the wild type at about two positions. Assuming a Poisson model for the distribution of deviations from the wild type, only 14% of the population was found to be wild type and the rest was accounted for by related mutants with up to 3–4 substitutions. Similarly, in the Hepatitis C virus, half of the RNA molecules were found to be identical and the rest one to four mutations away from each other [101]. In the case of HIV, the quasispecies concept has been used to explain the reappearance of the virus after the treatment with drugs that target only the wild type [102]. Many experiments, such as [103] on poliovirus, also show that RNA viruses operate close to the error threshold, since on a modest increase in mutation rate (through chemicals), the virus population was found to lose its genetic structure.

14.5.3 Dynamics of Microbial Evolution

The dynamics of adaptation have been studied in several long-term experiments on asexually reproducing microbes like viruses and bacteria. In experiments on *E. coli* [104, 105], several populations are derived from the same ancestor and allowed to replicate under identical conditions. The ancestor is engineered to have a selectively neutral marker so that it can be distinguished from the offspring colony. The process of evolution occurs because the progeny is grown in the presence of limited supply of glucose, unlike the ancestor.

To measure the fitness of the evolved type, the ancestor and the evolved progeny are made to compete for glucose by mixing them in equal amounts at time $t = 0$ and estimating their respective densities ρ_A and ρ_P at $t = 0$ and $t = 1$ where time is measured in days. Then the Malthusian fitness of the evolved type at any instant measured relative to the ancestor A is given by

$$w = \frac{\ln(\rho_P(1)/\rho_P(0))}{\ln(\rho_A(1)/\rho_A(0))}. \tag{14.76}$$

The experiments indicate that the fitness of all populations improves in time, but each of the replicate populations reaches a different fitness level at large times. This supports the picture of a rugged fitness landscape (Fig. 14.4) with several peaks in which the population, starting from the same initial point, reaches different local maxima via different evolutionary trajectories.

Initially, fitness changes rapidly but slows down considerably in the course of time. When the same experimental data is viewed at a finer scale, the best fit to the data is obtained if the fitness increases are assumed to occur in steps. The occurrence of punctuated evolution is associated with the selection of rare beneficial mutations [106]. Although a large number of advantageous mutations with small effects may have occurred, a few mutations with large effects quickly spread through the population and are responsible for the jumps in the fitness. For a review of other experiments with this bacterial population see [9,107].

The step-like nature of fitness trajectories, especially the properties of the first step, has been investigated in detail in other experiments as well. For instance, in [108], the distribution of the fitness conferred in the first step was measured in *E. coli*, which supports the above observation of the occurrence of few mutations with large benefits and many with small payoffs. Similar experiments have also been performed on the RNA virus ϕ_6 [109]. This study tracked the fitness recovery in a population, after a deleterious mutation has been induced by a population bottleneck, for about hundred generations. The fitness was seen to recover in steps but the number of steps (and the fitness benefit) was found to depend strongly on the population size. While large populations recovered in one large step, smaller populations required many steps each granting small favors. As discussed in detail in the chapter by Ester Lázaro, such population bottlenecks occur naturally in the life cycle of viruses, because the number of viral particles that are transmitted from one host to another is often very small.

Finally, we note that under certain conditions populations of RNA viruses display a linear increase or decrease of fitness with time [27, 110], which can be analyzed within the framework of the fitness space models as discussed in Sect. 14.4.3.

14.6 Conclusions

In this chapter, we have given an overview over a class of models of adaptive evolution, which include selection and mutation, but (due to their deterministic character) ignore effects of genetic drift in finite populations. A large body of work spread out over different scientific communities has been devoted to such models, and our survey must necessarily remain quite incomplete. We have, therefore, tried to focus on some general concepts – such as sequence space, fitness landscapes, error thresholds, and epistatic interactions – that we believe to be useful also beyond the specific biological situations in which the models apply.

Stochastic effects characteristic of finite populations are expected to be quantitatively and even qualitatively important for several of the phenomena we have described. Genetic drift induces a new mechanism of genetic degradation, *Muller's ratchet* [111], in which the fittest genotype is lost from the population because it is not sampled for reproduction. In the limit of infinite sequence length, this process is irreversible, and it generally contributes to the delocalization of the population from fitness peaks. Correspondingly, a common result of finite population studies in simple [23, 37, 48] as well as complex [42, 72] landscapes is a lowering of the error threshold mutation rate with decreasing population size. A comparison between Muller's ratchet and the error threshold in infinite population models can be found in [10, 50]. As described in the chapter by E. Lázaro, both mechanisms for genetic degradation are being considered as possible strategies for fighting viral infections.

The finite size of the population is also crucially important for the peak shifts in rugged landscapes as discussed in Sect. 14.4.2, because it imposes a cutoff on the tails of rare mutants, which are responsible for the communication between distant fitness peaks. Much of the analytic work on adaptive dynamics that takes stochastic aspects into account has considered the regime of low mutation rates,[17] where the population consists of a single genotype at most times and the generation and fixation of new mutations are rare events. In these studies, the geometrical constraints on the availability of new mutants in sequence space are usually ignored, and the timing and fitness effects of mutations are instead generated by a suitable stochastic process [112, 113]. An important task for the future will be to integrate the different theoretical approaches, with the ultimate goal of bringing them to bear on the experimental data that are becoming available.

Acknowledgments

This work was supported by DFG within SFB-TR12 *Symmetries and universality in mesoscopic systems*. JK is grateful to E. Ben-Naim, D. Krakauer, H. Levine, and T. Wiehe for useful discussions, and to the Laboratory of Physics of HUT for the kind hospitality during the completion of the chapter.

References

1. S. Wright, in *Proceedings of the 6th International Congress of Genetics* **I**, 356 (1932)
2. S. Gavrilets, *Fitness Landscapes and the Origin of Species* (Princeton University Press, 2004)

---

[17] The quantitative characterization of this regime is that the product of the population size and the mutation probability per site is small compared to unity [26].

3. J. Maynard Smith, Nature **225**, 563 (1970)
4. M. Eigen, Naturwissenschaften **58**, 465 (1971)
5. S. Kauffman, S. Levin, J. Theor. Biol. **128**, 11 (1987)
6. J.F. Crow, M. Kimura, *An Introduction to Population Genetics Theory* (Harper & Row, New York, 1970)
7. M. Travisano, Curr. Biol. **11**, R440 (2001)
8. H.A. Orr, Nat. Rev. Genet. **6**, 119 (2005)
9. S.F. Elena, R.E. Lenski, Nat. Rev. Genet. **4**, 457 (2003)
10. E. Baake, W. Gabriel, Annu. Rev. Comput. Phys. **7**, 203 (2000)
11. E. Baake, H. Wagner, Genet. Res. Camb. **78**, 93 (2001)
12. B. Drossel, Adv. Phys. **50**, 209 (2001)
13. L. Peliti, in *Fitness Landscapes and Evolution*, ed. by T. Riste, D. Sherrington. Physics of Biomaterials: Fluctuations, Self-Assembly and Evolution (Kluwer, Dordrecht, 1996), p. 267
14. L. Peliti, cond-mat/9712027
15. R. Bürger, Genetica **102/103**, 279 (1998)
16. U. Gerland, T. Hwa, J. Mol. Evol. **55**, 386 (2002)
17. A.S. Perelson, C.A. Macken, Proc. Natl. Acad. Sci. USA **92**, 9657 (1995)
18. C.O. Wilke, C. Adami, Trends Ecol. Evol. **17**, 528 (2002)
19. P.F. Stadler, in *Fitness landscapes*, ed. by M. Lässig, A. Valleriani. Biological Evolution and Statistical Physics (Springer, Berlin Heidelberg New York, 2002), p. 183
20. P. Schuster, in *A Testable Genotype–phenotype Map: Modeling Evolution of RNA Molecules*, ed. by M. Lässig, A. Valleriani. Biological Evolution and Statistical Physics (Springer, Berlin Heidelberg New York, 2002), p. 55
21. J.W. Drake, B. Charlesworth, D. Charlesworth, J.F. Crow, Genetics **148**, 1667 (1998)
22. Y. Fu, P.W. Anderson, J. Phys. A Math. Gen.**19**, 1605 (1986)
23. T. Wiehe, E. Baake, P. Schuster, J. Theor. Biol. **177**, 1 (1995)
24. D.B. Saakian, C.-K. Hu, Phys. Rev. E **69**, 046121 (2004)
25. E. Domingo, D. Sabo, T. Taniguchi, C. Weissmann, Cell **13**, 735 (1978)
26. L.M. Wahl, D.C. Krakauer, Genetics **156**, 1437 (2000)
27. I.M. Rouzine, J. Wakeley, J.M. Coffin, Proc. Natl. Acad. Sci. USA **100**, 587 (2003)
28. P.G. Higgs, Genet. Res. **63**, 63 (1994)
29. E. Baake, M. Baake, H. Wagner, Phys. Rev. Lett. **78**, 559 (1997)
30. H. Wagner, E. Baake, T. Gerisch, J. Stat. Phys. **92**, 1017 (1998)
31. J. Hermisson, H. Wagner, M. Baake, J. Stat. Phys. **102**, 315 (2001)
32. M. Eigen, P. Schuster, Naturwissenschaften **64**, 541 (1977)
33. M. Eigen, J. McCaskill, P. Schuster, J. Phys. Chem. **92**, 6881 (1988); ibid.: Adv. Chem. Phys. **75**, 149 (1989)
34. J.B. Kogut, Rev. Mod. Phys. **51**, 659 (1979)
35. I. Leuthäusser, J. Chem. Phys. **84**, 1884 (1986); J. Stat. Phys. **48**, 343 (1987)
36. P. Tarazona, Phys. Rev. A **45**, 6038 (1992)
37. M. Nowak, P. Schuster, J. Theor. Biol. **137**, 375 (1989)
38. D.S. Rumschitzki, J. Math. Biol. **24**, 667 (1987)
39. B. Derrida, L. Peliti, Bull. Math. Biol. **53**, 355 (1991)
40. T. Wiehe, Genet. Res. Camb. **69**, 127 (1997)
41. S. Franz, L. Peliti, J. Phys. A Math. Gen. **30**, 4481 (1997)

42. S. Bonhoeffer, P.F. Stadler, J. Theor. Biol. **164**, 359 (1993)
43. S. Galluccio, Phys. Rev. E **56**, 4526 (1997)
44. L. Peliti, Europhys. Lett. **57**, 745 (2002)
45. P. Schuster, J. Swetina, Bull. Math. Biol. **50**, 635 (1988)
46. W. Peng, U. Gerland, T. Hwa, H. Levine, Phys. Rev. Lett. **90**, 088103 (2003)
47. J. Berg, S. Willmann, M. Lässig, BMC Evol. Biol. **4**, 42 (2004)
48. G. Woodcock, P.G. Higgs, J. Theor. Biol. **179**, 61 (1996)
49. A.S. Kondrashov, Nature **336**, 435 (1988)
50. G.P. Wagner, P. Krall, J. Math. Biol. **32**, 33 (1993)
51. E. Baake, T. Wiehe, J. Math. Biol. **35**, 321 (1997)
52. E. Tannenbaum, E.J. Deeds, E.I. Shakhnovich, Phys. Rev. E **69**, 061916 (2004)
53. M. Nilsson, N. Snoad, Phys. Rev. Lett. **84**, 191 (2000)
54. C.O. Wilke, C. Ronnewinkel, T. Martinetz, Phys. Rep. **349**, 395 (2001)
55. M. Nilsson, N. Snoad, Phys. Rev. E **65**, 031901 (2002)
56. C.O. Wilke, Phys. Rev. Lett. **88**, 078101 (2002)
57. K. Aoki, M. Furusawa, Phys. Rev. E **68**, 031904 (2003)
58. P.W. Anderson, Proc. Natl. Acad. Sci. USA **80**, 3386 (1983)
59. J.S. McCaskill, J. Chem. Phys. **80**, 5194 (1984)
60. W. Fontana, P.F. Stadler, E.G. Bornberg-Bauer, T. Griesmacher, I.L. Hofacker, M. Tacker, P. Tarazona, E.D. Weinberger, P. Schuster, Phys. Rev. E **47**, 2083 (1993)
61. P. Schuster, W. Fontana, P.F. Stadler, I.L. Hofacker, Proc. Roy. Soc. Lond. B **255**, 279 (1994)
62. U. Bastolla, H.E. Roman, M. Vendruscolo, J. Theor. Biol. **200**, 49 (1999)
63. U. Bastolla, M. Porto, H.E. Roman, M. Vendruscolo, J. Mol. Evol. **56**, 243 (2003)
64. U. Bastolla, M. Porto, H.E. Roman, M. Vendruscolo, J. Mol. Evol. **57**, S103 (2003)
65. B. Derrida, Phys. Rev. B **24**, 2613 (1981)
66. C. Amitrano, L. Peliti, M. Saber, J. Mol. Evol. **29**, 513 (1989)
67. S. Franz, L. Peliti, M. Sellitto, J. Phys. A Math. Gen. **26**, L1195 (1993)
68. S.A. Kauffman, *The Origins of Order* (Oxford University Press, New York, 1993)
69. J.J. Welch, D. Waxman, J. Theor. Biol. **234**, 329 (2005)
70. E.D. Weinberger, Phys. Rev. A **44**, 6399 (1991)
71. P.R.A. Campos, C. Adami, C.O. Wilke, Physica A **304**, 495 (2002); Erratum **318**, 637 (2003)
72. C.O. Wilke, P.R.A. Campos, J.F. Fontanari, J. Exp. Zool. (Mol. Dev. Evol.) **294**, 274 (2002)
73. E. van Nimwegen, J.P. Crutchfield, M. Huynen, Proc. Natl. Acad. Sci. USA **96**, 9716 (1999)
74. A. Soshnikov, B. Sudakov, Commun. Math. Phys. **239**, 53 (2003)
75. U. Bastolla, M. Porto, H.E. Roman, M. Vendruscolo, Phys. Rev. Lett. **89**, 208101 (2002)
76. S. Gavrilets, J. Gravner, J. Theor. Biol. **184**, 51 (1997)
77. K. Jain, J. Krug, J. Stat. Mech. **P04008** (2005)
78. G.G. Simpson, *Tempo and Mode in Evolution* (Columbia University Press, New York, 1944)
79. C.M. Newman, J.E. Cohen, C. Kipnis, Nature **315**, 400 (1985)

80. N.H. Barton, S. Rouhani, J. Theor. Biol. **125**, 397 (1987)
81. N. Eldredge, *Macroevolutionary Dynamics* (McGraw-Hill, New York, 1989)
82. W. Fontana, P. Schuster, Science **280**, 1451 (1998)
83. E. van Nimwegen, J.P. Crutchfield, Bull. Math. Biol. **62**, 799 (2000)
84. D. Kim, W. Gill, J. Korean Phys. Soc. **44**, 973 (2004)
85. J. Krug, *Tempo and Mode in Quasispecies Evolution*, ed. by M. Lässig, A. Valleriani. Biological Evolution and Statistical Physics (Springer, Berlin Heidelberg New York, 2002), p. 205
86. J. Krug, C. Karl, Physica A **318**, 137 (2003)
87. J. Krug, K. Jain, Physica A **358**, 1 (2005)
88. E. Ben-Naim, P.L. Krapivsky, S. Redner, Phys. Rev. E **50**, 822 (1994)
89. L.S. Tsimring, H. Levine, D.A. Kessler, Phys. Rev. Lett. **76**, 4440 (1996)
90. D.A. Kessler, H. Levine, D. Ridgway, L. Tsimring, J. Stat. Phys. **87**, 519 (1997)
91. S.C. Manrubia, E. Lázaro, J. Pérez-Mercader, C. Escarmís, E. Domingo, Phys. Rev. Lett. **90**, 188102 (2003)
92. E. Cohen, D.A. Kessler, H. Levine, Phys. Rev. Lett. **94**, 098102 (2005)
93. D.R. Mills, R.L. Peterson, S. Spigelman, Proc. Natl. Acad. Sci. USA **58**, 217 (1967)
94. C.K. Biebricher, *Darwinian selection of self-replicating RNA molecules*, ed. by M. Hecht, B. Wallace, G.T. Prance. Evolutionary Biology, vol. 16 (Plenum, New York, 1983), pp. 1–52
95. C.K. Biebricher, W.C. Gardiner, Biophys. Chem. **66**, 179 (1997)
96. N. Rohde, H. Daum, C.K. Biebricher, J. Mol. Biol. **249**, 754 (1995)
97. E. Domingo, J.J. Holland, Annu. Rev. Microbiol. **51**, 151 (1997)
98. M.A. Nowak, R. May, *Virus Dynamics* (Oxford University Press, 2001)
99. M. Eigen, Proc. Natl. Acad. Sci. USA **99**, 13374 (2002)
100. A. Moya, E.C. Holmes, F. González-Candelas, Nat. Rev. Microbiol. **2**, 279 (2004)
101. M. Martell, J.I. Esteban, J. Quer, J. Genesca, A. Weiner, R. Esteban, J. Guardia, J. Gomez, J. Virol. **66**, 3225 (1992)
102. J.M. Coffin, Science **267**, 483 (1995)
103. S. Crotty, C.E. Cameron, R. Andino, Proc. Natl. Acad. Sci. USA **98**, 6895 (2001)
104. R.E. Lenski, M.R. Rose, S.C. Simpson, S.C. Tadler, Am. Nat. **138**, 1315 (1991)
105. R.E. Lenski, M. Travisano, Proc. Natl. Acad. Sci. USA **91**, 6808 (1994)
106. S.F. Elena, V.S. Cooper, R.E. Lenski, Science **272**, 1802 (1996)
107. R.E. Lenski, Plant Breed. Rev. **24**, 225 (2004)
108. M. Imhof, C. Schlötterer, Proc. Natl. Acad. Sci. USA **98**, 1113 (2001)
109. C.L. Burch, L. Chao, Genetics **151**, 921 (1999)
110. I.S. Novella, E.A. Duarte, S.F. Elena, A. Moya, E. Domingo, J.J. Holland, Proc. Natl. Acad. Sci. USA **92**, 5841 (1995)
111. J. Haigh, Theor. Popul. Biol. **14**, 251 (1978)
112. H.A. Orr, Genetics **155**, 961 (2000)
113. P. Gerrish, Nature **413**, 299 (2001)

15

Genetic Variability in RNA Viruses: Consequences in Epidemiology and in the Development of New Strategies for the Extinction of Infectivity

E. Lázaro

15.1 Introduction

Viruses constitute one of the simplest biological entities in nature. They possess some properties typical of life, such as the transmission of genetic information through generations, but lack a proper metabolism and a system to translate the genetic information into proteins. This ambivalence places them at the border between living and non-living matter. To reproduce themselves, viruses are forced to infect a host cell, behaving as intra-cellular parasites. Despite their simplicity, viruses have been able to develop a wide repertoire of infection mechanisms and replication strategies to adapt to the broad diversity of the cellular world in order to execute their genetic program.

All known viruses consist of one or several genomic nucleic acid molecules, covered by protective layers. Usually, there is one protein capside that can be enclosed by a lipid bilayer membrane proceeding from the host cell. The number of proteins encoded by the viral genomes is rather small, making their success to replicate and give rise to an offspring dependent on their ability to take advantage of the enzymatic activities provided by the host cell. After infection, cellular protein synthesis is stopped and most of the subcellular machinery is directed to produce copies of the viral nucleic acids and proteins. These newly synthesized viral components are assembled inside the cell into mature virions that can infect other cells of the same organism or establish transmission chains between different individuals. The stability of these chains strongly conditions the survival of the virus in nature. Outside an adequate host, viruses can still persist for some time in a latent state in which they are unable to replicate and exposed to irreversible damage by physical conditions of the environment. When they interact with the specific cellular receptors of a suitable host, they can initiate an infection that, if successfully transmitted in the population, can seriously compromise the survival of the host species.

In all cellular organisms, the genetic information is contained in the DNA. Before being translated into proteins (the molecules that execute the functions necessary for the performance of the cell), the DNA has to be copied to mRNA. But genetic information also needs to be maintained through generations, a process that takes place by means of DNA replication, in which many enzymatic activities are involved. In contrast to cells, viruses are more versatile and can use DNA or RNA to store the genetic information. DNA viruses can follow a scheme similar to that followed by cells to replicate their genomes and to synthesize their proteins. However, RNA viruses need another process, RNA replication, which is not among the functions carried out routinely in the cell [1]. Therefore, they have to encode and express the enzymatic activities necessary to copy their genomes. These enzymes are the RNA replicases and the reverse transcriptases, which in many cases are co-encapsidated with the nucleic acid during the assembly of the viral particles. In this way, they are available at the start of the infection.

All living systems must reach a compromise between the correct copy of the nucleotide sequence of their genomes and the ability to adapt to an environment that is continuously changing [2]. The generation of mutants upon replication provides the necessary diversity from which natural selection can choose the best adapted variants in a concrete environment. The observed divergences among mutation rates in different species suggest that possibly this character is selected depending on the variability of the environment [3]. Cellular systems are able to maintain a relative constancy in the intra-cytoplasmatic medium and because of that, they do not need a high genetic variability. Thus, evolution has emphasized the selection of a replicative machinery with several corrector activities that permit a high copying accuracy. However, even in the cellular world, mutation rate is not a fixed character that cannot be altered. It can be modified in response to environmental changes by selection of variants with higher or lower error rates. The isolation of hypermutator strains, which show deficiencies in some of the polymerase corrector activities, is frequent in conditions of environmental stress and is a proof of the versatility of mutability as a character that can be modified when the environment requires it [4, 5].

A relevant characteristic of RNA viruses is that they replicate their genomes with a copying fidelity several orders of magnitude lower than cellular DNA [6]. This fact has been interpreted as a consequence of the fluctuating environments that viruses have to face. High error prone replication, together with the short replication times and large population sizes typical of RNA viruses, instead of being a handicap for survival provides an extraordinary evolutionary advantage by permitting the generation of a wide reservoir of mutants with different phenotypic properties [7]. The high variability of RNA viruses facilitates their survival in presence of antibodies and other defence mechanisms produced by the immune system of the host. It also makes possible the acquisition of novel pathogenic properties that in occasions have allowed to cross species boundaries favouring the infection of alternative hosts [8, 9]. Finally, the heterogeneity of RNA virus populations makes it also difficult to eradicate diseases with antiviral drugs, due to the emergence of drug-resistant mutants [10], a problem that will be treated in more detail in the next sections. Whether the high genetic variability of RNA viruses is a selected character, necessary for survival in high fluctuating environments, or it is simply a consequence of the lack of corrector activities of RNA replicases and reverse transcriptases is a debated question. However, the fact that DNA organisms, which usually live in constant environments, have evolved corrector activities, whereas RNA viruses have not, suggests that replication with high error rates is a selected character that strongly favours viral adaptation to fast changing conditions.

15.2 Replication of RNA Viruses and Generation of Genetic Variability

The first requisite for the evolution of any population is the generation of a significantly wide genetic diversity on which natural selection and genetic drift can act to shape the properties of the new populations generated at subsequent generations.

The genetic variation attained by RNA viruses is mainly the result of mutation and recombination, two processes that are dependent on the properties of the enzymes that replicate their genomes. Genome segment reassortment occurs during encapsidation and can add extra variability in the case of viruses with segmented genomes.

The replication of RNA viruses takes place through two main mechanisms that involve the use of different enzymatic activities [1]. Riboviruses, including many prokaryotic RNA viruses as well as many animal and plant viruses (poliovirus, influenza virus, hepatitis A and C viruses, etc.), replicate their genomes using RNA replicases that catalyze the RNA-dependent RNA synthesis. The template RNA can be of positive polarity (it can work as mRNA) or negative polarity (it is the complementary strand that is translated into proteins). Viruses with positive polarity genomes deliver the nucleic acid directly to the cellular ribosomes and begin infections with translation. In contrast, viruses with negative polarity genomes begin infections with transcription to obtain mRNA molecules that can be translated. Retroviruses, HIV-1 (human immunodeficiency virus type 1) being the best known example, replicate their genomes through a different mechanism with an intermediary step that consists in the copy of the genomic RNA to DNA. This process is catalyzed by the enzyme reverse transcriptase, an RNA-dependent DNA polymerase carried by the viral particle. The DNA obtained in this way is integrated in the host chromosome, being transcribed by the cellular enzyme RNA polymerase II to produce transcripts that can function either as precursors of mRNAs or as genomic RNAs that can be assembled into progeny viruses.

The lack of corrector activities of both classes of enzymes RNA replicases and reverse transcriptases results in high mutation rates, which have been estimated in 10^{-4} to 10^{-5} misincorporations per nucleotide copied. For a virus with a genome length of 10,000 nucleotides, this amounts to the incorporation of one incorrect nucleotide per genome copied on the average [6,11]. Thus, each new viral genome differs from its parent at one or two nucleotide positions. The relative proportion of a specific mutant in the viral population depends on the rate at which the mutant is generated and on its fitness, which is defined as the ability to give rise to a progeny in competition with the rest of viruses replicating under certain environmental conditions [7]. The number of mutations that occur per time unit is also influenced by the number of replication rounds during that period, this is, the generation time. For viruses with similar error rate polymerases, the shorter the generation time, the larger the number of mutants that is produced in the same time interval.

Recombination takes place when a new genome is built from fragments belonging to different parental molecules. In RNA viruses, this process usually occurs by template switching during RNA or cDNA synthesis. Most studies suggest that recombination rates in RNA viruses are lower than in other organisms [12], although there are some notable exceptions, such as HIV-1 in which the recombination rate seems to be higher than the mutation rate [13]. Recombination can be a powerful mechanism to create advantageous genomes and to purge deleterious mutations in a very short time. However, the actual effects of recombination in RNA viruses have not been studied in detail, and it is not clear whether it is beneficial or it has a negative effect on fitness [14].

Genome segment reassortment occurs in viruses with segmented genomes and consists in the encapsidation in the same viral particle of genome segments proceeding from different parental viruses. Influenza viruses are the typical example in which this process has been responsible for antigenic shifts, probably resulting from combinations of segments of influenza virus of different specificity [15]. The natural reservoir of influenza is aquatic birds, although the virus can also infect domestic birds and mammals (human or pigs preferably). When a reassortant influenza virus emerges, its pathogenic potential can dramatically increase, because the infected host is not able to recognize the antigenic determinants of the new virus generated. These reassortant strains have been responsible for a number of pandemics through history and most studies suggest that a new influenza pandemic is unavoidable [16].

15.3 Structure of Viral Populations

The structure of viral populations results from the concerted action of the processes of mutation and selection acting in very large ensembles of replicating units. Population size fluctuations, which frequently take place during transmission of viruses in nature, constitute an additional and important factor influencing the extension of genetic diversity from which a new virus population will be generated.

The evolution and self-organization of heterogeneous populations composed by a large number of molecules subjected to error-prone replication and exposed to selection was first studied theoretically [17]. These studies showed that, for large population sizes and after long growth times in a constant environment, a steady-state is reached where each mutant represents a constant fraction of the total population. This equilibrium population was called quasi-species [18, 19]. The most frequently occurring molecular species, usually the one with the highest fitness, is called the master sequence. This sequence is accompanied by a mutant spectrum, composed by an ensemble of variants that differ in one or several nucleotide positions that can be responsible for fitness variations in individual mutants. The number of nucleotide differences between two sequences is called the Hamming distance. The consensus sequence is defined as the sequence of the most represented nucleotides at each genomic position in the ensemble of genomes constituting the population. The correspondence between fitness values and sequences (or between phenotypes and genotypes) reveals that fitness landscapes (a surface in the genotype space representing the fitness of each genotype as a point placed at a different height) are rather rugged, since relatively small sequence differences can cause great differences in fitness values.

Analysis of RNA virus populations, either at the phenotypic or genotypic level showed that these populations have a structure similar to the molecular quasi-species described theoretically [20, 21]. They present a master sequence surrounded by a mutant cloud and, in the absence of a cloning method to separate individual genomes, only the consensus sequence can be determined. However, two main differences have to be taken into account when comparing theoretical and viral quasi-species. The first one is that viral quasi-species usually are not equilibrium populations because viruses are continually confronted with many environmental perturbations that cause variations in the fitness distribution of the population. The second difference is that viral fitness is not only determined by the genomic replicative ability. In spite of its simplicity, a virus must complete successfully many processes to originate an infective progeny. These include recognition of the cellular receptors, uncoating and release of the nucleic acid inside the cell, interaction with many enzymes and cellular structures, correct assembly to give rise to new viruses and exit out of the cell. The ability of a virus to perform correctly all these processes, together with the replicative ability of its genome, is what determines its fitness value. These differences introduce uncertainties and additional complexity to viral evolution compared to molecular evolution described theoretically.

15.4 Viral Quasi-Species and Adaptation

Viral quasi-species constitute very dynamical structures in which the processes of generation of new mutants, selection of the best adapted and elimination of the less fit are continuously acting. Quasi-species replicating during a long time in a near-constant environment in the absence of large population size fluctuations can present a low rate of fixation of mutations in the consensus sequence, despite the continuous occurrence of mutants that is characteristic of the underlying dynamics of the population. In this case, the quasi-species is well adapted to the environment and can maintain a low rate of evolution, as determined by the stability of the consensus sequence. Many of the mutants generated are lethal or very deleterious and are eliminated or maintained at low frequency by the action of negative selection. In contrast, mutants with a selective advantage can be present at high frequency, even if they are produced at a low rate [20]. Neutral mutations are thought to be very restricted in RNA viruses because of their highly compact genomes [22]. Low rates of evolution have been described for viruses well adapted to their animal reservoir (the host in which the virus is usually maintained in nature) as influenza in birds or hantavirus in rodents. The same happens in the laboratory, where viruses are usually cultivated during years in the same cellular type. In both cases an almost invariant consensus sequence can be found, although the dynamics of the quasi-species is always dominated by the processes of mutation and selection acting in close concert.

The factors that promote the fixation of mutations in the consensus sequence are usually environmental changes that favour the selection of the best adapted genomes in the new conditions or drastic reductions in the number of individuals that will originate a new population, what is called population bottlenecks. The occurrence of genetic alterations with an adaptive advantage in the absence of environmental perturbations is also possible, although it happens more rarely. In this section, we will focus on some features that favour the action of positive selection and amplification of advantageous mutants. We will mention three examples that make enormously difficult virus eradication:

1. *Treatment with antiviral drugs.* When a viral infection is treated with an antiviral agent the usual outcome is that, after a short time of success, the treatment loses its efficacy. The failure is generally due to the presence in the mutant spectrum of some genomes able to resist the action of the drug. Usually, these genomes have lower fitness in the absence of the drug and they are maintained at low frequencies by the action of negative selection. The presence of the drug inhibits the replication of the sensitive genomes, but not of the resistant ones, which are selected and amplified. It can also occur that, at the beginning of the treatment, no drug-resistant genomes are present in the population. However, the high mutation rates and large population sizes of RNA viruses make highly probable that, after a variable time lag, a resistant mutant appears, which in a short time can dominate the population. Maybe, the most dramatic example of drug-induced resistance occurs in patients infected with HIV-1 in which variants resistant to all currently used drugs have been isolated [23, 24]. At present, the most effective treatment to control HIV-1 infection is the so called highly active anti-retroviral therapy (HAART), which involves a combination of several drugs, aimed at preventing the emergence of variants with mutations conferring resistance to all the drugs at the same time.

2. *Antigenic drift.* There are considerable differences in the nature and duration of the immune response elicited by different viruses [25]. Some human viruses, such as measles or chicken pox, can only infect once, because their antigenic determinants have very slow evolution rate and the immunological response of the memory cells continues being effective along the whole life of the individual. In contrast, there are other viruses, influenza being the paradigmatic example, that can infect the same organism repeated times. Most experimental evidence leads to the conclusion that the tolerance of a virus to accept immune-escape mutations is limited by the restriction of conserving the cell tropism [26]. Modifications in the capside antigen domains of measles virus seem to have very deleterious effects, possibly because they affect the recognition of the cellular receptors. In contrast, influenza can experience a continuous change in the antigenic properties of the two main surface proteins involved in the entry of the virus inside the cell, the hemagglutinin and the neuraminidase. The evolution of the virus seems to be strongly influenced by selection of new antigenic variants to escape the immune system at the same time that the capacity of interaction with the cellular receptor is preserved [15]. In the case of the hemagglutinin gene, 18 codon sites have been identified in which non-synonymous nucleotide substitutions are much more frequent than synonymous [27, 28]. The remaining sites show the more common pattern of synonymous substitutions, indicating that possibly they are subjected to stronger evolutionary constraints.
3. *Change of tropism.* Many viruses are maintained in nature in animal reservoirs that do not manifest symptoms of disease. This probably occurs because the relation virus–host is very old (hundreds or thousands of years) and both species have had enough time to co-evolve, meaning that the virus has attenuated its virulence and the host also has acquired some properties that permit coexistence with the pathogen. The long time of evolution in the same host has permitted to these viruses to be close to the equilibrium between mutation and selection processes and to maintain a high stability in the consensus sequence. Occasionally, a virus well-adapted for replication in a particular host can cross the species boundaries and infect a new host. This can be facilitated by genetic changes in the virus and/or by ecological factors that involve alterations in the relationships established among different species in nature [29, 30]. The infection of a new host constitutes a sudden change in the environment in which viral replication takes place, usually with the consequence of a drastic decrease in the average fitness of the virus population, which prevents further transmission. The success of a virus to establish as a new infectious agent in the new host relies largely on two features (a) its ability to interact with a cellular receptor that permits the entry inside the cells and (b) the acquisition and fixation of mutations that allow efficient replication and capacity of transmission between organisms. Most recent virus emergences in humans include HIV-1, whose closest animal ancestor seems

to be the simian immunodeficiency virus found in a particular species of chimpanzees (SIVcpz) [31], the coronavirus causing SARS (severe acute respiratory syndrome) [32] and the influenza virus H5N1, an avian virus strain that can infect directly humans without further human-to-human transmission [33].

15.5 Population Dynamics of Host–Pathogen Interactions

Virus dynamics in nature cannot be separated from host population dynamics, constituting two processes in continuous interaction [25, 34, 35]. Factors such as the transmission mode, the basic reproductive number (R_0), the duration of the infectious period, the renewal of susceptible hosts and the durability of the immune response contribute to shape the genetic heterogeneity of viruses and the quasi-species structure. They also strongly condition the evolution of the pathogen along the time, adding a great complexity to the epidemiological and phylogenetic studies on RNA viruses.

An important factor in viral evolution that takes place at the inter-host level is the number of viral particles that are transmitted from one host to the other [20]. When this number is very small, a population bottleneck takes place. Then, only one or few individuals originate a new population, resulting in a strong reduction in the genetic diversity. The consequence is that any mutation present in the founder genomes will have a high probability of being transmitted to the progeny, accelerating in this way the rate of fixation of mutations (Fig. 15.1). Since most mutations are deleterious, the expected effect of their accumulation through repeated bottlenecks is a decrease in the

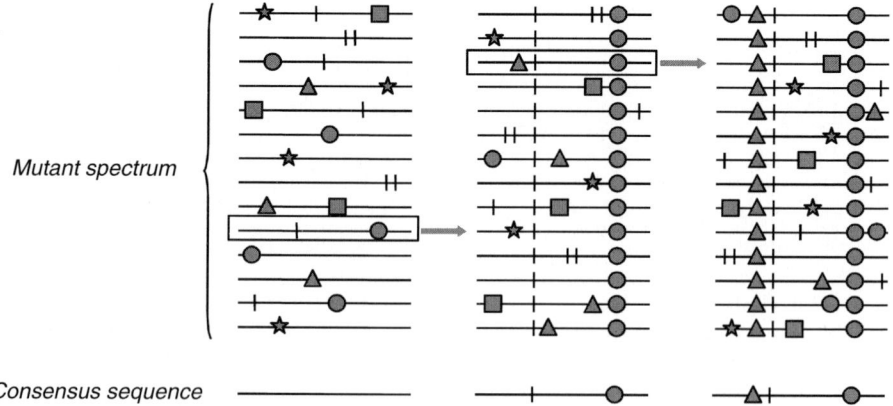

Fig. 15.1. Accumulation of mutations in the consensus sequence of a heterogeneous virus population when a single genome (*in the box*) is selected to found a new population. All the mutations carried by this genome are transmitted to the progeny, and consequently they will be fixed in the consensus sequence

average fitness that eventually could lead to the extinction of the population. Bottlenecks are very frequent in nature, during the inter-organ or inter-host transmission of many viruses. Thus, in each new infected organism, the quasispecies must be rebuilt from one or a few founder genomes, a fact that could lead to a wide diversity in diseases in which the usual form of transmission is mediated through bottlenecks. The persistence of viruses in nature and the limited number of circulating strains in diseases such as influenza, despite the frequent occurrence of bottlenecks, is paradoxical [36]. It is believed that the inter-host competition that can induce stochastic losses of the less fit variants, together with the action of previous immune responses on genetically related virus variants (the so called cross-immunity) are factors that restrict the strain diversity. The intra-host competition that takes place after each transmission event is an additional factor that favours the optimization of the viral population inside each infected individual and also contributing to the resistance to extinction of viruses transmitted through bottlenecks.

Grenfell et al. [37] have classified RNA viruses in four phylodynamic categories, according to factors pertaining to the host–pathogen interactions (mainly the duration of the infection and the nature and strength of the immune response). They are briefly described:

1. *Short infections with strong cross-immunity.* The best known viruses included in this category belong to the family of morbilliviruses (measles being a well studied example). In these viruses, epidemic cycles are mainly determined by the lifelong immunity elicited by the pathogen, which causes that the renewal of susceptible hosts takes place only at the birth of new individuals. The existence of a strong immune response that is powerful against all circulating strains (strain-transcending immunity) would prevent the action of selection. In these viruses, the burden of many different strains seems to be limited by spatio-temporal parameters of the dynamics of the epidemic process.
2. *Short infections with partial cross-immunity.* The best example of this category is influenza A virus. The high mutation rate characteristic of RNA viruses, together with the transmission of influenza through bottleneck events, opposes to the limited variability within lineages. In contrast to measles, cross-immunity against virus variants is only partial and the replacement of susceptible individuals takes place, not only through the birth of new hosts but also through generation of new influenza strains that may affect individuals previously exposed to the virus. Evolution of influenza and its epidemic dynamics have been modelled in several studies, trying to reproduce the strong seasonality of infections and the replacement of strains at each epidemic. The most successful models reproduce the behaviour of influenza epidemics when a short-lived strain-transcending immunity (in contrast to the long-lived immunity characteristic of viruses in the previous category) is included as an essential factor limiting viral diversity in the host population [36]. However, the role

of within-host dynamics after each bottleneck mediating transmission remains to be added to the epidemic model.
3. *Infections with immune enhancement.* These are infections with the possibility of antibody-dependent enhancement (ADE). An example is dengue virus that comprises four serotypes co-circulating in tropical regions. ADE causes that secondary infections produced by a different virus serotype usually curse with more severe symptoms than primary infections.
4. *Persistent infections.* In this category are included viruses such as HIV and HCV (hepatitis C virus) that can persist in their host during long times periods. For these viruses, inter-host dynamics is slow, being more important and faster the intra-host period of evolution that is driven by continuous and strong immune pressure.

15.6 The Limit of the Error Rate

Since the high genetic heterogeneity of RNA viruses provides an enormous adaptive capacity, it could be naively expected that additional increases in the replication error rate makes evolutionary adaptation even more efficient. However, there are many theoretical and experimental evidences showing that RNA viruses have selected the maximal error rate, which is compatible with the preservation of their genetic information.

Theoretical studies on molecular evolution postulate that the higher the error rate and the genome length, the smaller is the probability of obtaining a progeny identical to the parental genome and to conserve the master sequence in the population [17, 19]. There is a sharp limit, called error threshold, which cannot be crossed without catastrophic consequences for the survival of the population (see the chapter by Jain and Krug in this book). Below this limit, the quasi-species can maintain a large genetic variability from which the best adapted molecules are selected. When the threshold is crossed, the dispersing force of mutation cannot be compensated by selection of the best adapted phenotypes and the genetic information melts away in a process with the physical characteristics of a first order phase transition, as the melting of a solid. The transition takes place in an 'information space' that is multidimensional, comprising 4^N sequences of length N [38].

The error rate that can be maintained is related to the genome length according to this relation:

$$N_{\max} < \ln s_0/(1-q). \tag{15.1}$$

Here $N_{\max}$ is the maximal length of the genome and it is inversely proportional to the error rate per nucleotide $(1-q)$. The factor s_0 indicates the selective advantage of the master sequence in relation to the mutant spectrum. Measurements of the chain lengths and the replication error rates of RNA viruses show that the genome lengths of RNA viruses are close to the maximum that can be maintained at the error rates of their replication. Moreover, phylogenetic analysis of RNA viruses reveals a negative correlation between rates of nucleotide substitution and genome size [39]. As a direct consequence, all viral functions must be encoded within a limited genomic space (10–15 kb on average for most RNA viruses), meaning that certain regions of the genome will often have to participate in several functions at the same time, resulting in restrictions to the capacity of RNA virus to alter their nucleotide sequences. The most frequent evolutionary constraints identified are the following [22]:

1. Usually, the antigenic determinants of a virus are domains of the same proteins involved in the recognition of the cellular receptor [26]. This fact restricts the possibilities of immune escape to the occurrence of mutations in domains that are not crucial for penetration of the virus inside the cell.
2. Genomic coding regions can also be involved in the interaction with enzymes or cellular structures and in the regulation of the correct synthesis and assembly of the viral components to constitute mature particles.
3. Synonymous mutations may be not silent and have effect on fitness because they can affect the secondary structure of RNA domains critical for keeping the stability and functionality of the molecule [40].
4. Sometimes the same genomic region can encode several proteins through the use of overlapping reading frames.

In most RNA viruses, a high amount of particles is not infectious, suggesting that viral populations operate near the error threshold and most mutations are not easily tolerated, possibly due to the above-mentioned constraints. The large population sizes constituted by RNA viruses seem to be necessary to avoid stochastic extinctions that could happen due to the generation of many deleterious mutants.

Given the high mutation rate of RNA viruses, and the increased fraction of deleterious mutations over advantageous ones that occur when a population is well adapted to the environment, one can think of two alternative strategies for driving a viral population to extinction. Both of them involve an increase in the number of mutations in individual viral genomes, which can be related, although not necessarily, to changes in the consensus sequence of the population.

The first pathway is the classical one described by molecular evolution error catastrophe theories. It consists in the increase of the replication error rate, usually through the use of mutagens. The new populations generated exhibit larger complexity than the initial ones. Advantageous mutations, even if they occur, would be spoiled by the continuous generation of deleterious mutations, before they can be fixed by natural selection. In this case, it is the strong dispersing force of mutation what dominates the dynamics of the population.

The second pathway consists in the application of successive bottlenecks to the population. After each bottleneck, the founder genomes give rise to a new population through a limited number of replication rounds. The larger the number of generations between bottlenecks, the closer is the new population to the equilibrium between mutation and selection [41]. The resulting populations have two essential characteristics. The first one is their low complexity, because the low number of copy rounds taking place between bottlenecks does not permit to generate a large genetic diversity. The second one is an increased rate in the fixation of mutations in the consensus sequence, since most mutations present in the founder genomes are transmitted to the descendants (Fig. 15.1). Given the high amount of deleterious mutations, the expected result of their accumulation is a progressive reduction in the average fitness of the population that could lead to the extinction of infectivity.

The structure of the viral populations generated through the two pathways described here have different evolutionary consequences that have been explored experimentally by several groups. Next sections contain a review of the main results published in this field.

15.6.1 Increases in the Error Rate of Replication. Lethal Mutagenesis As a New Antiviral Strategy

There are many experimental evidences documenting extinction of RNA viruses experiencing an increased mutation rate due to the action of mutagens [42–47]. The mutagens most currently used are 5-fluorouracil (FU), 5-azacytidine (AZC), azidothymidine (AZT), ribavirin and 7-hydroxyurea. Some of them are nucleoside analogues that, in addition to increasing the rate of erroneous incorporation of nucleotides, can also interfere with other cellular or viral processes, such as endogenous nucleotide metabolism, viral replication or transcription.

Foot-and-mouth disease virus (FMDV), poliovirus, HIV-1 and lymphocytic choriomeningitis virus (LCMV) are some examples of RNA viruses in which successful extinctions of infectivity have been documented. The results agree with molecular evolution theories that postulate that viral replication operates very close to an error threshold that cannot be crossed without compromising the transmission of genetic information and the existence of the population (reviewed in [48]). Although many studies have been devoted to the characterization of the mutant spectrum of pre-extinction populations [43, 49, 50], it is not clear how the quasi-species looses its infective capacity. It is not known whether all the genomes are carrying lethal mutations and therefore are unable to replicate or it is the disorganization of the mutant spectrum what makes the quasi-species to be non-infective. In the last case, the quasi-species could still conserve some viable genomes that, in the absence of the interfering mutants, could initiate the development of an infective population.

Mutagenized populations of FMDV treated with AZC and FU could be efficiently extinguished [47]. As expected, the characterization of the RNA genomes composing the pre-extinction populations did not show mutations in the consensus sequences, but displayed an increase in the complexity of the mutant spectrum (reviewed in [51]). The maximum increases in complexity occurred in the polymerase gene, which usually is well conserved. Other studies have also demonstrated the invariance of the consensus sequence, despite the occurrence of a high number of mutations in individual genomes [43].

The same mutagenic agent can behave differently in different viruses. As an example, LCMV was systematically extinguished after only two or three passages in the presence of FU [43,52], whereas extinction of FMDV was stochastic and required a larger number of passages in the presence of similar amounts of mutagenic agent. Differences in the susceptibility of a virus to a mutagen can be explained by a number of factors including different affinity of the polymerase for the mutagen, effect of the mutagen in other viral or cellular processes, type of mutations preferentially induced by the mutagen that can affect viral functions differently depending on the nucleotide composition of the virus genomes, etc., [53]. The influence of variations in the mutation rate of different virus polymerases in the capacity of mutagens to extinguish infections is not well known. In principle, it should be expected that the closer is the virus to the error threshold, the easier should be its extinction by increased mutagenesis. However, the error rate of the polymerase is very difficult to estimate, and it can change depending on environmental factors and the region of the genome sequenced. Mutation rates are usually obtained from measurements of mutation frequency, a procedure that can lead to underestimation of the true mutation rates, because only replicating genomes are abundant enough to be detected. The isolation of a poliovirus mutant with a high fidelity polymerase [54,55] that is resistant to the action of ribavirin and other mutagens clearly indicates that the error rate of a particular polymerase is a relevant factor contributing to the efficiency of increased mutagenesis to extinguish viral infections.

Studies in both riboviruses and retroviruses suggest that host enzymes also represent a potential source of variation by RNA editing [56]. There are some cellular enzymes able to produce hypermutation in the viral genomes, which occurs as clusters of specific base substitutions. A documented example is the enzyme APOBEC3G, which has been shown to generate G→A hypermutations in HIV-1. Enzymes of this type could act as a natural strategy for limiting viral infection by increasing mutagenesis above the error threshold. The discovery of these host factors constitutes an alternative for the development of agents that specifically enhance the natural antiviral activity of cells.

Recently, extinction by lethal mutagenesis has been shown to involve more complex mechanisms than those affecting only the replicative ability of genomes [44]. It is well known that in a normal infection a variable amount of the viruses produced are non-infective because they are unable to code for all functional proteins [57]. However, inside the cell, it is plausible that many of these non-infective genomes behave as parasites and replicate using the proteins produced by other viruses. When the mutation rate is kept below a critical threshold, defective mutants maintain an equilibrium with viable genomes. The increase in the mutation rate forces the appearance of a larger amount of defective genomes that, beyond a critical fraction, can exhaust the

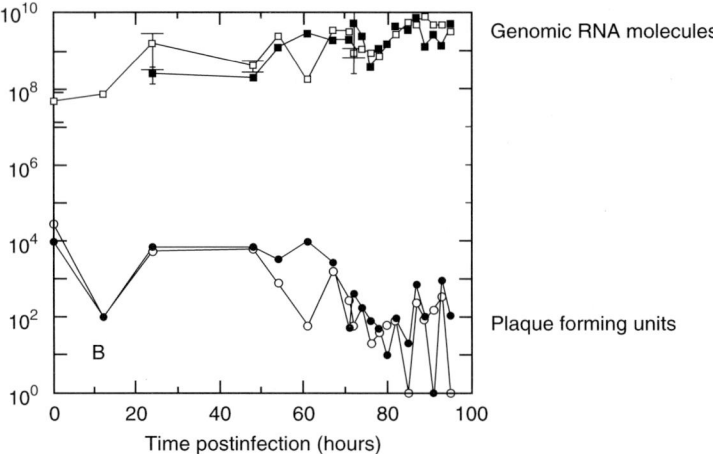

Fig. 15.2. Quantification of the viral genomic RNA and infectivity in supernatants and cell fractions of LCMV incubated with 100 µg ml^{-1} of 5-FU. Although a high amount of RNA is still present in the samples, infectivity declines until undetectable levels, indicating that replicative ability does not disappear simultaneously with infectivity. Further details of this experiment can be found in [44]. *Open symbols* correspond to the intra-cellular fraction. *Filled symbols* correspond to the supernatant fraction

resources necessary for viral replication, becoming an additional force that can promote extinction.

This conceptual framework derives from several 'in vitro' experiments with LCMV [44]. Infective viruses and RNA genomic molecules were monitored during a virological steady-state persistent infection of BHK-21 cells by LCMV in the absence and presence of 5-FU (Fig. 15.2). In the course of the infection, there is a clear increase in the number of genomic RNA molecules, both in the intra-cellular fraction and in supernatants of control and mutagenized virus. However, in FU-treated virus cultures, infectivity declines and falls below detection, despite the high number of genomic RNA molecules. The number of infective units per RNA molecule as a function of the mutation frequency yields a curve with a sharp decay when the mutation frequency overcomes a critical threshold. The sudden loss of infectivity takes place through a transition analogous to that predicted by error catastrophe theories. However, the unexpected outcome of the experiment was the presence of large numbers of RNA molecules, revealing that the replicative ability does not disappear simultaneously with infectivity. Similar results have been found with poliovirus and Hantaan virus where decreases in infectivity preceded decreases in viral RNA levels [42, 58, 59].

Lethal mutagenesis probably presents many of the same difficulties as conventional antiviral therapy. An important problem takes place in viruses, such as retroviruses, that can stay in a latent state during a long time in cellular or anatomical reservoirs. Activation of these latent viruses can contribute to the resurgence of the disease after interruption of drug treatment in vivo [60]. However, the strongest obstacle to antiviral mutagenesis is the appearance of drug-resistant mutants due to the presence of enhanced fidelity polymerases. Possibly these mutants are less able to generate resistances to other antiviral drugs, due to the diminished ability of adaptation that results from the reduction of the genetic diversity because of the higher fidelity of the polymerase. Therefore, combined therapies consisting of lethal mutagenesis and other antivirals could be a promising strategy for the treatment of viral infections [54].

15.6.2 Evolution of Viral Populations Through Successive Bottlenecks

The probability of extinction of small asexual populations due to the accumulation of mutations was first studied by Muller several decades ago [61]. He predicted that the genomes with the lowest mutational load could be stochastically lost due to population fluctuations through a mechanism similar to the clicks of a ratchet. When the ratchet clicks the first time, this means that the genomes with no mutations are lost and the least loaded class corresponds to individuals carrying one mutation. In the next click, the one-mutation class disappears by a similar mechanism, and the least mutated class corresponds now to genomes with two mutations and so on. At that time it was believed that the least mutated genomes were the best adapted and that reversions were the only mechanism able to recover fitness. Thus, this process, which is particularly effective at high mutation rates, as it happens in RNA viruses, should inevitably imply a progressive fitness loss that can lead populations to extinction.

The experimental study on the transmission of RNA viruses through successive bottlenecks usually is carried out making serial plaque-to-plaque transfers (Fig. 15.3). At each transfer, the viral population is plated at low multiplicity of infection to get well-isolated lytic plaques that are the result of the infection by a single virus, which after several replication rounds gives rise to a progeny. Since at each transfer the effective population size is reduced to one individual, this constitutes the most extreme form of bottleneck. The population contained in a randomly chosen plaque is isolated, properly diluted and plated again in a process that is serially repeated. The consequences on fitness of successive repetitions of this process have been analyzed with several RNA viruses including bacteriophages MS2 [62] and Phi 6 [63], vesicular estomatitis virus (VSV) [64–66], FMDV [67–69] and HIV-1 [70]. In all these studies, progressive fitness declines were found, although extinctions of infectivity were only observed in the case of HIV-1.

The most complete study on the effect that the accumulation of mutations through plaque-to-plaque transfers has on fitness evolution has been carried out with FMDV [67–69, 71].

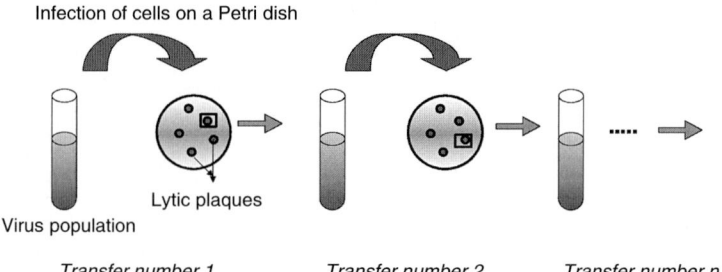

Fig. 15.3. Schematic representation of the experimental procedure of plaque-to-plaque transfers. The starting viral population is plated to isolate individual lytic plaques. The virus contained in a single plaque (in the box) is diluted for titration of infectious particles and to be used in the successive plaque transfer. The process is repeated as many times as desired

In this study, the titer of the plaques (determined as the number of infectious units per plaque or pfu) at each transfer was taken as a measure of fitness. The mutations that accumulated along the process were identified by determining the consensus sequence of the viral population isolated from single plaques at different transfers. The expected result of the experiment was a progressive decrease in fitness accompanied by an increase in the number of mutations fixed in the consensus sequence. After a certain number of transfers, extinctions of infectivity were expected. In contrast to these expectations, a biphasic dynamics of fitness decrease was observed. There was an initial period of roughly exponential fitness loss, but after a variable number of passages, a statistically stationary state of fitness with large fluctuations around a mean constant value was reached (Fig. 15.4). In this state, the virus exhibits a great resistance to extinction, since when it reaches a very low fitness value, the usual outcome at the next passage is a sudden fitness recovery. A detailed statistical analysis of the viral titers at the stationary state showed that fluctuations in the viral yield followed a Weibull distribution [69]. This distribution is indicative of an underlying dynamics with two main features (a) an exponential amplification of the founder genomes during the development of each plaque, which makes that small fitness differences are considerably amplified and (b) large variations in the initial state of the system at each transfer, which is determined by the stochastic nature of the sampling process.

Strikingly, mutations accumulated at the same rate in the phase of fitness decrease and in the stationary state [68]. This might indicate that the nature and effects of mutations can vary with the transfer number, depending on the restrictions imposed by the selection of the genomes able to form plaques. When the population is well adapted to the environment, as it happens at the beginning of the experiment, deleterious mutations are well tolerated. However, as the population is getting more debilitated, less deleterious mutations can be accepted and possibly there are many extinctions of individual genomes

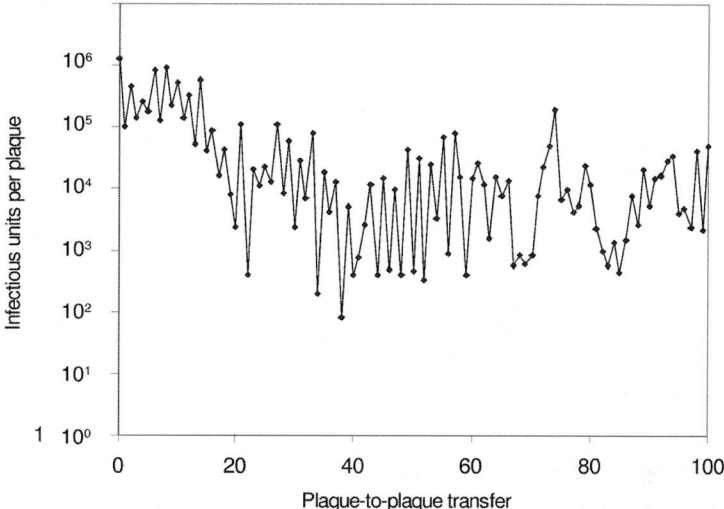

Fig. 15.4. Infectious units per plaque produced along the process of plaque-to-plaque transfers experienced by the viral clone C10. After an exponential decay of infectivity, a statistical stationary state with strong fluctuations is attained

that become unable to replicate. Nevertheless, a fraction of the genomes contained in a plaque can possess advantageous mutations, in some occasions because the mutation has a positive effect 'per se' and in others because it has a compensatory effect in a concrete genome carrying a particular combination of mutations. In the stationary state, where average fitness values possibly are the lowest ones compatible with virus survival, advantageous mutations would be more easily selected, because only the genomes carrying them can form plaques and be chosen for the next transfer. Each advantageous mutation produces a fitness increase that moves the genome to a different position in the fitness landscape. This permits the acceptance of additional deleterious mutations, originating the fluctuating pattern of infectivity that is observed in the experiments.

An interesting result is the preferential accumulation of mutations in certain genomic regions that present a mutation frequency significantly higher than the average obtained considering the whole genome [68] (Fig. 15.5). An unusual distribution of mutations has also been found in bottlenecked HIV-1 clones in which there was a higher accumulation of mutations in the gene gag and the first third of the genome, compared to the gene env, which is less conserved in natural populations of the virus [73]. Bottlenecked VSV clones also accumulated a high number of mutations in the N open reading frame, contrasting with the conservation of this region in natural isolates [66]. All these results suggest that bottlenecks permit the isolation of genomes that

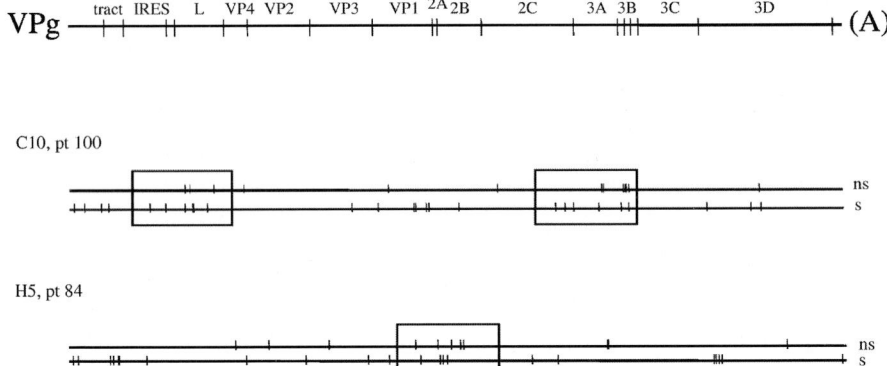

Fig. 15.5. Location of the mutations found in the genome of FMDV clones C10 and H5 subjected to 100 and 84 plaque transfers respectively. The *top horizontal line* is a scheme of the FMDV genome showing the main regulatory regions and the encoded proteins [72]. The *lines* below the genome indicate the non-synonymous (ns) and synonymous (s) mutations present in the virus. The *boxes* indicate the genomic regions where the number of mutations is significantly higher than the average for the whole genome

otherwise would be eliminated under the action of positive selection that dominates virus optimization. Genomic regions that seem to be much conserved might mutate with the same mutation rate as the rest of the genome, although subjected to stronger constraints. Nevertheless, the evolutionary relevance and the molecular mechanism by which the mutation clusters observed in the bottlenecked FMDV clones are generated is unknown and further experiments are in progress to answer this question.

A numerical model of evolution through bottlenecks was developed with the aim of identifying the parameters that are responsible for the biphasic dynamics of fitness loss [74,75]. The main features of the model are the occurrence with low probability of advantageous mutations and the presence of an extinction threshold, which means that genomes reaching the minimal allowed fitness value are eliminated. The results of the simulations were very similar to those observed in the experiments: a biphasic dynamics of fitness decrease and large fluctuations in the fitness values attained at the stationary state. Moreover, the statistical analysis of fitness values reveals that, similarly to the experimental results, they follow a Weibull distribution, strongly supporting that the underlying dynamics must be the same in both the simulations and the experiments. The elimination of individuals as their fitness falls below the extinction threshold and the probability of selecting for the subsequent transfer genomes with compensatory mutations constitute two factors acting in close concert to avoid extinctions due to an excessive accumulation of deleterious mutations. The occurrence of compensatory, advantageous mutations was not introduced in most models of Muller's ratchet that considered that back mutations were the only mechanism to revert the negative effect of deleterious mutations [76,77]. However, compensatory mutations are much more frequent than reversions as a mechanism to increase fitness, as it has been demonstrated in several theoretical and experimental studies [78, 79]. Accordingly, during the process of fitness recovery of FMDV and VSV bottlenecked clones upon large population passages, both reversions and compensatory mutations were found to be responsible for the observed fitness increases [64,66,71]. None of the recovered strains reverted to a wild type sequence, confirming that bottlenecks move the quasi-species through the fitness landscape towards regions where the adaptive value of mutations can be drastically altered.

The results of all these studies show that there are different mechanisms able to modulate the adaptive value of mutations. When the environment is altered, a new fitness landscape appears where the effect of particular mutations varies. In a similar way, even if the fitness landscape is not modified, bottlenecks constitute an effective way to explore new regions, where the selective value of mutations can differ from that present in the initial quasi-species. This means that the effect of mutations can vary depending on the mutations previously accumulated in the genome, a fact that points to epistatic interactions. Sanjuán et al. have studied the effect of pair of mutations in the VSV genome, compared to their effects as single mutations [80]. They found mainly antagonistic interactions between deleterious mutations (the effect of both mutations appearing together is smaller than the sum of the separate effect of each mutation). This finding can partially explain the non-linear dynamics of fitness loss observed in the FMDV clones. Some theoretical studies also show that antagonistic epistasis can reduce the speed of the ratchet.

A relevant question concerns the effect that the high mutational load of viral populations with a long history of bottlenecks has on their adaptability. The studies of Novella [81] have shown that bottlenecked viruses, even if they have recovered fitness through massive passages, always loss in competition experiments with the wild type, meaning that they have lower adaptability. It would be quite interesting to investigate if the high number of mutations accumulated in bottlenecked viruses also has negative consequences for adaptation to a new environment with a different fitness landscape. These studies can be carried out with viruses carrying different combinations of mutations and having the same fitness value, as those obtained at different transfer number in the stationary state attained by FMDV bottlenecked clones. The results of experiments of this type would allow to get more insight in the alternative adaptive solutions that can be explored by RNA virus populations differing in the consensus sequence.

15.7 Conclusions

Most of the viruses that are important human pathogens have RNA as genetic material. All of them share high mutability and a great potential for adaptation that makes their eradication enormously difficult. The isolation of drug-resistant mutants, the emergence of new diseases in humans caused by viruses that usually are maintained in animal reservoirs or the appearance of viral variants able to resist the action of the immune system of the host constitute important challenges for research in this century. One of the most promising strategies for the control of viral diseases consists in the increase of the error rate of viral replication above the threshold that prevents further transmission of genetic information. The difficulty to apply lethal mutagenesis to the treatment of viral infections largely rely, as it happens with other antiviral drugs, on the emergence of resistant mutants, which in this case would probably be those carrying high fidelity polymerases. The knowledge of the exact mechanisms leading a population to error catastrophe implies a detailed study of the composition and structure of the mutant spectrum of the quasi-species. In this sense, the comparison with the structure of bottlenecked populations that have accumulated a large number of mutations still compatible with survival can help to design new strategies for the extinction of infectivity.

Acknowledgments

The author thanks S.C. Manrubia, C. Escarmís, A. Grande-Pérez, J.Pérez-Mercader and E. Domingo for valuable suggestions and their participation in many of the studies detailed in this chapter. J.E. González-Pastor is also acknowledged for critical reading of the manuscript. Work at Centro de Astrobiología has been supported by INTA (Instituto Nacional de Técnica Aeroespacial).

References

1. L.A. Ball, in *Fields Virology*, 4th edn., ed. by D.M. Knipe, P.M. Howley (Lippincott Williams and Wilkins, 2001)
2. D.J. Earl, M.W. Deem, Proc. Natl. Acad. Sci. USA **101**, 11531 (2004)
3. J.W. Drake, B. Charlesworth, D. Charlesworth, J.F. Crow, Genetics **148**, 1667 (1998)
4. P.L. Foster, Mutat. Res. **569**, 3 (2005)
5. M.M. Tanaka, C.T. Bergstrom, B.R. Levin, Genetics **164**, 843 (2003)
6. J.W. Drake, J.J. Holland, Proc. Natl. Acad. Sci. USA **96**, 13910 (1999)
7. E. Domingo, J.J. Holland, Annu. Rev. Microbiol. **51**, 151 (1997)
8. S. Cleaveland, M.K. Laurenson, L.H. Taylor, Philos. Trans. R. Soc. Lond. B Biol. Sci. **356**, 991 (2001)
9. M.E.J. Woolhouse, L.H. Taylor, D.T. Haydon, Science **292**, 1109 (2001)
10. R. Chen, M.E. Quinones-Mateu, L.M. Mansky, Curr. Pharm. Des. **10**, 4065 (2004)
11. E. Batschelet, E. Domingo, C. Weissmann, Gene **1**, 27 (1976)
12. D. Posada, K.A. Crandall, E.C. Holmes, Ann. Rev. Genet. **36**, 75 (2002)
13. T. Rhodes, H. Wargo, W.S. Hu, J. Virol. **77**, 11193 (2003)

14. M.C. Boerlijst, S. Bonhoeffer, M.A. Nowak, Proc. R. Soc. Lond. B Biol. Sci. **263**, 1577 (1996)
15. D.J. Earn, J. Dushoff, S.A. Levin, Trends Ecol. Evol. **17**, 334 (2002)
16. R.J. Webby, R.G. Webster, Science **302**, 1519 (2003)
17. M. Eigen, Naturwissenschaften **58**, 465 (1971)
18. M. Eigen, C.K. Biebricher, in *Variability of RNA Genomes*, vol. III, ed. by E. Domingo, J.J. Holland, P. Ahlquist (CRC, Boca Raton, FL, 1998) pp. 211–245
19. M. Eigen, P. Schuster, Naturwissenschaften **64**, 541 (1977)
20. E. Domingo, C. Biebricher, M. Eigen, J.J. Holland, *Quasispecies and RNA Virus Evolution: Principles and Consequences* (Landes Bioscience, Austin, 2001)
21. E. Domingo, D.L. Sabo, T. Taniguchi, C. Weissmann, Cell **13**, 735 (1978)
22. E.C. Holmes, Trends Microbiol. **11**, 543 (2003)
23. U. Parikh, C. Calef, B. Larder, R. Schinazi, J.W. Mellors, in *HIV-1 Sequence Compendium* ed. by C. Kuiken, B. Foley, B. Hahn, P. Marx, F. McCutchan, J. Mellors, S. Wolinski, B. Korber (Theoretical Biology and Biophysics Group, Los Alamos National Laboratory, Los Alamos, 2001) p. 191
24. R.W. Shafer, Clin. Microbiol. Rev. **15**, 247 (2002)
25. R.M. Anderson, R.M. May, *Infectious Diseases of Humans: Dynamics and Control* (Oxford University Press, Oxford, 1991)
26. E. Baranowski, C.M. Ruiz-Jarabo, E. Domingo, Science **292**, 1102 (2001)
27. R.M. Bush, W.M. Fitch, C.A. Bender, N.J. Cox, Mol. Biol. Evol. **16**, 1457 (1999)
28. Y. Ina, T. Gojobori, Proc. Natl. Acad. Sci. USA **91**, 8388 (1994)
29. P. Daszak, A.A. Cunningham, A.D. Hyatt, Science **287**, 443 (2000)
30. S.S. Morse, Emerg. Infect. Dis. **1**, 7 (1995)
31. F. Gao, E. Bailes, D.L. Robertson, Y. Chen, C.M. Rodenburg, S.F. Michael, L.B. Cummins, L.O. Arthur, M. Peeters, G.M. Shaw, P.M. Sharp, B.H. Hahn, Nature **397**, 436 (1999)
32. P.A. Rota, M.S. Oberste, S.S. Monroe, W.A. Nix, R. Campagnoli, J.P. Icenogle, S. Peñaranda, B. Bankamp, K. Maher, M.H. Chen, S. Tong, A. Tamin, L. Lowe, M. Frace, J.L. DeRisi, Q. Chen, D. Wang, D.D. Erdman, T.C. Peret, C. Burns, T.G. Ksiazek, P.E. Rollin, A. Sanchez, S. Liffick, B. Holloway, J. Limor, K. McCaustland, M. Olsen-Rasmussen, R. Fouchier, S. Gunther, A.D. Osterhaus, C. Drosten, M.A. Pallansch, L.J. Anderson, W.J. Bellini, Science **300**, 1394 (2003)
33. R.G. Webster, D.J. Hulse, Rev. Sci. Tech. **23**, 453 (2004)
34. E. Domingo, Curr. Opin. Microbiol. **6**, 383 (2003)
35. A. Moya, E.C. Holmes, F. González-Candelas, Nat. Rev. **2**, 279 (2004)
36. N.M. Ferguson, A.P. Galvani, R.M. Bush, Nature **422**, 428 (2003)
37. B.T. Grenfell, O.G. Pybus, J.R. Gog, J.L.N. Wood, J.M. Daly, J.A. Mumford, E.C. Holmes, Science **303**, 327 (2004)
38. M. Eigen, Proc. Natl. Acad. Sci. USA **99**, 13374 (2002)
39. G.M. Jenkins, A. Rambaut, O.G. Pybus, E.C. Holmes, J. Mol. Evol. **54**, 156 (2002)
40. P. Simmonds, D.B. Smith, J. Virol. **73**, 5787 (1999)
41. S.C. Manrubia, C. Escarmís, E. Domingo, E. Lázaro, Gene **347**, 273 (2005)
42. S. Crotty, C.E. Cameron, R. Andino, Proc. Natl. Acad. Sci. USA **98**, 6895 (2001)

43. A. Grande-Pérez, S. Sierra, M.G. Castro, E. Domingo, P.R. Lowenstein, Proc. Natl. Acad. Sci. USA **99**, 12938 (2002)
44. A. Grande-Pérez, E. Lázaro, P. Lowenstein, E. Domingo, S.C. Manrubia, Proc. Natl. Acad. Sci. USA **102**, 4448 (2005)
45. L.A. Loeb, J.M. Essigmann, F. Kazazi, J. Zhang, K.D. Rose, J.L. Mullins, Proc. Natl. Acad. Sci. USA **96**, 1492 (1999)
46. L.A. Loeb, J.I. Mullins, AIDS Res. Hum. Retroviruses **16**, 1 (2000)
47. S. Sierra, M. Dávila, P.R. Lowenstein, E. Domingo, J. Virol. **74**, 8316 (2000)
48. C.K. Biebricher, M. Eigen, Virus Res. **107**, 117 (2005)
49. C. González-López, A. Arias, N. Pariente, G. Gómez-Mariano, E. Domingo, J. Virol. **78**, 3319 (2004)
50. N. Pariente, S. Sierra, P.R. Lowenstein, E. Domingo, J. Virol. **75**, 9723 (2001)
51. N. Pariente, S. Sierra, A. Airaksinen, Virus Res. **107**, 183 (2005).
52. C.M. Ruiz-Jarabo, C. Ly, E. Domingo, J.C. de la Torre, Virology **308**, 37 (2003)
53. E. Domingo, C. Escarmís, E. Lázaro, S.C. Manrubia, Virus Res. **107**, 129 (2005)
54. J.J. Arnold, M. Vignuzzi, J.K. Stone, R. Andino, C.E. Cameron, J. Biol. Chem. **280**, 25706 (2005)
55. J.K. Pfeiffer, K. Kirkegaad, Proc. Natl. Acad. Sci. USA **100**, 7289 (2003)
56. J.P. Vartanian, P. Sommer, S. Wain-Hobson, Trends Mol. Med. **9**, 409 (2003)
57. N. Pariente, A. Airaksinen, E. Domingo, J. Virol. **77**, 7131 (2003)
58. S. Crotty, R. Andino, Microbes Infect. **4**, 1301 (2002)
59. W.E. Severson, C.S. Schmaljohn, A. Javadian, C.B. Jonsson, J. Virol. **77**, 481 (2003)
60. T. Pierson, J. McArthur, R.F. Siliciano, Annu. Rev. Immunol. **18**, 665 (2000)
61. H.J. Muller, Mut. Res. **1**, 2 (1964)
62. M. de la Peña, S.F. Elena, A. Moya, Evolution **54**, 686 (2000)
63. L. Chao, Nature **348**, 454 (1990)
64. D. Clarke, E. Duarte, A. Moya, S. Elena, E. Domingo, J.J. Holland, J. Virol. **67**, 222 (1993)
65. E. Duarte, D. Clarke, A. Moya, E. Domingo, J. Holland, Proc. Natl. Acad. Sci. USA **89**, 6015 (1992)
66. I.S. Novella, B.E. Ebendick-Corpus, J. Mol. Biol. **342**, 1423 (2004)
67. C. Escarmís, M. Dávila, N. Charpentier, A. Bracho, A. Moya, E. Domingo, J. Mol. Biol. **264**, 255 (1996)
68. C. Escarmís, G. Gómez-Mariano, M. Dávila, E. Lázaro, E. Domingo, J. Mol. Biol. **315**, 647 (2002).
69. E. Lázaro, C. Escarmís, J. Pérez-Mercader, S.C. Manrubia, E. Domingo, Proc. Natl. Acad. Sci. USA **100**, 10830 (2003)
70. E. Yuste, S. Sánchez-Palomino, C. Casado, E. Domingo, C. López-Galíndez, J. Virol. **73**, 2745 (1999)
71. C. Escarmís, M. Dávila, E. Domingo, J. Mol. Biol. **28**, 495 (1999)
72. G.J. Belsham, Curr. Top. Microbiol. Immunol. **288**, 43 (2005)
73. E. Yuste, C. López-Galíndez, E. Domingo, J. Virol. **74**, 9546 (2000)
74. E. Lázaro, C. Escarmís, E. Domingo, S.C. Manrubia, J. Virol. **76**, 8675 (2002)
75. S.C. Manrubia, E. Lázaro, J. Pérez-Mercader, C. Escarmís, E. Domingo, Phys. Rev. Lett. **90**, 188102 (2003)
76. I. Gordo, B. Charlesworth, Genetics **154**, 1379 (2000)
77. A.S. Kondrashov, Genetics **136**, 1469 (1994)
78. S. Maisnier-Patin, D.I. Andersson, Res. Microbiol. **155**, 360 (2004)
79. G.P. Wagner, W. Gabriel, Evolution **44**, 715 (1990).
80. R. Sanjuán, A. Moya, S.F. Elena, Proc. Natl. Acad. Sci. USA **101**, 15376 (2004)
81. I.S. Novella, J. Mol. Biol. **336**, 61 (2004)

Index

ACNUC, 207–231
adaptation, 299, 301, 304, 327, 334, 345
adaptive landscape, 299
ADDA, 42
adjacency matrix, 6, 19, 326
amino acid, 220–231
   composition, 157
antigenic drift, 346
antiviral drugs, 346
archea, 39
AT content, 153–158

bacteriocyte, 149–150
Baumannia cicadellinicola, 149–162
Bayesian inference, 170, 185
beta distribution, 186, 187
binding cooperativity, 279, 280
binomial distribution, 184
blind-ant regime, 76, 77, 82
Blochmannia floridanus, 149–162
breakpoint distances, 160
Buchnera aphidicola, 149–162

capside, 341
Carsonella ruddii, 149–162
CATH, 37, 41
   domains, 41
cellular computation, 130
chaperones, 158
chemokines, 197
chromosomal rearrangements, 153–162
CLUSTAL format, 214
codon, 221–231
   usage, 231
   usage bias, 153, 156

combinatorics, 4, 20
commensalism, 149
compatibility, 8, 10, 12, 20, 21
compensatory substitutions, 293
conjugate prior distribution, 186
consensus sequence, 310, 344
cooption, 295, 296
correspondence analysis, 225
CRAN, 207–231

DALI, 65
data storage medium, 129
databases, 207–231
defective genomes, 353
delta method, 172–175
designability, 104
detailed balance, 78, 107, 273
disintegration rate, 159
distance-based method, 178, 179, 181, 184
divergence
   expression, 235, 240, 241, 243, 246, 247
   sequence, 236, 246, 247
DNA, 207–231
   loss, 158–159
   methylation, 185
   uptake, 159
dotchart, 222
dynamic programming, 264

electrophoresis, 288
EMBL, 207–231
endocytobiosis, 149

endosymbiosis, 149
environments
  simple vs diverse, 292
epistasis, 299, 314–316, 325
equilibrium frequencies, 172, 174
equilog, 116
error threshold, 299, 307–310, 312–320, 323, 324, 327, 334–336, 350, 351, 353
evolution
  convergent, 113–124
  divergent, 113–124
evolutionary trajectory, 91, 93, 328, 330–332
  correlations, 91, 93
  fluctuations, 91, 93
expression
  conservation, 250
  divergence, 235, 240, 241, 243, 246, 247
    natural selection, 248
    neutral changes, 247, 248
    probe-to-gene mapping, 241, 242
  level, 243
    clustering, 245, 246
  pattern, 235
  polymorphism, 249
  profile, 244
extinction, 348, 352

FASTA format, 213–231
FastME, 178, 184
Fibonacci numbers, 22
Fisher's model, 294
Fitch–Margoliash, 178, 184
fitness, 236, 238, 239, 246, 247, 300, 301, 303, 306, 308, 314–317, 321–324, 332–336, 343
  effects of mutations, 292
  landscape, 299–301, 303, 304, 306–308, 310, 311, 313–317, 319, 321–328, 332–336, 344
  Malthusian, 300, 303, 316, 317, 321, 323, 324, 332–334
  step-wise increase, 327, 328, 335
  Wrightian, 301, 303, 306, 308, 314–317, 321, 322, 324
fixation, 76
  probability, 77, 291

FM criterion, 178–180, 185
foot and mouth disease virus, 352–359
frequency distribution of the fitness effects, 293
functional
  annotation, 113–122
  conservation, 84
functional flexibility score, 118

gamma-distributed rates, 180
GC content, 153–158
GenBank, 207–231
gene
  duplication, 31, 32, 191, 193
  inactivation, 155–163
  interaction, 254, 280
  loss, 158
  networks, 280
  ontology, 118
  regulation, 253, 254, 259, 260, 263–271, 273–280
    Bayesian model, 263–265, 277
    binding cooperativity, 279, 280
    evolution, 266–271, 273–276
    evolutionary equilibria, 270, 271, 274, 276
    evolvability, 260
    neutral evolution, 273
    programmability, 260
    score-fitness relation, 274
    selection, 275
Gene3D, 37, 41
genealogy, 27
genetic
  background effects, 296
  code, 211, 231
  distance, 170–172, 174, 176, 178, 180
  diversity, 343
  drift, 299, 328, 335, 343
  load, 287
  regulatory network, 286
    bottom-up evolution, 295
genome
  changes, 162
  complexity, 294
  evolution, 135, 136
  formatting, 131
  rearrangement rates, 160
  reduction, 158–163

segment reassortment, 343, 344
  stasis, 160
genotype-phenotype map, 301, 322
globins, 57
  secondary and tertiary structure, 59

haem group, 59
haemoglobin, 57
Hamming distance, 302, 344
homolog, 32
  identification, 236
horizontal gene transfer, 159, 160
human immunodeficiency virus, 343–359
hypermutation, 342, 353

identification
  ortholog, 236
immune enhancement, 350
cross-immunity, 348, 349
strain-trascending immunity, 349
indel bias, 159
influenza, 344–359
informative site, 181, 182
insertion sequence, 163
interaction systems, 287
interactivity parameters, 98
intersection theorem, 21, 22
inversion distances, 160
Ising model, 304, 306, 307, 309, 314

JC69 model, 171, 174, 183

K80 model, 173, 174

lattice structure graph, 119
likelihood, 169, 170, 177, 178, 181–189
long-branch
  attraction, 180, 182
  repulsion, 180
LS criterion, 180
lupin leghaemoglobin, 68

Markov chain, 171, 176, 186
MASE format, 214
master sequence, 310, 312–316, 318–320, 323, 332, 333, 344
matching, 6, 8, 15, 19, 21
ME criterion, 179, 180
metazoa, 58

methylation, 185
mobile genetic elements, 131, 136
modular structures, 285, 287
molecular
  clock, 75, 287, 288
  phylogenetics, 169, 170, 176, 187
monomorphic site, 181
MSF format, 214
Muller's ratchet, 156, 336, 355, 358
multidomain, 38, 39
multigene families, 191
multiple hits, 170
multivariate analyses, 225–231
mutagenesis, 351
  lethal, 352
mutant spectrum, 78, 344
  advantageous mutations, 78
  nearly neutral mutations, 78, 79
  neutral mutations, 78–80
  strongly deleterious mutations, 78
mutation, 76, 343
  advantageous, 351
  compensatory, 359
  deleterious, 344–359
  rate, 76, 303, 342, 343
  synonymous, 351
mutation and selection
  dynamics, 299, 300, 302, 303, 308, 317, 328
  mean-field model, 107, 108
mutualism, 149, 152
myoglobin, 57

narrow range of heterozygosity, 288
natural selection, 77, 236, 238, 239, 246, 247, 285, 342
nearly neutral theory, 285, 288, 294
  mutant substitutions, 289
  spreading, 296
neutral
  connectivity, 286
  evolution, 273
  mutations, 80
  network, 15–18, 28–30, 91, 322, 326
    blind-ant regime, 81
    quasispecies regime, 81, 82
    SCN model, 91
  path, 17, 18
  space, 286

substitutions, 80
　theory, 75, 288
neutrality, 15, 17, 18, 326
　degree of, 16
new gene expression, 295
NewFam, 44
　domains, 41
newly arisen enhancers, 295
NK model, 324, 325
nonsynonymous substitutions, 230–231, 290

operational taxonomic unit, 178
ortholog, 185
　identification, 236

paralogue, 31, 32
parasitism, 149
parsimony, 180–182
partition function, 8, 13, 18, 19
pathogens, 37
percolation, 327
persistent infections, 350
PFAM, 37, 41
PFscape, 41
PHYLIP format, 214
phylogenetics, 30, 31, 191
plasmid, 160
　variability, 160
polymorphism, 181
　of gene expression, 294
polyploidization, 192
population
　bottleneck, 335, 346, 348
　genetics, 76
　size, 342, 344
　　fluctuations, 344
ProDom, 43
protein
　domain, 37, 113
　evolution, 37, 113
　folding, 83
　sequence, 99, 100
　　vectorial representation, 98
　stability, 83
　structure, 99, 100
　　representation, 88
　　stability against missfolding, 89–91
　　stability against unfolding, 88
　　vectorial representation, 99
　thermodynamics, 83
Protein Domain Universe Graph, 116–124
proteome, 45
ProtoNet, 42
pseudogene, 31, 158–159
　half-life of, 159
pseudoknot, 4–6, 19
punctuated evolution, 327, 328, 335

quantum spin Hamiltonian, 304, 306, 307, 314, 316
quasi-species, 344
quasispecies, 299, 305, 308, 309, 311–313, 318, 319, 326, 328–331, 333, 334
　regime, 76, 77, 82

R, 207–231
random energy model, 90, 322, 323
random genetic drift, 285
random graph, 15, 16
recombination, 153–163, 344
　rate, 344
reformation of regulatory networks, 295
relay series, 27, 28
repetitive DNA, 129, 131, 132, 138
replication, 214
　origin, 214
　rounds, 344
　terminus, 214
reproductive number, 348
drug-induced resistance, 346
reverse transcriptase, 342
ridges-into-grooves model, 63
RNA
　evolution, 3, 18, 23, 25–32, 333
　kinetic folding, 3, 11, 13, 23–25
　noncoding, 3, 23, 30
　replicase, 333, 342
　switch, 13

Saccharomyces cerevisiae, 192
scale-free networks, 117, 118
SCN model, 76, 87
selection, 344
　negative, 345
　positive, 346

selection coefficient
  distribution of, 291
seqinR, 207–231
sequence, 207–231
  alignment, 237
  consensus, 310
  conservation, 104
  distance, 237, 239, 240
  divergence, 236, 246, 247
  entropy, 273
  master, 310, 312–316, 318–320, 323, 332, 333
  space, 302–304, 308, 310–312, 314, 321, 322, 326–329, 335, 336
site-specific amino acid distributions, 97, 101–106
slightly deleterious mutation theory, 288
  amino acid substitutions, 290
  mutant substitutions, 293
Sodalis glossinidius, 149–162
SOPE (primary endosymbiont of the weevil *Sitophilus oryzae*), 149–162
Spearman coefficient, 46, 47
spin glass, 322–324
  random energy model, 322, 323
  Sherrington–Kirkpatrick model, 323
spurious transcription binding sites, 294
stepwise increase of fitness, 296
structural approach, 75
structural conservation, 84–87, 100
subfunctionalization, 194
subneofunctionalization, 195
substitution
  acceleration of rate, 153, 156
  matrices, 104
  models, 169–171, 176, 177, 182
  nonsynonymous, 238, 239

process, 93, 95, 96
saturation, 179, 182
synonymous, 238, 239
superfamily, 46, 49
  evolution, 48
SWISS-PROT, 207–231
symbiosis, 149
synonymous, 230–231
  mutation, 185
  substitutions, 290
systems engineering, 141
SYSTERS, 42

Thermotoga maritima, 202
TIM barrel, 199, 204
TN93 model, 171, 175
transcription, 253
  factor, 254–263, 266, 276, 278, 279
    binding energy, 254–259, 261, 262, 266
    search kinetics, 258, 259
  regulation, 37
transversion, 173–176, 185
tree of life, 115
tropism, 346, 347

UPGMA, 178

variance of the evolutionary rate, 290
vectorial representation
  protein sequence, 98
  protein structure, 99
virus, 341–360
  emergence, 348

Wigglesworthia glossinidia, 149–162
Wright's shifting balance theory, 294

Printing: Krips bv, Meppel
Binding: Stürtz, Würzburg